The Antibiotic Era

The Antibiotic Era

Reform, Resistance, and the Pursuit of a Rational Therapeutics

SCOTT H. PODOLSKY

Johns Hopkins University Press
Baltimore

KH

This book was brought to publication with the generous assistance of the Department of Global Health and Social Medicine, Harvard Medical School.

Printed in the United States of America on acid-free paper
9 8 7 6 5 4 3 2 1

Johns Hopkins University Press
2715 North Charles Street
Baltimore, Maryland 21218-4363
www.press.jhu.edu

Library of Congress Cataloging-in-Publication Data
Podolsky, Scott H., author.
The antibiotic era : reform, resistance, and the pursuit of a rational therapeutics / Scott H. Podolsky.
p. ; cm.
Includes bibliographical references and index.
ISBN 978-1-4214-1593-2 (hardcover : alk. paper) — ISBN 1-4214-1593-3 (hardcover : alk. paper) — ISBN 978-1-4214-1594-9 (electronic) — ISBN 1-4214-1594-1 (electronic)
I. Title.
[DNLM: 1. Anti-Bacterial Agents—history—United States. 2. Drug Industry—history—United States. 3. Drug Resistance—United States. 4. History, 20th Century—United States. 5. Legislation, Drug—history—United States. QV 11 AA1]
RM267
615.7'922—dc23 2014014551

A catalog record for this book is available from the British Library.

Johns Hopkins University Press uses environmentally friendly book materials, including recycled text paper that is composed of at least 30 percent post-consumer waste, whenever possible.

9/15/17

For Roz and Leon Fink,

and in memory of

Lorna and Jack Podolsky

CONTENTS

ACKNOWLEDGMENTS

In many ways, this book grew out of my prior book project, *Pneumonia Before Antibiotics*, as I've continued to engage in the clinic and as a historian with issues of therapeutic autonomy, education, regulation, and innovation. Indeed, Jeremy Greene has quipped that I should have titled this book "Pneumonia After Antibiotics." Since the publication of *Pneumonia*, I've had the great fortune to become director of the Center for the History of Medicine at the Francis A. Countway Library of Medicine, to teach with and learn from fellow members of Harvard Medical School's Department of Global Health and Social Medicine, to continue to provide primary care to my patients at Massachusetts General Hospital, and to engage with colleagues worldwide attempting to define and promote the rational production and delivery of antibiotics. I've learned a great deal in the process, and it's a privilege to be able to give thanks in this space.

At the Countway, Kathryn Hammond Baker has been a brilliant collaborator, continually positioning our center to support scholarship, while our remarkable staff members always make us look good through their dedication and consistent excellence. Special thanks go to Jack Eckert and Jess Murphy for facilitating (and tolerating) my extensive research on Max Finland in particular; I appreciate them all the more, knowing they work as diligently for all of our center's users. Zak Kohane and Alexa McCray, the Countway's director and deputy director, have been supportive of both our center and my own scholarship, as have Jay Jayasankar and Roz Vogel, president and administrative head of the Boston Medical Library. It's an embarrassment of riches to work not only amid the books and journals of the Countway but also amid its remarkable librarians and staff, and I'm especially grateful to Elizabeth Bueso, Betsy Eggleston, David Osterbur, and Julia Whalen for their ongoing wisdom and availability.

At Harvard Medical School's Department of Global Health and Social Medicine, it's impossible not to be inspired and influenced by the examples of our leaders, Paul Farmer and Anne Becker. I'm personally grateful for their support, as well as that of Jennifer Puccetti and Rebecca Grow, in particular. My esteemed colleagues Allan Brandt and David Jones continue to shape my thinking and support my efforts in many ways, as they have for many years. I am truly fortunate and grateful. Leon Eisenberg provided wise counsel at the start of this book, and he's greatly missed by all who were lucky enough to know him and learn from him. At HMS itself, Dean Jeffrey Flier has created an environment where history is a means not only of celebration but also of critical inquiry. Across the river, I'm proud to be affiliated with Harvard's History of Science Department and exposed from the one side to such role models as Charles Rosenberg and to fantastic students from the other side. Again, this is an embarrassment of riches.

At the Massachusetts General Medical Group, I continue to learn from both my colleagues and my patients. I'm grateful to everyone in the practice, while special thanks go to our group's directors during this project, Steve Levisohn, David Finn, and Amy Schoenbaum, to my office roommate Patty Gibbons, and to my "team" of Anne Drake, Angela McCaul, Tina Rosado, and Fred Rose, who take such wonderful care of our patients. MGH and its Division of General Internal Medicine provided critical funding at an early stage of the project and have been consistently supportive of my work.

My thinking about the issues examined in this book has evolved considerably over the past decade, and I'm honored to thank colleagues who have provided both encouragement and critique. I'm especially indebted to my frequent collaborator Jeremy Greene, who generously read through several versions of the manuscript, by which time he had already provided countless leads and shaped my understanding of the history of therapeutics more broadly. Ed Dwyer and Fred Tauber, both long-term mentors, read through the entire manuscript, as did Robert Guidos, Calvin Kunin, John Powers, and Dominique Tobbell, who have been engaged with the issues described here for many years. At Harvard, I've benefited from discussions with Jerry Avorn, Daniel Carpenter, Yonatan Grad, Ted Kaptchuk, Aaron Kesselheim, and Peter Tishler, while beyond Harvard I've received valuable input and insights from Robert Bud, Iain Chalmers, Arthur Daemmrich, David Herzberg, Greg Higby, Suzanne Junod, Claas Kirchelle, Jerry Klein, John Lesch, Stuart Levy, Nick Rasmussen, John Swann, Ulrike Thoms, Jo Tricker, Elizabeth Watkins, and the late Mark Finlay. The enduring influence of Harry Marks will be apparent throughout.

I extend particular thanks to the European Science Foundation's Drugs Networking Programme and to the hosts of a series of conferences concerning the history (and future) of antibiotics and pharmaceuticals more generally. Christoph Gradmann, Flurin Condreau, Anne Kveim Lie, and María

Jesús Santesmases not only individually influenced my thinking, but they also provided environments where scholars could converse across national and disciplinary boundaries. If the intention of such conferences was to broaden the thinking of participants, then they were wildly successful in my case. I'm also grateful to the hosts and audiences at the American Association for the History of Medicine, Beth Israel Deaconess Medical Center's Division of Pulmonary Medicine, the Boston Colloquium on the History of Psychiatry and Medicine, Brigham and Women's Hospital's Division of Pharmacoepidemiology, Clark University's Department of Biology, Dartmouth Medical School's Pathology Research and Review Seminar Series, the Food and Drug Administration, the Massachusetts Infectious Disease Society, multiple venues at Massachusetts General Hospital, the New England Tuberculosis Symposium, and Yale University's Program in the History of Science and Medicine for close attention and feedback.

Now that I live on both sides of the archival fence, I more fully appreciate just how much work goes into the acquisition and curation of archival collections. This book could not have been completed without the efforts of the archivists and librarians at the locations listed at the end of this book. Special thanks to Bill Davis, Janice Goldblum, and Stephen Greenberg for making going over and beyond the call of duty seem easy.

I am particularly grateful as well to Mark Leasure, Robert Guidos, and Kathy Cortez at the Infectious Diseases Society of America and to Stuart Levy at Tufts University School of Medicine for making their private collections available to me without in any way attempting to censor my conclusions. In both cases, their engagement with issues around antibiotic resistance was matched by their generosity and open-mindedness.

I'd like to acknowledge Oxford University Press for permission to draw from my article "Antibiotics and the Social History of the Controlled Clinical Trial, 1950–1970," *Journal of the History of Medicine and Allied Sciences* 65 (2010): 327–67, and to acknowledge Johns Hopkins University Press (JHUP) for permission to revise and reprint excerpts from "Chapter 2: Pharmacological Restraints: Antibiotics and the Limits of Physician Autonomy," in Jeremy A. Greene and Elizabeth Siegel Watkins, eds., *Prescribed: Writing, Filling, Using, and Abusing the Prescription in Modern America* (Baltimore: Johns Hopkins University Press, 2012), pp. 46–65. My debt to JHUP only starts there, however. Executive editor Jackie Wehmueller has again provided consistent support, good humor, and wise counsel in equal measure, while I'm grateful to Glenn Perkins for his careful copy editing and to Courtney Bond for her thoughtful production editing.

Finally, it's always a privilege to be able to thank my family. Neither my skepticism nor my enthusiasm fall far from the tree, and I'm grateful for the ongoing enthusiasm of my family—from my parents, Ellen and Steve Podolsky, to the treasured clans of Does, Finks, Gottholds, MacWrights, McGuires, Mirons and Shapiros—for my work. At home, my wife Amy remains my best friend and chief support, and the love of my life. Our boys, Josh and Danny, are our pride and joy, even as they are now old enough to be both amused and slightly bemused by my own enthusiasms.

The Antibiotic Era

Introduction

In March 1994, as both experts and the media increasingly turned their attention to the "crisis" of antibiotic resistance, *Newsweek* posed a question: Is this "The End of Antibiotics?" Three weeks later, *Newsweek* asked again, this time as its cover story, "Antibiotics: The End of Miracle Drugs?" Offering "plenty of blame to go around," the story's writers pointed to the host of actors responsible for antibiotic overuse and the "medical disaster in the making": greedy pharmaceutical marketers, frightened patients, and harried physicians at the mercy of both, unwilling to stand up for a rational therapeutics. In an era of medical surfeit, the use of antibiotics seemed a matter of "too much of a good thing," an ironic denouement to the twentieth-century therapeutic revolution itself.[1]

More than two decades later, while *Newsweek* has gone in and out of print form, antibiotics are still here. Yet so are concerns about the enduring utility of such miracle drugs. Whether in academic journals, popular media, or the blogosphere, we remain awash in ever-escalating antibiotic-resistance jeremiads, captured in such terms as *superbugs*, *antibiotic crisis*, and the *post-antibiotic era*. And antibiotic use continues to be generalized to stand for something more, to serve as a commentary on how we handle—or have mishandled—the therapeutic resources at our command.

Antibiotics had served as the leading edge of the post–World War II wonder drug revolution, allowing medicine to powerfully rebrand itself while such drugs indelibly altered the very practice of medicine.[2] Lederle's Aureomycin (chlortetracycline), released in 1948 as the first broad-spectrum antibiotic, was aptly named.[3] This was indeed the "golden age" of medicine.[4] But Pfizer's Terramycin (oxytetracycline), released two years later and the first antibiotic actually termed a "broad-spectrum antibiotic," was likewise aptly (if unintentionally, in this respect) named. From the beginning, reformers sought to bring antibiotics back down to earth, serving as a leading edge for those expressing anxiety and skepticism about the use and potential misuse of the wonder drugs more broadly.

Over the ensuing six-plus decades, antibiotics have continued to embody the power and limitations of the wonder-drug armamentarium, even if they entail special ecological considerations all their own. Perhaps no domain in conventional medicine has been subjected to such ongoing scrutiny—or so frequently labeled as "irrational"—as the development and prescribing of antibiotics. Such attention to therapeutic irrationality—with all its professional, epistemological, moral, legal, economic, and ecological implications—has led, inevitably, to discussions regarding therapeutic autonomy and the relative roles of education versus regulation in promoting or enforcing a "rational" therapeutics.

There are many books that could be—and have been—written about antibiotics and their history.[5] While this book traces the history of antibiotic development and usage, it focuses on antibiotic *reformers*—those who would change how we develop and prescribe antibiotics—and their institutional origins, motivations, strategies, struggles, successes, and limitations. It is about their evolving notions of therapeutic rationality. It is about the collaborations they have formed and the resistances they have faced at the crowded intersection of bugs and drugs, front-line clinicians and patients, medical academia and industry, Congress and the media, government regulatory and public health agencies, and national and international organizations advocating for a particular form of rational (or prudent or appropriate) therapeutics. It is about what their apparent successes and failures say about the practice and regulation of medicine more broadly.

In 1986, legendary antibiotic researcher (and penicillin biographer) Gladys Hobby wrote to Harvard's Edward Kass about a young historian, Harry Marks, who had written to her wanting to know more about the antibiotics that "came after penicillin."[6] Harry Marks, as usual, was on to something, and he would give extended attention in his *Progress of Experiment* to the evolution of twentieth-century reform groups and those who would aspire to inculcate a "rational" therapeutics.[7] Shortly before, James Whorton had delivered a fascinating talk, "'Antibiotic Abandon': The Resurgence of Therapeutic Rationalism," focusing on the emergence in the 1950s of antibiotic reformers as a group of self-described "therapeutic rationalists."[8] In many ways, my book took its methodological starting point at the intersection of *Progress of Experiment* and "Antibiotic Abandon," tracing over the course of five chapters and six-plus decades the progress—or lack of progress—of attempts to rein in antibiotic abandon.

Chapter 1 sets the stage for the remainder of the book. Today, we turn to infectious disease experts in the context of spreading bugs and an apparent shortage of drugs. But infectious disease experts emerged in the United States in the post–World War II era as a distinct specialty in response to receding bugs (the experts were among the few who had seen certain declining diseases) and an emerging array of drugs. They were asked to evaluate which drugs were useful both for individual patients and broadly. Soon, a subset of such *experts* would coalesce into the first generation of antibiotic *reformers*. This chapter concentrates on the introduction, in the late 1940s and early 1950s, of the broad-spectrum antibiotics. In part, this is to acknowledge the work that Jack Lesch and Robert Bud have performed in demonstrating the degree to which first the sulfa drugs in the 1930s, and then penicillin in the 1940s, transformed the nature and self-image of medical practice.[9] But this focus also accentuates how the broad-spectrum antibiotics in particular epitomized the widespread advent of the patented, branded, and heavily marketed postwar "wonder drugs" that would soon come to include steroids, minor tranquilizers, antipsychotics, and antihypertensives (among other drug classes).[10] In the eyes of a gathering band of infectious disease–based therapeutic reformers, headed by Harvard's Maxwell ("Max") Finland, such antibiotic development and marketing proliferated amid a regulatory vacuum left by the American Medical Association (AMA) and the Food and Drug Administration (FDA, which at the time could formally adjudicate only drug safety, rather than efficacy), threatening the very basis of a "rational" therapeutics. In response, this generation of experts would set in motion over six decades of ongoing antibiotic reform.

Chapter 2 details how such reform efforts were catalyzed in the 1950s by the advent of a relatively forgotten therapeutic class, the "fixed-dose combination antibiotics." Developed as the medical profession confronted widespread antibiotic resistance for the first time, the fixed-dose combination antibiotics were what they sounded like, set combinations of more than one antibiotic in a single pill. Widely promoted on the basis of the "synergy" (more than additive impact) of their components, their wide range of antimicrobial coverage, and their capacity to overcome individual antibiotic resistance, they were described as a "third era" in antibiotic evolution, after such "narrow"-spectrum antibiotics as penicillin and streptomycin and then the broad-spectrum antibiotics. However, reformers like Max Finland,

Harry Dowling at the University of Illinois, and Ernest Jawetz at the University of California–San Francisco (UCSF) not only deemed the drugs no better (and at times worse, in the setting of drug antagonism or combined or conflated side effects) than their component parts, but they also worried that the combinations represented portents of a future in which marketing would supersede substance, bringing down the entire and increasingly interconnected edifice of medicine and the pharmaceutical industry. Such reformers would be moved to argue that the "testimonials" of the pharmaceutical companies in support of such remedies would need to be countered by the *controlled clinical trial* in particular; in the process, the reformers contributed to both the uptake and regulatory encoding of the controlled clinical trial in twentieth-century medicine. At the same time, the gaze of reformers—focused on the pharmaceutical industry—turned away from individual prescribers at a key moment of therapeutic reform.

Chapter 3 serves as the dividing line of the story. By the late 1950s, intraprofessional antibiotic anxieties intersected with FDA, media, and congressional reform efforts to help shape Senator Estes Kefauver's hearings on the pharmaceutical industry between 1959 and 1962. Kefauver, the liberal from Tennessee who had taken on the mob in the early 1950s, had initially sought to tackle monopolistic tendencies in the pharmaceutical industry. But as the investigation and proceedings developed, he and his staff increasingly turned their attention to the pharmaceutical marketing of drugs, which in turn shone a powerful light on the FDA's inability to explicitly rule on drug efficacy prior to new drug approval. The investigation and hearings ultimately resulted in the Kefauver-Harris amendments of 1962, which mandated that novel drugs needed to be proven efficacious by controlled clinical studies prior to new drug approval, setting the foundation for the subsequent half-century of pharmaceutical testing and marketing.[11] Daniel Carpenter has carefully demonstrated how such an apparent FDA revolution in many ways represented a natural evolution from normal science activity gaining force within the FDA throughout the 1950s.[12] But the fixed-dose combination antibiotic story highlights the importance of the process by which such academic reformers as Max Finland and his colleagues—motivated by dystopic visions of "irrational" medicines stemming from the convergence of shoddy clinical investigation and overmarketing—united with FDA and congressional representatives to contribute to the construction of the amendments.

Both within and beyond the domain of antibiotic development and prescribing, the results would be profound. The Kefauver-Harris amendments permitted the FDA, through the resulting Drug Efficacy Study and Implementation (DESI) process, to remove *existing* inefficacious drugs from the market; and emblematically, the FDA removed every existing fixed-dose combination antibiotic, arguing that they had not been proven, via controlled clinical trials, to be more efficacious than their component parts.[13] Such FDA empowerment would be contested by Upjohn with respect to Panalba—its fixed-dose combination of tetracycline and novobiocin—as epitomizing government encroachment on physician prescribing autonomy, and the contest over Panalba would be taken all the way to the U.S. Supreme Court in 1970. Yet the judiciary found in favor of the FDA, and broadly, the Panalba rulings undergirded a crucial era of FDA empowerment.[14]

However, as chapter 4 relates, such a top-down "victory," as pertaining to antibiotics, would be a pyrrhic one. The limits to government encroachment on the prescribing of antibiotics in the United States would be reached with Panalba and the fixed-dose combination antibiotics. While the FDA had been empowered to remove seemingly "irrational" drugs from the marketplace, no one had been empowered to rein in the seemingly inappropriate *prescribing* of appropriate drugs. The 1970s would witness ongoing professional and government attention given to the increasingly quantified prevalence of "irrational" antibiotic prescribing and its consequences, and such attention would in fact lead to attempts to restrain such prescribing through both educational and regulatory measures. The DESI process, though, had generated a vocal backlash against centralized attempts to further delimit individual antibiotic prescribing behavior in the United States, resulting in generally failed attempts to control prescribing at local, let alone regional or national, levels in the United States.

This is a legacy we are left with today, and chapter 5 carries the story forward through a detailed history of the responses to antibiotic resistance in the United States and globally, from the decades covered in the first four chapters through today. Contemporary discussions of antibiotic resistance focus on both the "supply" side of the equation (the need for industry to develop novel antibiotics) and the "demand" side (how to prevent inappropriate prescribing so as to lessen the degree of antibiotic resistance). Examining the relevant co-evolution of such organizations as the AMA, the FDA, the World Health Organization (WHO), the Alliance for the Prudent Use of Antibiotics (APUA), the Centers for Disease Control and Prevention (CDC), the National Institute of Allergy and Infectious Diseases (NIAID), and the Infectious Diseases Society of America (IDSA), this chapter demonstrates the degree to which such contemporary discussion and efforts must be understood against the backdrop of the historical forces detailed in the preceding four chapters. It complicates our narratives of antibiotic "reform" itself by demonstrating the heterogeneity—and at times, even the collision—of such reform efforts. And in doing so, it reveals the degree to which efforts to forestall the "end of antibiotics" remain subject to the complex interactions of bugs, drugs, clinicians, patients, investigators, reformers, institutions, and the structures that govern the worldwide usage of antibiotics themselves.

All histories represent particular perspectives, and all historians make choices. I have six major caveats or points of orientation to declare before diving into the history. First, this is primarily an American story. As the conclusion relates, this perspective in part reflects my stance that all such histories are "local" case studies to some degree, contextualized within (and in turn affecting) particular social, cultural, and political settings. Robert Bud, for example, depicted the politicization of antibiotic resistance in Great Britain, and I expect that similarly detailed analyses pertaining to "local" (regional, national, etc.) settings will be valuable both individually and collectively.[15] However, as chapter 5 most clearly demonstrates, the distinction between the "local" and the "global" inevitably collapses, in historically situated fashion. With respect to antibiotic resistance, this process became self-evident starting in the late 1970s, though of course was present to some degree from the dawn of the antibiotic era.[16]

Second, within the American story, I devote a great deal of attention to Harvard's Max Finland (1902–1987). I initially feared that the richness of Finland's archival collection at the Francis A. Countway Library of Medicine could distort my overall perception of events (despite my relying on more than two dozen other archival collections), but I believe this initial fear of a misshapen archival homunculus was unfounded. Barely over five feet tall and literally living at Boston City Hospital for most of his career, Finland spent much of the twentieth century as the country's foremost and most influential clinical investigator of antimicrobials. By any measure, his impact was enormous. Paul Beeson, in his 1976 "honor roll" of the twelve most important American contributors to the understanding and conquest of infectious disease over the prior two centuries, placed Finland alongside Oswald Avery, John Enders, Walter Reed, Albert Sabin, Jonas Salk, and Selman Waksman (among other luminaries).[17] Finland trained over one hundred fellows in infectious diseases, including seven future presidents of the IDSA (he served as its first president in 1963) and many others who would go on to chair infectious disease departments and lead training programs of their own.[18] As his Harvard colleagues remembered, "There were relatively few places in which to train in infectious diseases in the years immediately following the cessation of World War II, and expansion of training programs basically had to wait until his former fellows had risen to prominence."[19] Finland published more than 800 papers, many of them classics, and was easily the most cited infectious disease expert of his generation.[20] And more than any other individual, he would shape the contours of antibiotic reform throughout the latter half of the twentieth century, both directly and through his impact on his fellows and colleagues (including Harry Dowling, his first fellow, and Calvin Kunin, his self-labeled longest-tenured fellow).

Third, I focus little attention on the use of antibiotics in animal husbandry. This topic, of course, has important ecological, economic, political, and regulatory components and consequences and is a crucial one for those interested in countering antibiotic resistance and forestalling the "end of antibiotics." But this volume is primarily about the use of antibiotics in human clinical practice, and I thus only refer to the agribusiness aspects of antibiotics when related to the story at hand (especially in chapter 5). Those interested in the topic would do well to start with Robert Bud's excellent account of the British experience.[21] I expect that treatments concerning other regions will follow as scholars working at the interface of environmental history, the history of medicine, and political science investigate this pivotal subject.[22]

Fourth, the voices of patients are rarely heard in this volume. From the 1990s onward (as described in chapter 5), clinicians and sociologists have sought to capture what patients think about antibiotics and how patients (or their parents) contribute to apparent antibiotic overuse. But these are relatively recent efforts and analyses. There has been minimal patient activism regarding antibiotic usage—whether regarding overuse or underuse—in this country or elsewhere. In fact, Brad Spellberg of the University of California–Los Angeles concluded his 2009 book, *Rising Plague*, with a plea to patients afflicted by antibiotic-resistant pathogens to communicate their stories to their political representatives in an effort to lend support to efforts to forestall the "postantibiotic era."[23] Rather, as Tanya Stivers explains in her 2007 book on pediatric antibiotic prescribing, patient (or in this case, parental) pressure on physicians occurs more subtly, privately, and diffusely, in an uncoordinated yet surprisingly consistent orchestration of key phrases and negotiations that generally promote antibiotic prescribing in uncertain clinical circumstances.[24] As such, absent direct patient perspectives or extended individual clinician chart reviews, I have for the most part focused on clinicians' and reformers' publicly stated perceptions of the influence of patients on prescribing patterns from the 1940s onward.

Fifth, as noted earlier, my focus on self-styled "therapeutic reformers" owes an enormous debt to the work of Harry Marks. The primary lineage of antibiotic reformers described in this book indeed represents a strain of therapeutic conservatism and skepticism in the attempt to use science to rationalize therapeutics. At times such an ethos seems timeless and ideological—and one can trace this ethos from Oliver Wendell Holmes through William Osler to Max Finland—while at other times it appears clearly networked and/or lineal. One of the goals of this book is to situate this particular strain of therapeutic reform during a fruitful, if ultimately contested, period of its existence. Yet "therapeutic reformer" is a loose and loaded term. Who is to say that those advocating for the more widespread use of antibiotics in the 1950s, or for the loosening of FDA regulations today, are not advocating a particular model of antibiotic reform? Dominique Tobbell has detailed the collaborations formed between the pharmaceutical industry and members of academic medicine throughout the same era, as they attempted to forestall what they feared would be further government interference with the practice of medicine.[25] Another goal of this book is thus to demonstrate the possibility of the *collision* of reform efforts, and of particular therapeutic utopias and dystopias, and to show that such conflicting reform efforts must themselves be historically situated to be appreciated and examined.

Sixth, and finally, such antibiotic reform efforts have been mobilized throughout much of the past six decades on behalf of a rational therapeutics, even if the term *rational* has been replaced gradually by seemingly more neutral terms like *prudent*, *appropriate*, or, more generally, *evidence-based*. And if *reformer* is a loaded category, then *rational therapeutics* is even more so. The notion of a rational therapeutics has long conceptual and rhetorical histories, from the ancient Greeks through the complicated therapeutic meanderings of the nineteenth century through early-twentieth-century efforts to adjudicate the merits of therapies emanating from an expanding drug industry.[26] However, the second half of the twentieth century, as the wonder drug revolution was both embraced and questioned, represented a unique era in the history of attempts to define and inculcate an explicitly "rational" therapeutics. Antibiotics served, and continue to serve, as a key focus of such attempts, as the notion of the right drug (whether defined mechanistically or empirically) for the right bug (or disease) for the right patient at the right time at the right price has been juxtaposed with the factors—patients, the marketing efforts of the pharmaceutical industry, and the cultures and structures of practice and drug delivery—that supposedly confound such practice.

This book, at its core, represents an attempt to situate such notions of rational therapeutics, the debates over their construction, and the nature of—and resistance to—the measures proposed and at times taken in their names. Antibiotic researcher and therapeutic rationalist Ernest Jawetz wondered aloud in 1957 whether it was "asking for too much that in a few areas man behave as a rational being," while pharmaceutical executive W. Clarke Wescoe shot back fourteen years later that "'rational' . . . almost invariably serves as the label for the *opinion* of those who happen to be in authority."[27] The tensions between Jawetz's aspirations and Wescoe's resistance—and between data and action, town and gown, autonomy and accountability, education and regulation, and private practice and public health—have played out throughout the antibiotic era. As we contemplate a post-antibiotic era, we would do well to consider their historical foundations.

CHAPTER ONE

The Origins of Antibiotic Reform

Is it asking for too much that in a few areas man behave as a rational being? —ERNEST JAWETZ, "Patient, Doctor, Bug, and Drug" (1957)

THE BIRTH OF ANTIBIOTIC REFORM in the United States predated our present near-exclusive focus on antibiotic resistance, emerging bugs, and receding drugs. Such 1950s reform efforts were not even a response to the well-publicized adverse effects of Parke-Davis's Chloromycetin, though such adverse effects are detailed in this chapter, and certainly contributed to later antibiotic reform efforts in the 1960s and 1970s.[1] Rather, antibiotic reform initially grew out of a series of transformations taking place from the late 1940s through the mid-1950s, as infectious disease experts themselves came into existence as a distinct group of clinician investigators at the interface of receding bugs and emerging drugs. In the context of the revolutionary transformation of pharmaceutical branding attendant to the advent of the broad-spectrum antibiotics, taking place amid a perceived regulatory vacuum in which no single authority was adjudicating the appropriate place of such drugs in the clinician's armamentarium, these recently emergent infectious disease experts would transform into a group of would-be reformers attempting to inculcate a "rational" therapeutics. This chapter sets the stage for such reform, while the next chapter examines the particular form it would take in the latter half of the 1950s.

The Emergence of Infectious Disease Experts

From the earliest discovery of disease-causing microbes in the 1870s, scientists, clinicians, public health departments, and commercial entities sought the means to combat them.[2] During the first decades of the twentieth century, a host of immunological, microbial, and chemotherapeutic agents—from passive serotherapy and active vaccination, through bacteriophage and lactobacillus therapies, to salvarsan and antiparasitics—were administered and advertised in the name of medical science and commerce.[3] The bacteriophage remedies and specific antibodies of Sinclair Lewis's *Arrowsmith* were far from science fiction when the book was published in 1925.[4] And as I have described elsewhere, antipneumococcal serotherapy for pneumonia during this time epitomized the approach of scientific medicine and efforts to ensure its equitable distribution.[5]

Such achievements, though, would be dwarfed by the revolutionary impact of first the sulfa drugs, introduced in the mid-1930s, and then penicillin, introduced in America during World War II. The sulfa drugs were indeed the first "miracle drugs," stimulating equally powerful changes in the organization of pharmaceutical research and the expectations of physicians and their patients. Merck, the first U.S. company to sell the antipneumococcic sulfapyridine, grew dramatically in the decade after the introduction of the sulfa drugs, pouring its profits back into an expanding research base.[6] Penicillin quickly superseded the sulfa drugs in its demonstration of the power of pharmaceutical know-how (and even cooperation) and its impact on patient care, becoming the symbol of modern medicine.[7]

Nevertheless, throughout the first half of the twentieth century, there was no unified American "infectious disease" community or specialty of the kind we know today. Infectious disease laboratories had been established, from William Welch's laboratory at Johns Hopkins University to the Hygienic Laboratory founded at the Marine Hospital on Staten Island in 1887 (which moved to the nation's capital in 1891, evolving into the National Institute of Health in 1930).[8] There was Chicago's Memorial Institute for Infectious Diseases, analogous in many ways to the Rockefeller Institute, and from which the *Journal of Infectious Diseases* would emanate beginning in 1904.[9] And yet the study of "infectious diseases" in the United States remained nosologically and institutionally heterogeneous. Nosologically, American textbooks devoted to "infectious diseases" as a grouped entity would certainly increase in number throughout the first decades of the twentieth century, and infectious diseases considered broadly attracted literary and popular attention; medical approaches to such diseases and groupings as tuberculosis, syphilis, diphtheria, and tropical medicine, however, were more often organized in isolation from one another, in separate textbooks, clinics, and hospital services.[10] Institutionally, the study of infectious diseases would be embedded in public health departments and their associated contagious disease hospitals, the military (from the pioneering efforts of Surgeon General George Sternberg, the "Father of American Bacteriology," onward), pharmaceutical firms, research institutes like the Memorial Institute and the Rockefeller, and a wide array of academic medical departments.[11]

In other words, there were indeed those who focused on infectious diseases from public health bases, as well as those who focused on infectious disease from within academic medicine. Unlike cardiology or gastroenterology, however, there was no clinical "infectious disease" association, no board certification, and no organizing force to bring the field together under a single clinical banner.[12] Today's American Society for Microbiology was initially formed in 1899 as the Society of American Bacteriologists, a self-consciously heterogeneous grouping of fundamental scientists (including industrial and agricultural), public health workers, and the occasional clinician; the members amiably resolved at their 1910 meeting "that, while the purpose of this Society is primarily for the advancement of microbiology as a pure science, this must not be interpreted as excluding papers of applied microbiology."[13] The Infectious Diseases Society of America would not be formed until 1963, in large part owing to the events described in this book's first three chapters, and formal infectious disease subspecialization, under the aegis of the American Board of Internal Medicine, would not begin until 1972.[14]

Still, for all their heterogeneity, when taken as a whole the clinical investigators who studied the effects of antimicrobial agents in the first decades of the twentieth century differentiated themselves from those in related fields in at least one respect. In the context of therapeutic uncertainty, the commercial proliferation of remedies, and hard methodological endpoints (e.g., mortality), they served at the vanguard of those attempting to use the empirical clinical trial to differentiate effective from ineffective remedies. Infectious disease concerns, such as diphtheria in the 1890s and syphilis in the 1920s, had served as the foci of some of the most ambitious "cooperative" investigations (in which large groups of clinicians attempted to standardize and pool their results) of therapeutic remedies in American medicine.[15]

More important to the story at hand, those studying the effects of antimicrobial remedies also led in the conduct of alternate allocation studies, in which alternated patients (e.g., every other patient or patients admitted to a hospital every other day) were allocated to treatment versus control groups, followed by a statistical comparison of treated versus untreated patient outcomes. Indeed, in the half-century between the publication of Johannes Fibiger's famous use of alternate allocation to evaluate anti-diphtheria antitoxin in 1898 and the British Medical Research Council's use of the randomized controlled study to assess the efficacy of streptomycin in tuberculosis in 1948, approximately 130 mentions of alternate allocation studies appeared in the pages of the *Journal of the American Medical Association* (*JAMA*) alone. Of these, 103 concerned antimicrobial agents, ranging from serotherapy to the sulfa drugs to the first antibiotics.[16] When placed against the broad denominator of clinical investigation and care, this represented an admittedly scattered sample.[17] Yet this sample illustrates how combatting infectious diseases lent itself to early attempts to inculcate a rational therapy more generally via the controlled clinical trial in the first half of the twentieth century. And several of the key figures who would shape the antibiotic era—especially Harvard's Max Finland and the University of Illinois's Harry Dowling (who served as Finland's first fellow from 1933 to 1934)—got their methodological feet wet evaluating earlier antimicrobial modalities via such alternate allocation studies throughout this era.[18]

By the 1940s, against this background, those who emerged as infectious disease experts precipitated out of the epidemiological transition (from infectious diseases to cardiovascular disease, cancer, and other chronic diseases as the leading causes of death in this country) on the one hand and the advent of the wonder drug era on the other. As Edward Kass has related, as diphtheria, smallpox, typhoid and their ilk began to recede in America, infectious disease experts emerged as the collective repository of those who could identify and advise regarding the treatment of such ever-rarer clinical entities.[19] At the same time, as additional varieties of first the sulfa drugs and then novel antibiotics emerged in the late 1930s and 1940s, infectious disease experts could advise the government, the pharmaceutical industry, and their colleagues regarding the evaluation and application of such novel remedies.

From an investigative standpoint, and with respect to the government, sulfa drugs and antimalarials were the subjects of intensive military scrutiny throughout World War II (and the Armed Forces Epidemiological Board would remain an important locus of antimicrobial investigation for decades thereafter).[20] And the unpatented (or inexpensively licensed) penicillin and streptomycin, discovered during the war, were subjected to large-scale, government-initiated "cooperative" studies to determine their appropriate application in such conditions as syphilis and tuberculosis.[21] In parallel, after the 1937 Elixir Sulfanilamide fiasco, in which more than 100 children died as a result of taking sulfanilamide (the first of the sulfa drugs) to which diethylene glycol had been added to improve palatability, the Food and

Drug Administration (FDA) was empowered to ensure that novel remedies were indeed "safe" prior to their interstate sale.[22] This would mandate, if not alternate allocation studies, at the very least more intensive clinical investigation of emerging remedies prior to their release and marketing.[23]

In this context, what would come to be known as "clinical pharmacology" —the study of the action of drugs in patients—began to emerge as a formal field, and those studying the effects of antimicrobial remedies would serve as leaders in the field.[24] And by the late 1940s, an increasingly competitive pharmaceutical industry began to test a series of antibiotics—Lederle's Aureomycin (chlortetracycline), Parke-Davis's Chloromycetin (chloramphenicol), and Pfizer's Terramycin (oxytetracycline)—with a seemingly broader range of activity than that exhibited by penicillin and streptomycin. Peter Temin and Walsh McDermott have related how the advent of such patented "broad-spectrum" antibiotics represented a significant change from the wartime clinical investigation of penicillin and streptomycin, as the clinical investigation of antibiotics shifted to a dynamic interaction between academic investigators and the pharmaceutical industry, with direct government involvement in the testing process significantly reduced.[25] As John Swann and Nick Rasmussen have demonstrated, however, characterizing this moment as a powerful break obscures continuities from the interwar era concerning the relationships between such investigators and the pharmaceutical industry.[26]

Rasmussen has described three types of interwar physician collaborator: the "free-lancer," unpaid by industry and instead garnering free drugs to study in anticipation of earning recognition and status; the "efficient," funded by industry to study its remedies (usually with more company control over study design); and the "friendly expert," funded by industry but serving a more directive role, whether in the design of studies or with respect to giving broader advice to industry colleagues. This typology is a useful analytic, but across such a horizontal axis of clinical researcher "types" may be erected a vertical axis indicating investigators' explicit and self-conscious stances as either "enthusiasts" or "skeptics." The tension between therapeutic enthusiasm and skepticism is as old as the American medical profession.[27] And such categories are not absolute, as clinicians can be arrayed across a spectrum and can change stances under varying circumstances. Yet this tension and categorization would play out in particularly visible fashion with the emergence and study of novel antibiotic classes beginning in the late 1940s.

The case of Lederle's Aureomycin makes an illustrative starting point. Streptomycin, discovered by microbiologist Selman Waksman (who coined the term "antibiotic" in 1942, and who would garner the Nobel Prize in Physiology or Medicine in 1952) as a result of extensive soil-sifting and analysis, had served as the prototype for private industry to discover its own antibiotic soil derivatives.[28] Aureomycin, derived from a soil sample sent to Lederle from the University of Missouri in 1945, appeared to possess a seemingly wider range of application than that achieved by penicillin or streptomycin, with in vitro and experimental in vivo activity found not only against both "gram-positive" and "gram-negative" bacteria but seemingly against rickettsial and certain large "viral" organisms as well.[29] By 1948, it was time for clinical investigation in humans.

For remarkable Harlem Hospital surgeon Louis Tompkins Wright, himself a famous "first" in so many ways—the first African American physician appointed to the staff of a New York hospital, the first African American presiding over the medical board of such a hospital, and the first African American admitted to the American College of Surgeons since its founding—Aureomycin became an opportunity to garner further "firsts."[30] With his surgical interest in lymphogranuloma venereum (LGV, a painful sexually transmitted disease) converging with Aureomycin's apparent activity against large viral organisms such as the disease's causative agent, Wright became the self-described "first" to test the drug in human patients, successfully treating a series of LGV-afflicted patients.[31] When Lederle effectively launched the drug at a symposium at the New York Academy of Sciences in July 1948, Wright and his colleagues stated, "Apparently we [are] dealing with an antibiotic that could cure this virus infection [LGV] in humans."[32] In January 1949, Wright gained still more publicity through reportedly saving the life of a British peeress by personally sending Aureomycin to London to "cure" her viral pneumonia, echoing the famous treatment of Winston Churchill's pneumococcal pneumonia with sulfapyridine five years previously.[33] Six months later, in its account of a "possible wonder drug with curative powers beyond the wildest dreams of the average doctor," *Ebony* enthusiastically reflected that "a Negro doctor was the first to use the drug and a Negro patient was the first volunteer cured by its magic. . . . Before Dr. Wright's team made its history-making experiments, no scientist had ever administered aureomycin to humans."[34] Wright may have been a "free-lancer" in Rasmussen's typology, but he was an enthusiastic one, earning additional fame and status through his association with Aureomycin.[35]

At the other end of the spectrum, Max Finland likewise had served as a presenter at the Aureomycin launch; in fact, his group presented two papers, with another presented by Harry Dowling and his group.[36] Finland enjoyed a relationship with Lederle vice president and general manager Wil-

The Aureomycin Research Group at Harlem Hospital, New York City, 1949. Louis Tomkins Wright (1891–1952), leader of the group, is in the front row, center. Box 22, ff 17, Louis Tomkins Wright Papers (H MS c56), Harvard Medical Library in the Francis A. Countway Library of Medicine. Reproduced with the permission of Black Star Photography Archive.

bur Malcolm that had originated in the antipneumococcal antiserum era and that would ultimately span nearly four decades.[37] With his laboratory generously supported by Lederle funding, and serving as adviser to Lederle's developing extramural fellowship program, Finland embodied Rasmussen's notion of the "friendly expert."[38] Yet Finland, carrying forth a stance that dated back to Oliver Wendell Holmes and William Osler, served as a Skeptical Friendly Expert.[39] It was a cautious, conservative therapeutic stance, one he had first developed during the 1930s, when it appeared the sulfa drugs, on the strength of seemingly meager studies, would threaten to supplant the established antipneumococcal antisera in the treatment of pneumonia. At the time, Finland (along with Harlem Hospital's Jesse Bullowa and Bellevue's Norman Plummer) had written: "If evaluation in experimental animals

The Thorndike Memorial Laboratory staff, 1950. Maxwell Finland (1902–1987, *first row, third from the left*) had just been named the associate director of the Thorndike. Director William Castle is seated to Finland's left. Two other future presidents of the Infectious Diseases Society of America are pictured: Edward Kass (*second row, far left*) and George Gee Jackson (*back row, far left*). Finland would ultimately publish over 800 papers in infectious diseases and train over 100 fellows, including seven future presidents of the Infectious Diseases Society of America. (He would serve as its first president in 1963.) Box 29, ff 66, Maxwell Finland Papers (H MS c153), Harvard Medical Library in the Francis A. Countway Library of Medicine.

under standard and controlled conditions is difficult, it is all the more reason for extreme caution in reporting results in human beings. . . . It would be unfortunate if the appearance of a new therapy, no matter how promising, were to cause the abandonment of agents whose curative efficacy and life-saving qualities have become established."[40] His sensibilities honed during years of research concerning the treatment of pneumonia, Finland carried such caution forward to his investigation of antimicrobials more broadly beginning in the 1940s.[41]

When it came to Aureomycin, Finland was certainly optimistic regarding its wide utility, concluding in *JAMA* in 1948 that "aureomycin has definite

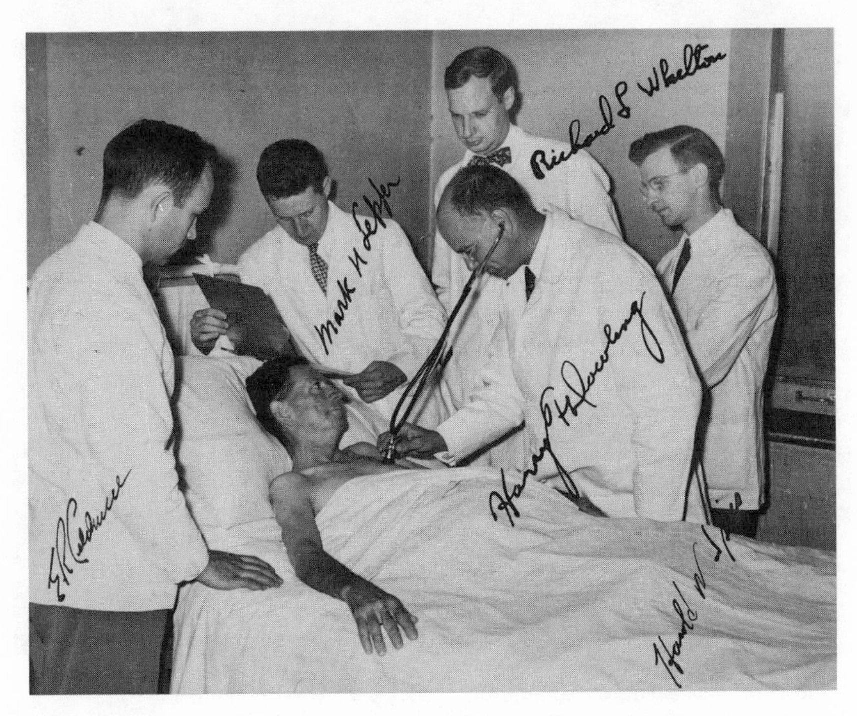

Harry Dowling (1904–2000), with members of his own Aureomycin research team, at Gallinger Municipal Hospital in 1949. Box 1, ff 1, Harry F. Dowling Papers (MS C 372), Modern Manuscripts Collection, History of Medicine Division, National Library of Medicine.

antibacterial activity against many micro-organisms, including coccic and bacillary forms."[42] However, in noting Aureomycin's equivocal effect on several severe gram-negative infections—seemingly beneficial in some cases, and less so in others (and dependent, in Finland's opinion, as much on the severity of illness and host characteristics as on the in vitro efficacy of the antibiotic)—he warned that "the present cases indicate quite clearly the need for caution in interpreting the results of the use of any new therapeutic agent in many infectious diseases."[43] Finland maintained this skeptical, conservative stance in writing to his friends at Lederle. He cautioned Benjamin Carey, Lederle's director of laboratories, that Louis Tomkins Wright's favorable interpretation of results obtained with the use of Aureomycin for ulcerative colitis were "tinged with enthusiasm."[44]

By the mid-1960s, therapeutic conservatives such as Finland and Harry Dowling would find themselves classified as "academic prigs" in an industry representative's own typology, to be contrasted with industry "friends" and "the indifferent."[45] Throughout the 1950s, however, it would be the marketing excesses of industry itself that would unsettle Finland, Dowling, and an emerging faction of infectious disease experts evaluating such remedies. Appearing within the context of ever-growing antibiotic usage, a perceived regulatory vacuum, and the increasingly recognized dangers of antibiotics, the enthusiasm/skepticism dynamic took on much larger significance in the era of the wonder drug.

Broad-Spectrum Branding

The era between 1948 and 1956, which witnessed the advent of the "broad-spectrum" antibiotics, represents a watershed not only in the history of antibiotic usage, but in the history of wonder-drug marketing and sales themselves.[46] The broad-spectrum antibiotics would come along at—indeed, epitomize—the very moment that a shift from "proprietary" to "ethical" pharmaceuticals, increasingly based on unique brand-name drugs offered by each company, impelled manufacturers to expand their sales forces (whose members would "detail" physicians) and refine their campaign strategies.[47] As of 1948, two forms of antibiotics—penicillin and streptomycin—accounted for 99.7 percent of U.S. antibiotic output.[48] Not only did they exhibit apparent limitations to their scope of therapeutic application, but given that neither was exclusively patented by a single company, they also exhibited obvious limitations to their profitability.[49] Pharmaceutical companies stimulated by the World War II efforts to produce penicillin, such as Lederle, Parke-Davis, and Pfizer, thus launched worldwide searches for novel—and hence, patentable—alternatives, ideally with wider therapeutic application.[50]

Lederle struck first, offering Aureomycin for interstate sale in December 1948. Promoted as "the most versatile antibiotic yet discovered, with a wider range of activity than any other known remedy," its sales would epitomize the literal change in scale ushered in by the broad-spectrum antibiotics.[51] Lederle's sales of the patented drug were driven by marketing on a likewise unprecedented scale. As *Fortune* magazine reported a decade later: "The company originated the now familiar 'blitz' technique of marketing new drugs; to introduce its aureomycin, the first broad-spectrum antibiotic, in 1948, for example, Lederle shipped ten carloads of samples to about 142,000 doctors, at an estimated cost for the product alone of about $2 million."[52] Beginning in January 1950, Lederle published and sent to physicians the near-monthly *Aureomycin Digest*, replete with a bibliography that grew from 321 titles in January to 1,859 by December.

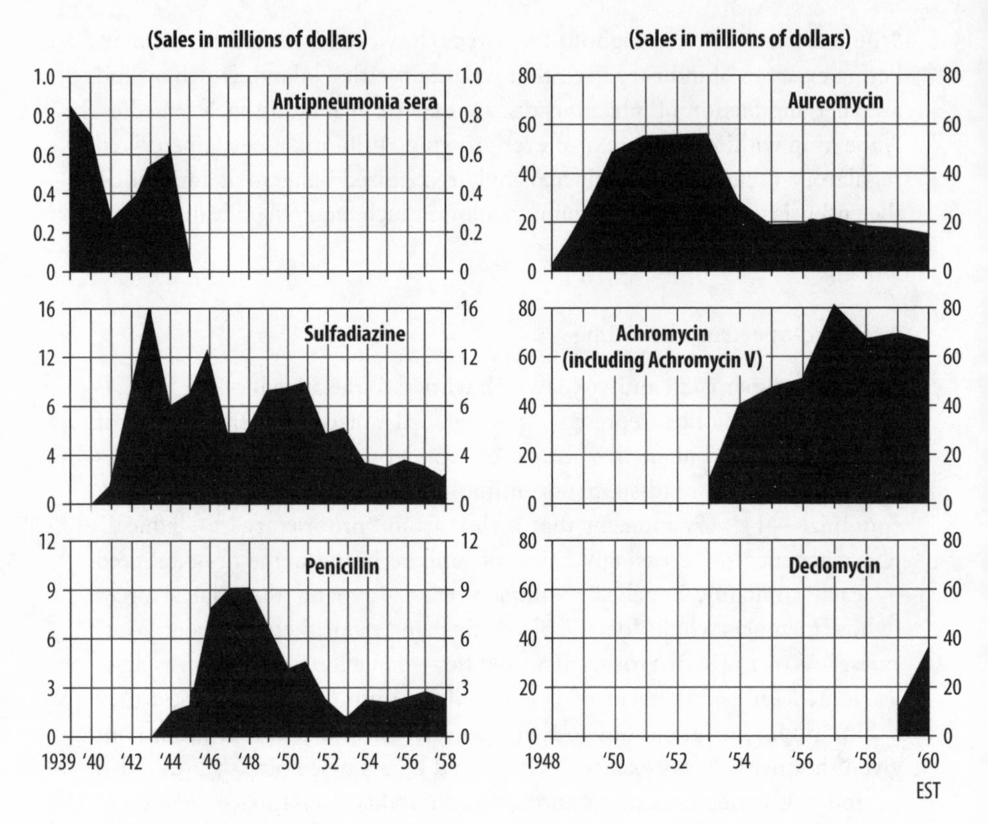

Selected Lederle antimicrobial sales figures, 1939–1960, as presented by Wilbur Malcolm, president of American Cyanamid (parent company of Lederle Laboratories), at the Kefauver hearings in 1960. Ostensibly illustrating that "Lederle's Antibacterial Drugs Have a High Rate of Obsolescence" (hence both demonstrating pharmaceutical risk and justifying drug profit margins), the graph also shows the enormous change in revenue ushered in by the broad-spectrum antibiotics. "Statement of Dr. W. G. Malcolm, President, American Cyanamid Co.," U.S. Congress, Senate Committee on the Judiciary, Subcommittee on Antitrust and Monopoly, *Administered Prices in the Drug Industry*, 86th Congress, 2nd session, 1960, Part 24, pp. 13635, 13637.

Within five years, Lederle would be the first to introduce tetracycline, marketed as Achromycin. And the launch for Achromycin would be supported by its own $2.5 million "blanketing" advertising campaign, entailing over $1 million spent on detailing, $851,000 on direct mail (105 mailings to each physician in the country), $470,000 on medical journal advertis-

"Doctor, why is Aureomycin best?" As Lederle's 1952 advertisement answered: "There are many reasons, which may be summed up by describing this versatile drug as at present the nearest approach to a 'perfect' antibiotic. . . . In many cases, it has proved successful where other antibiotics have failed. Experience has shown that AUREOMYCIN achieves better results with lower dosages . . . without serious side reactions . . . and with less likelihood that disease germs will build up immunity to its health-restoring powers." Author's collection.

ing, and $100,000 on exhibits at medical meetings. "Sales promotion devices," including pens, tongue depressors, and brushes, were to be included as well.[53] Lederle had good reason to be aggressive. In 1955, antibiotics accounted for $81 million out of $123.2 million in Lederle drug sales. And while Lederle's average profitability on its drugs was 20 percent, the profitability on its antibiotics was 35 percent (with profitability on non-antibiotic drugs being just 3%), despite being weighed down by actual losses on its sales of sulfa drugs and penicillin.[54]

Venerable pharmaceutical house Parke-Davis—maker of such drugs as adrenaline, Dilantin, and Benadryl—would follow Lederle in March 1949 with its own broad-spectrum antibiotic, Chloromycetin. Thomas Maeder has exhaustively documented the tremendous (if ultimately, mixed) impact of Chloromycetin on Parke-Davis's fortunes. By 1951, Chloromycetin accounted for $52.4 million out of an industry-leading $138.1 million in Parke-Davis drug sales, and the company would ride the roller-coaster fortunes of Chloromycetin throughout the 1950s.[55] Parke-Davis purported to be above the fray with respect to aggressive marketing, with an executive stating in *Fortune* that it was more like staid Midwestern counterparts such as Eli Lilly, "not like that eastern bunch."[56]

Yet while Parke-Davis's initial approach was indeed more modest than that of Lederle or (as will be seen) Pfizer, the executive nevertheless protested too much.[57] Chloromycetin made a grand entrance in Parke-Davis's house organ, *Therapeutic Notes*, in the summer of 1949, presented in a "Special Chloromyctein Issue" as "originally characterized as the greatest antibiotic since penicillin."[58] By the end of 1950, Chloromycetin carried the moniker of "described as the greatest of antibiotics" in the journal's ads, with articles on Chloromyctein appearing in every issue of *Therapeutic Notes* from the summer of 1949 through the summer of 1951. At the same time, Parke-Davis increased the size of its detailing force by 50 percent, from 500 to 750 men.[59] With respect to diagnostic specificity (or lack thereof), a lead editorial in the summer 1950 "Special Chloromyctein issue" noted the drug's utility "especially in cases where specific laboratory diagnosis is inadvisable or impracticable and in febrile patients clinically classified as 'PUO' or pyrexia of unknown origin."[60]

By 1952, such nonspecific usage would carry still more ominous connotations, as the use of Chloromycetin appeared possibly linked to the rare development of fatal aplastic anemia, or the destruction of the bone marrow's capacity to generate new blood cells. Even as concern about the relationship between Chloromycetin and aplastic anemia developed, no mention of the potential link would appear in the pages of *Therapeutic Notes* throughout

1952.[61] Instead, in another "Special Chloromycetin Issue," in January 1953, an exonerating article on "The Facts on Chloromycetin" noted, "If certain precautionary measures are taken—judicious use, alert observation of the patient, and adequate blood studies—the benefits of treatment with Chloromycetin can be utilized to their fullest extent."[62] As researchers over the next two decades would document, though, and as Maeder has recounted, such "precautionary measures" were seldom taken in the outpatient setting, as marketing features concerning this "greatest of all antibiotics" outweighed more cautionary reminders.

However, no company would be so transformed by the broad-spectrum antibiotics as would Chas. Pfizer & Company, and reciprocally, no company would so alter the marketing of antibiotics—and, critically, stimulate attention given to the marketing of antibiotics—as would Pfizer. The company had been formed in 1849 by two German immigrants, chemist Charles Pfizer and confectioner Charles Erhart. For nearly the first century of its existence, the company did not market pharmaceuticals but instead centered on the production and refinement of such chemicals as citric acid, of which Pfizer was the world's largest producer by World War II.[63] But the fermentation techniques used in the production of citric acid would render Pfizer a leader in American efforts to mass-produce penicillin during World War II, and Pfizer soon became one of the largest producers of the equally nonexclusive streptomycin and dihidrostreptomycin, selling to other companies to distribute.[64]

As prices for these wonder drugs began to plummet, Pfizer president John McKeen uttered to the New York Society of Security Analysts in March 1950 the well-cited warning, "If you want to lose your shirt in a hurry, start making penicillin and streptomycin."[65] Instead, he proposed that "from a profit point of view, the only realistic solution to this (antibiotic) problem lies in the development of new and exclusive antibiotic specialties."[66] Five months previously, Pfizer had applied for a patent on the production of Terramycin. The drug was ostensibly named for the Terre Haute, Indiana, site of its origin in a clod of soil, but conveniently, as depicted in contemporary accounts of Pfizer's efforts, "terra" encapsulated the entire post-streptomycin soil-sifting model of antibiotic discovery, linking globally dispersed soil-gatherers—"from Alaska to Australia, from the banks of the Amazon to the shores of the Ganges, from the swamps of Florida and from the Swiss Alps"—with antiseptically enshrouded mycologists, bacteriologists, and engineers involved in the scanning, screening, testing, and production of novel antibiotic agents.[67]

One month prior to McKeen's speech, Pfizer's board of directors had "voted to change its traditional marketing policy, and commence selling directly to retailers, wholesalers, and hospitals, notwithstanding the fact

A

B

C

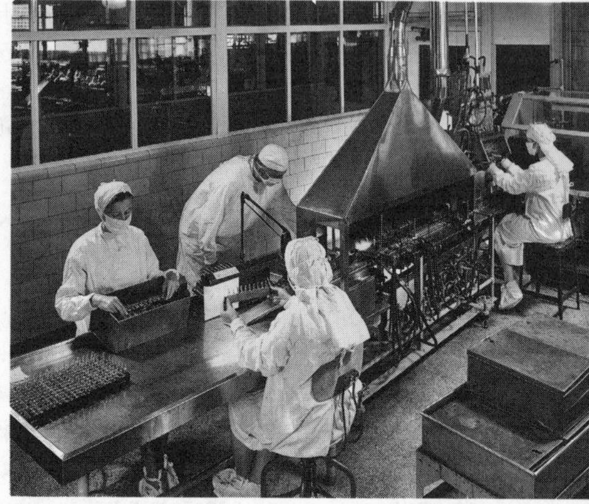

D

Pfizer wasn't the only pharmaceutical company advancing the soil-to-bedside model of antibiotic production. In *A*, we see Eli Lilly's portrayal of Nobelist soil microbiologist Selman Waksman, discoverer of streptomycin. In *B–D* we see photographs of "antibiotic development" commissioned by Parke-Davis, likely at the time of Chloromycetin's launch. Lilly advertisement featuring Selman Waksman located in Box 31, ff 37, Selman A. Waksman Papers, Rutgers University Libraries, Special Collections and University Archives. Parke-Davis images from Box 52, "Photographs [numbers 19, 34, and 45]," Parke, Davis Research Laboratory Records, Archives Center, National Museum of American History, Smithsonian Institution.

that Pfizer had had no experience in selling to these groups and had no sales force adequate to undertake the promotion of ethical drug products."[68] And the initial Terramycin campaign, as well as Pfizer's follow-up efforts, would indelibly alter the entire pharmaceutical marketing landscape, as "Pfizer exploded into the pharmaceutical industry, shattering precedents, records and taboos."[69] Pfizer's initial efforts, after hiring the William Douglas McAdams advertising agency, would serve as the very case study for the Federal Trade Commission's later evaluation of the rise of aggressive antibiotic marketing.[70] McAdams was led by Arthur Sackler, medical psychiatrist, defender of free enterprise and the beneficial role of pharmaceutical marketing in the dissemination of medical information, eventual target of federal inquiry at the Kefauver hearings on the pharmaceutical industry (as detailed in chapter 3), and benefactor of a series of art museums and institutions that bear his name.[71] And between the Herculean efforts of a group of only eight newly hired detail men and a terrifically managed campaign, Pfizer was initially able to "blitz" (as *Business Week* put it) wholesalers and physicians in 1950 despite its recent entrance into pharmaceutical sales, setting "something of a speed record among antibiotics for the trip from laboratory to wide clinical use."[72] By 1952, working with future Nobel Prize–winning chemist Robert Burns Woodward of Harvard, Pfizer had reported the molecular structure of Terramycin, boasting to physicians of its theriac-like qualities as "unusual, differing markedly from the structure of other antibiotics, . . . one of the most complex structures ever to be found in nature."[73] Within eighteen months of Terramycin's release, Pfizer had increased its sales force to 300 detail men (including 70 medical students); by 1957, it would boast 2,000.[74]

Pfizer's John McKeen declared in a presentation to the New York Security Analysts' Society in 1953, "It has been possible to expand the sales of the Antibiotic Division rapidly only by using vigorous promotional techniques."[75] And as fellow medical advertisers would write of impresario Arthur Sackler nearly half a century later:

> No single individual did more to shape the character of medical advertising than the multi-talented Dr. Arthur Sackler. His seminal contribution was bringing the full power of advertising and promotion to pharmaceutical marketing. Until the early 1950s, "ethical drug" promotion had been a low-key "trade" exercise relying principally on sales calls to physicians. The campaign conducted by Sackler's agency, William Douglas McAdams, for the antibiotic Terramycin (Pfizer) forever changed the Rx industry's marketing model. He showed how intelligently written, strikingly illustrated/designed advertising used in volume could greatly influence the success of a product.[76]

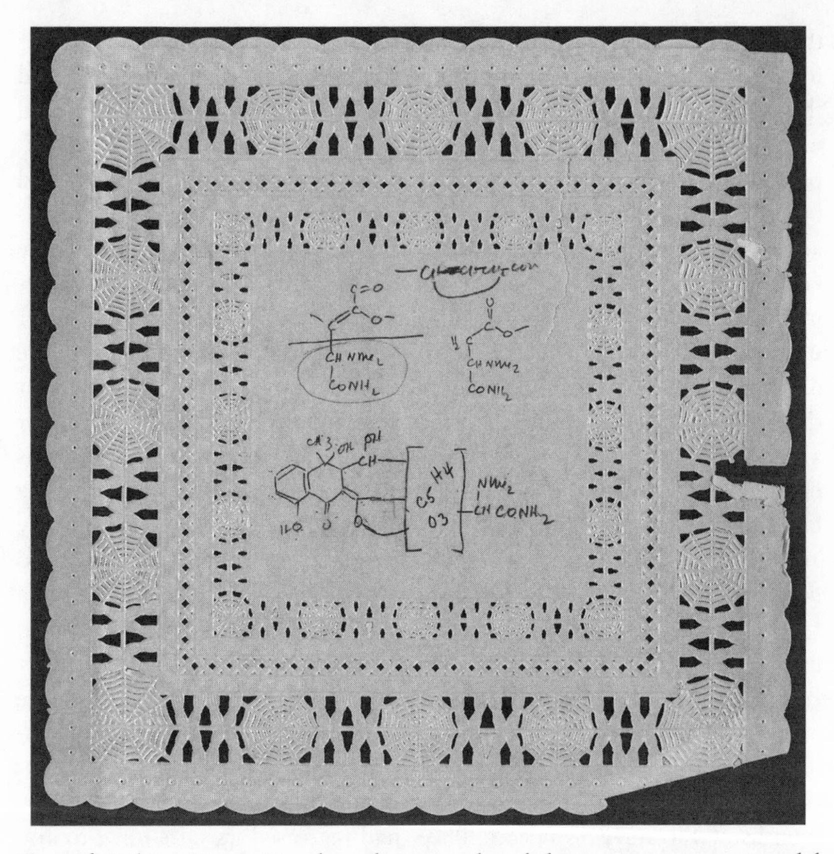

One of Robert Burns Woodward's many hand-drawn attempts to model Terramycin, sketched on a doily in the early 1950s. Pfizer boasted of Terramycin as having "one of the most complex structures ever to be found in nature." Woodward, winner of the Nobel Prize in Chemistry in 1965, has been considered the greatest organic chemist of the twentieth century. Series 68.10, Box 46, Robert Burns Woodward Papers (HUGFP 68), Harvard University Archives.

As the advertisers acknowledged, however, and as I discuss below, the "change in Rx promotion did not sit well with some physicians, academics, and consumerists of the day."[77]

Pfizer and its Terramycin directly changed the nature of journal advertising, as exemplified by advertising in *JAMA*. Antibiotic advertising in *JAMA* prior to 1950 had been modest in extent and typified by a lack of attention to trademarks or brand names.[78] Over the ensuing seven years, though, the brand-name advertising of antibiotics (and especially broad-spectrum antibiotics) in the journal increased seven-fold.[79] Pfizer had already been *JAMA*'s leading broad-spectrum antibiotic advertiser between 1950 and 1952, accounting for 68 percent of the total broad-spectrum advertising pages.[80] But the real transformation took place between June 1952 and April 1956, when nearly every issue of *JAMA* came to physicians with the appropriately named, Terramycin-touting Pfizer house organ, *Spectrum*, tucked within its pages.[81]

The change in promotion applied to drug detailing as well, exemplified still more by Terramycin's successor, Tetracyn (Pfizer's brand of tetracycline).[82] Lederle may have been credited with inventing the "blitz," but Pfizer perfected it, as it were. Detailing was organized with military precision and likewise described in the language of combat. Driving across the South, for example, Pfizer's forces simultaneously "blitzed" Memphis and Birmingham with tetracycline detailing the same day in mid-1955; countermeasures by competitors such as Lederle were described in similar terms ("Holding the line in hospitals is a 24-hour ulcer. By and large we're in good shape, but the sniping is terrific. . . . Men in good fighting shape.").[83] Tactics and armaments included an escalating process of intelligence-gathering from doctors and pharmacists, event sponsorship, and the distribution of reprints, gadgets (regarding one ping-pong ball device, it was noted that "doctors are amused by this device and usually want more of them"), and free samples, lots of free samples.[84]

What would eventually become commonplace was taking shape in real time, driven by the competitive crucible of the broad-spectrum antibiotic market. Even the sales forces themselves could be astounded—and at times taken aback—by the transformation they were engendering. As one Lederle sales manager reported in early 1955:

> In the doctors' offices, sampling has become a scandal. Packages of 16's [describing the lot-size of antibiotic courses included] are given lavishly. As a result of one detail a doctor may find himself with as much as $20 or $30 in material. A strong temptation for him to sell,—he can't possibly use it all. This gives the peddlers another inning. . . . It's fast getting out of hand. The level of business conduct has fallen to that of the cheapest chiseler. Always, there has to be a gimmick.[85]

But Lederle (like Parke-Davis before it) protested too much, attempting to match Pfizer sample-for-sample. As one member of Squibb's sales force lamented later that same year of hospital antibiotic sales: "Neither Pfizer nor Lederle will knowingly be undersold by the other. The end result is a state of chaos where not only does the hospital get the 'deal' but also a load of free ointment, drops, suspension etc. to give one company an 'edge' over the other."[86]

Ascribable to an admittedly unknown degree to such promotion and detailing, antibiotic prescribing exhibited phenomenal growth throughout the 1950s. In 1948, U.S. antibiotic output totaled 240,332 pounds; by 1956, it had increased to 3.081 million pounds.[87] Between 1950 and 1956, consumption nationwide likewise increased from 139.8 to 645.2 metric tons, as antibiotics "ranked as the leading ingredient of prescriptions in virtually every fortnightly survey made by the *American Druggist* magazine since 1952."[88] Moreover, despite the decidedly non-zero-sum situation in which penicillin production generally continued to rise, it was soon surpassed by broad-spectrum production and sales.[89] In terms of production, in 1948, penicillin had represented 64.9 percent of United States output (with streptomycin accounting for 34.8%); by 1956, the broad-spectrum antibiotics (including erythromycin, introduced by Eli Lilly and Abbott in 1952) made up 40.9 percent of such output, and penicillin only 34.4 percent.[90] In terms of sales, as net operating profit margins for broad-spectrum agents ranged from a remarkable 35.1 to 52.1 percent each year, the gap was even wider, with the broad-spectrum agents accounting for $183 million in sales in 1956, as compared to $67 million for penicillin.[91]

Pfizer, the self-described "world's largest producer of antibiotics," was equally transformed by the broad-spectrum revolution.[92] In his 1953 report to the New York Security Analysts' Society, McKeen related that "sales of Pfizer-labelled antibiotics through our own direct sales organizations account for almost 55 per cent of the Company's consolidated volume."[93] By 1956, Pfizer accounted for 26 percent of the United States antibiotic output, rivaled only by Lederle's 23.1 percent.[94] That same year, an industry-leading 39.4 percent of Pfizer's consolidated net domestic sales were still accounted for by antibiotics (compared to an industry average of 13.9% among antibiotic producers), and Terramycin continued to make up more than half of Pfizer's antibiotic sales.[95]

Such numbers, finally, reflected an expansive vision of the role of antibiotics in the growth of industry and the care of patients. With respect to industry, Pfizer's Thomas Winn, sales manager of the antibiotics division, was already emphasizing to retailers in 1950 the degree to which antibiotics would transform traditional drugstore sales:

> All of us can easily visualize the bigness of the so-called oral hygiene products' market—the tooth pastes, tooth powders, liquid dentifrices, tooth

> brushes, mouth washes and gargles, false teeth adhesives, false teeth cleansers, and dental floss, without which we cannot even imagine a drug store— . . . add all of these products together and you have slightly less than half the estimated normal 1950 market for antibiotics, and antibiotics are dispensed only on prescription! Add together *all* the vitamin sales to all outlets (except hospitals) and its [*sic*] not anywhere near as much as the 1950 sales of antibiotics will be. Add together all sales of hormones, botanicals and sulfa drugs—antibiotic sales are larger! . . . Antibiotics, it is estimated by our company's market research people, will account for nearly $1.00 out of every $4.00 spent for prescription drugs this year![96]

Moreover, as antibiotics continued to drive the ethical drug trade, and as pharmaceutical sales continued to dramatically outpace the national growth of disposable income, Winn would feel free to cite former *JAMA* editor Morris Fishbein's prophecy that "a few years hence, with the wider spread of health care and the increased productivity in new research medicines, the drug industry may easily be the number one industry in dollars, in the United States."[97] Pfizer annual sales would rise from $39 million in 1947 to $254 million in 1959.[98]

With respect to medical practice, the broad-spectrum manufacturers expected to have an equal impact. Pfizer's Terramycin campaign, geared toward a wide array of individual specialties, was explicitly framed to demonstrate the wide range of applicability—the broad spectrum of utility—of its wonder drug.[99] In 1950, Pfizer president John McKeen estimated that "antibiotics can be used with fair to excellent results in 30 to 50 per cent of the cases requiring the attention of the physician."[100] Moreover, after enumerating a list of disease categories in which antibiotics held therapeutic utility—including in such "general respiratory infections" as the common cold, for the sake of "preventing respiratory complications"—McKeen reported that looking to the future, "if the gaps in antibiotic effectiveness were filled in, antibiotic usage could easily double or even triple."[101] It was a wide "vista," culminating in the prediction that "it is not impossible that the present broad [*sic*] and energetic search for new antibiotics will lead within the next few years to the discovery of microbial antagonists capable of hobbling all infectious disease."[102]

However, in the view of a key cohort of infectious disease experts, the traditional agencies relied on at least to examine, if not to curtail, such exuberance appeared to have neglected or abdicated their traditional responsibilities. In particular, by the mid-1950s, those looking for advocates for restraint in either the marketing or the prescribing of antibiotics appeared to find them in neither the American Medical Association (AMA) nor the federal government.

The Perceived Regulatory Vacuum

The perceived regulatory vacuum left by the AMA and the federal government would draw out the anxieties of therapeutic skeptics like Max Finland and Harry Dowling throughout the decade. With respect to the AMA, it should be noted that the 1906 federal Pure Food and Drugs Act, passed in the context of alcohol-laden proprietary medications and Upton Sinclair's *The Jungle*, had solely mandated truth in labeling. The 1938 federal Food, Drug, and Cosmetic Act, passed in the wake of the elixir sulfanilamide tragedy, had solely mandated Food and Drug Administration (FDA) proof of drug *safety* prior to release for interstate marketing.[103] Thus, while FDA evaluations of drug safety would be conducted in the implicit context of relative drug efficacy, as of the early 1950s, the primary source of authority in the United States concerning the overt evaluation of drug efficacy lay with the AMA.[104] The AMA's efforts had emerged during the Progressive era, as a self-consciously reforming group of clinicians and pharmacologists had sought to promote a "rational therapeutics" free from commercial influence.[105] Not only had the AMA advocated for the passage of the 1906 federal Pure Food and Drugs Act, but the year before it had formed a Council on Pharmacy and Chemistry, intending, through the pages of *JAMA*, to educate clinicians regarding the therapeutics emerging from an expanding pharmaceutical industry. By 1929, the AMA had granted teeth to the council in the form of a Seal of Acceptance program, according to which only council-approved products could appear in advertisements in *JAMA* and its affiliated specialty organs.[106]

By the late 1940s and early 1950s, however, the AMA was adjusting to the era of the wonder drug. While it continued to maintain its Council on Pharmacy and Chemistry, in 1946 the AMA created a formal Business Division, in part to augment *JAMA*'s marketing efforts. Pfizer's *Spectrum* inserts were a component of a 48.3 percent increase in *JAMA* advertising revenue between 1949 and 1953.[107] And such manifestations reflected still-deeper aspirations by the novel division to curry favor with the pharmaceutical industry.

The early 1950s were notable for the first efforts by sociologists, pharmaceutical companies, and medical journals to understand the various influences on physician prescribing habits.[108] And in September 1952, the AMA contacted the management consulting firm Ben Gaffin and Associates "to uncover fundamental thinking of advertisers and physicians regarding basic advertising problems in general, and the peculiar problems of medical advertising in particular."[109] On the one hand, as reported by Gaffin, it was sincerely felt that such efforts would "enable the AMA to raise the standards of medical advertising" and consequently to enhance the educational function of such advertising.[110] On the other hand, it was well understood at the time that such efforts were likewise directed toward "increasing [the AMA's] advertising revenue."[111] Gaffin's subsequent 1953 "Survey of Advertisers," based on "extensive personal interviews with 92 executives of 78 representative companies," was underscored by two mutually related concerns: first, that "the advertisers, in general, feel that the AMA, especially through the Councils, distrusts them and views them as potential crooks who would become actively unethical if not constantly watched," and second, that in such an environment, the AMA was losing quantifiable advertising revenue to such throwaway journals as *Medical Economics* and *Modern Medicine*.[112]

Consequently, beginning in late 1953, the AMA's Business Division began mailing to pharmaceutical companies (approximately monthly, and continuing through mid-1955) a series of "marketing" reports regarding physician prescribing habits. Topics ranged from the origins of physicians' information regarding new products, to physicians' reactions to direct mailings and detailing efforts and were informed by a Gaffin poll of 500 physicians spread across the country.[113] Explicitly framed as a service to the industry by a mutual partner "in promoting products to physicians," the mailings at the same time emphasized that "in relation to dollars spent, medical journal advertising does double its share in acquainting physicians with new drugs," and that *JAMA* was more extensively read than the nemesis (or model) throwaway journals.[114] Regardless of its impact—and *JAMA* advertising revenue would climb another 19 percent between 1953 and 1955—such courting certainly reflected an empowering of the Business Division of the AMA in relation to the more skeptical members of the Council on Pharmacy and Chemistry.[115] By 1956, as Gaffin presented to the AMA the better-known follow-up survey—"The Fond du Lac Study," perhaps the most ambitious and intensive study of the impact of pharmaceutical marketing conducted to that point—it was clear to the AMA's business representatives that "the purpose of this study was to aid pharmaceutical manufacturers in evaluating promotional methods in selling their products [while at the same time we would] put it to good use as a sales promotional tool for the *Journal of the American Medical Association*."[116]

By that time, moreover, a critical parallel shift in AMA policy—the dropping of its Seal of Acceptance program and the conversion of the regulatory Council on Pharmacy and Chemistry into an educational Council on Drugs—came to most visibly represent such overall movement. In later government investigations and hearings, the dropping of the program would be characterized as a craven capitulation to advertising interests.[117] Yet the move, while not driven solely by a desire to expand the horizon of the council (as the AMA reported in *JAMA* at the time and in later congressional testimony), represented a reaction to a complex of forces all pointing in the same direction.[118] As both AMA representatives and Harry Dowling later identified, by the era of the wonder drug, it had become difficult for the Council on Pharmacy and Chemistry to keep up with a seemingly endless procession of new products.[119] At the same time, not only were the negative advertising revenue implications of the seal program clearly recognized, but the legal implications of the program also had become manifest. After motions were raised before the AMA Board of Trustees in October 1953 for "a proposed reorganization of the method of procedure, now in effect, for the acceptance of products by the Council on Pharmacy and Chemistry," an ad hoc committee was convened to discuss potential revisions to the program.[120] A Chicago legal firm was contacted, and four potential sources of seal-related litigation were identified: patients harmed by council-accepted products, companies harmed by council disapproval, and the possibility that either physicians or the Federal Trade Commission would contest AMA authority in the first place.[121] In this setting, by late 1954, the ad hoc committee and Board of Trustees moved to replace the regulatory seal program with a flexible educational mission grounded in articles in *JAMA* and monographs; and on February 4, 1955, the nearly three-decades-old program officially came to an end.[122]

The forces behind the seal program's disbanding thus may not have been as single-minded as later described in Congress, and the AMA's advertising supervisors were certainly more willing to push back at their industry counterparts than depicted in such testimony.[123] Nevertheless, the process reflected an apparent shift in overall AMA ethos, an overture toward the pharmaceutical industry that would continue throughout the decade. In 1956, a member of the AMA's House of Delegates "introduced a resolution that the AMA officially go on record as condemning certain of the promotional practices of the pharmaceutical industry as being unethical." Not only did the motion fail to carry, but the AMA instead appointed a joint medicine-industry liaison committee to promote still "better understanding between the medical profession and the drug manufacturers."[124] In the spring of 1958, "A Study of Drug Sampling" was convened by the AMA "as a service to the pharmaceutical industry," followed a year later by a Gaffin study concerning "Attitudes of U.S. Physicians toward the American Pharmaceutical Industry."[125]

Finally, and most important to this narrative, whatever the AMA's (or its Business Division's) intentions, in the eyes of an emerging cadre of academic reformers, the dropping of the seal program signaled its abdication of traditional responsibilities and the emergence of a regulatory vacuum.[126] Max Finland reported to Harry Dowling in November 1956, "the fact that the Councils of the A.M.A. have relinquished their position as leaders in this field."[127] Five months later, Finland lamented to the National Academy of Sciences–National Research Council's Thomas Bradley:

> The major agencies dealing with this matter [of drug evaluation] have either relaxed or relinquished their role as guardians of the consuming public and as sources for reliable information to practicing physicians. This role, in turn, has been taken up in a large measure by the drug manufacturers through literature which they create and control and by their direct representatives who pass on infirmation [*sic*] which, while in some instances may be accurate and honest, is usually slanted and often untrue and unreliable as well as misleading.[128]

The American Dental Association, it should be noted, had retained its own Seal of Acceptance program, begun in 1930; but the AMA, it seemed to such emerging reformers as Finland and Dowling, had lost its teeth.

Aside from the AMA, the institution capable of overseeing antibiotic utilization and marketing seemed to be the federal government, whether through the U.S. Public Health Service (USPHS) or the FDA. With respect to the USPHS, such infectious disease experts as Max Finland and Harry Dowling had served in the pre-sulfa era at the forefront of the treatment of pneumonia with antipneumococcal antiserum, an expensive and logistically complicated modality. In this context, between 1937 and 1945, more than two-thirds of the states had boasted USPHS-funded "pneumonia control programs," constructed for the provision and, at times, oversight of anti-pneumococcal therapeutics.[129] But with the widespread advent of first effective anti-pneumococcal sulfa drugs and then penicillin, such pneumonia control programs had collapsed completely by the end of World War II, leaving the antimicrobial treatment of patients by individual clinicians largely beyond the control of the public health system.[130]

If the role of the public health system was notable chiefly for its absence, it would be the actions and entanglements of the FDA's Division of Antibiotics that would incense the emerging reformers. The division was headed by Henry Welch, who had acquired his Ph.D. in bacteriology before coming to the FDA in 1938. By 1945, he had been placed in charge of the Division of Penicillin Control and Immunology to ensure the adequacy of each batch of penicillin produced in the country, and in 1951, the division was expanded into a Division of Antibiotics, with Welch as chief.[131] In 1950, Welch had agreed to helm "an authoritative journal dealing with the subject of antibiotics and chemotherapy" and to co-author a textbook on antibiotics, obtaining permission to do so from his superiors in the federal government.[132] When, by 1952, the publisher of the two ventures (*Antibiotics and Chemotherapy* and *Antibiotic Therapy*) had fallen into "financial difficulties," Welch joined with Félix Martí-Ibañez to form two parent organizations—M.D. Publications and Medical Encyclopedia Inc.—to take over the respective journal and monograph publishing duties.[133]

Martí-Ibañez was in possession of a truly remarkable CV by 1952.[134] Born in Cartagena, Spain, in 1911 and trained as a psychiatrist, he had served as under-secretary of public health and social service in Spain prior to the victory of Franco and the Nationalists in 1939. Wounded during the Spanish Civil War, Martí-Ibañez emigrated to the United States. After serving as medical adviser for overseas sales to Hoffman-LaRoche, he would be appointed "medical director" in charge of Latin American sales at Winthrop (1942–1946) and then Squibb (1946–1950), just as medical marketing began its own postwar revolution.[135] Once he left Squibb, in 1950, the multitasking Martí-Ibañez would soon be seeing patients in New York City in private practice, teaching and writing medical history (by 1956 he would become the director of the history of medicine department at New York Medical College, Flower and Fifth Hospitals), creating his joint publishing enterprise with the nation's reigning antibiotic tsar, and working closely with fellow psychiatrist and "dear friend" Arthur Sackler at William Douglas McAdams, the emerging marketing powerhouse behind Pfizer's Terramycin release.[136]

In mid-1953, Welch proposed to the commissioner of Health, Education, and Welfare (HEW) the convening of an international symposium on antibiotics, the proceedings of which were to be published by M.D. Encyclopedia as *Antibiotics Annual*.[137] The symposia were held each year for the next seven years and were attended by more than 600 scientists and clinicians by 1954.[138] Originally requested by pharmaceutical industry members themselves, they also maintained an appearance of heavy industry representation.[139] And the antibiotic ethos put forth by Welch—and by de facto co-host Martí-Ibañez, especially—would serve as a lightning rod for criticism from the emerging cohort of would-be antibiotic reformers.

Maxwell Finland (*left*), FDA commissioner George Larrick (*center*), and FDA Division of Antibiotics director Henry Welch (*right*) in happier times, at Welch's inaugural antibiotics symposium in 1953. By 1960, owing in large measure to Finland's efforts, all three men would find themselves enmeshed in debate over the adjudication of antibiotics by the FDA. (See chapters 2 and 3.) Box 29, ff 100, Maxwell Finland Papers (H MS c153), Harvard Medical Library in the Francis A. Countway Library of Medicine.

Henry Welch, Sir Howard Florey (who had shared the Nobel Prize in Physiology or Medicine in 1945 with Alexander Fleming and Ernst Chain for his role in the development of penicillin), and Félix Martí-Ibañez, reading greetings from President Dwight D. Eisenhower to participants at Welch's sixth annual antibiotics symposium in 1958. *Antibiotics Annual* (1958–1959): xviii.

In his self-appointed role as official historian-philosopher of the "Era of Antibiotics," Martí-Ibañez would lead off each annual symposium with a grandiose view of the state of antibiotic therapy. Constructing (and at times, mixing) elaborate metaphors and analogies, Martí-Ibañez predicated his speeches on the notion that "in the history of medicine there is perhaps no other event as revolutionary as the discovery of antibiotics," which was causing radical, beneficial changes in "our way of thinking in medicine."[140] It is ironic that Martí-Ibañez has been wrongly cited as one who first called for caution with respect to antibiotic resistance, when instead he consistently called for an ever-escalating race between germs and science.[141] As he stated before the second annual symposium in 1954:

> The physician is beginning to understand that in his struggle against infection two factors are involved: his own science, aided by chemotherapeutic resources, and the microbe itself, and that between the two a fascinating game of chess is being played in which the human body is the chessboard and the life of the patient at stake. As in a ballet *pas de deux*, it is necessary for each participant to anticipate the reactions of the other if the dance is to reach a successful end. The microbe defends itself by resisting the drugs and developing new forms of attack. Hence the aphorism: "Be quick to use a remedy while it is still effective."[142]

He could point forward—in the very presentation (inaugurating the third symposium) in which he was said to have warned against the emergence of resistance—to still "more prophylactic and less therapeutic ends" and, still further off, to prophesy that "by the year 2000 the diseases caused by bacteria, protozoa, and perhaps viruses will be considered by the medical student as exotic curiosities of mere historical interest."[143]

To facilitate such a revolution, a second, parallel revolution was required: one concerning *medical communication*.[144] And in defining such ideal communication, Martí-Ibañez hoped for "the widest diffusion of the maximum amount of practical knowledge on antibiotic medicine to the greatest number of physicians in the shortest time possible."[145] Speed of dissemination, rather than quality control, was the key. According to Martí-Ibañez's vision, while such journals as his were "merchants of light," even "true university chairs on paper," he could look forward to *real* university chairs of antibiotic medicine, as well as an "International Institute of Antibiotics dedicated exclusively to organizing knowledge in this field, a center to act as a universal brain for receiving, classifying, synthesizing, and making available all printed materials on antibiotics."[146]

Who, in such a model, was to finance this institution and foster such communication? The pharmaceutical industry, the "scientific subconscious" of the nation—not an id to be subdued but a beneficent source of creative progress spurred on by free competition.[147] Martí-Ibañez summarized his overall ethos before the fourth annual symposium in 1956:

> "*Public health is purchasable; within natural limitations any community can determine its own death rate.*" So also might we say that medical communication can be financed, and within certain limits each medical community can determine the degree of knowledge its members may attain. Who better than the pharmaceutical industry could organize, coordinate, and integrate on an international scale the vast and increasing knowledge on antibiotics?[148]

However, the irony of transforming Herman Biggs's public-health dictum into a call for industry control of the education of physicians would not be lost on Max Finland, Harry Dowling, and their fellow infectious disease–based reformers who begged to differ.[149]

The Emergence of the "Therapeutic Rationalists" and the Mirage of Reform

Twentieth-century clinical reform has depended recurrently on self-appointed groups of academically based therapeutic reformers arising in particular therapeutic contexts.[150] More than thirty years ago, James Whorton first drew attention to the "Antibiotic Abandon" of 1950s America and the consequent emergence of a cohort of infectious disease–based "therapeutic rationalists."[151] And while I differ with Wharton concerning the primary focus of the rationalists' gaze and activities (as I describe below), I fully agree that by the mid-1950s, the perceived excesses of the wonder drug era and the anxieties they engendered provided the setting in which a cohort of the nation's leading infectious disease experts—self-styled therapeutic skeptics—would coalesce into a reforming group of "therapeutic rationalists."

Immediately prior to this time, as noted above, the nation's infectious disease experts were emerging as the repository of knowledge regarding infectious diseases and the investigators of novel remedies deriving from the pharmaceutical industry. In 1948, Hobart Reimann reported in his fourteenth annual review in the *Archives of Internal Medicine* of significant publications in infectious diseases on the "decline of importance of infectious disease, . . . [as] nothing at present . . . threatens to stem the downward trend to . . . an irreducible minimum [of mortality]."[152] The chief problem engendered by such a seemingly cheerful state—exacerbated, as it were, by the advent and proliferation of the broad-spectrum antibiotics and ever more formulations of each antibiotic—would be deciding which agent to choose in each therapeutic instance. Enter the "new specialist, the antibioticist," happy to help the general practitioner (or, at the hospital level, proposed committees on chemotherapeutics) render such decisions.[153]

Yet the placidity of the era—when the most frightening situation a clinician would apparently find himself in was as a "fielder trying to catch a dozen fly balls [pharmaceutical options] at once"—would be shaken by the mid-1950s by several phenomena: the seemingly unfettered proliferation and marketing of the wonder drugs, the increasingly recognized costs and adverse effects (allergic and otherwise) of antibiotics, novel syndromes of "superinfection" resulting from alterations in patients' microbial florae, apparently correlated diagnostic sloppiness, and finally, increasingly documented antibiotic resistance, particularly among the feared staphylococci (discussed at length in chapter 5).[154] It is in this setting that the refocused attention of the would-be antibioticists—toward a collective attempt to inculcate a "rational" application of the antimicrobials—must be understood.[155]

The term *rational therapy*, as invoked in the first half of the 1950s, chiefly meant achieving an appropriate balance between the risks and benefits of a particular drug (the evolving and increasing rhetorical and regulatory invocation of "rational" and "irrational" therapy by the late 1960s and early 1970s is detailed in chapter 4). At an epidemiological level, Ernest Jawetz of the University of California–San Francisco—who publicly and explicitly invoked the notion of rational therapy more than any other infectious disease–based reformer throughout the 1950s—described the phases of overenthusiasm and disappointment that typically preceded a final equilibrium appropriate to new drugs, when "some sort of balance is reached between the real usefulness and the potential harm of the drug, and a tentative, rational place in the medical armamentarium can be assigned to it, the 'phase of final stabilization.'"[156] However, according to Jawetz, as of 1954, 95–99 percent of contemporary antimicrobials were administered inappropriately, with the era of rational antibiotic usage off on the horizon. Three years later, at Henry Welch's annual antibiotics symposium, Jawetz would still lament, "Is it asking for too much that in a few areas man behave as a rational being?"[157]

At the individual level, others were willing to acknowledge the rationale of those who seemingly overprescribed antibiotics. As Wendell Hall reported at a symposium on the "Abuses of New Therapeutic Agents" at the annual meeting of the Minnesota State Medical Association in May 1952:

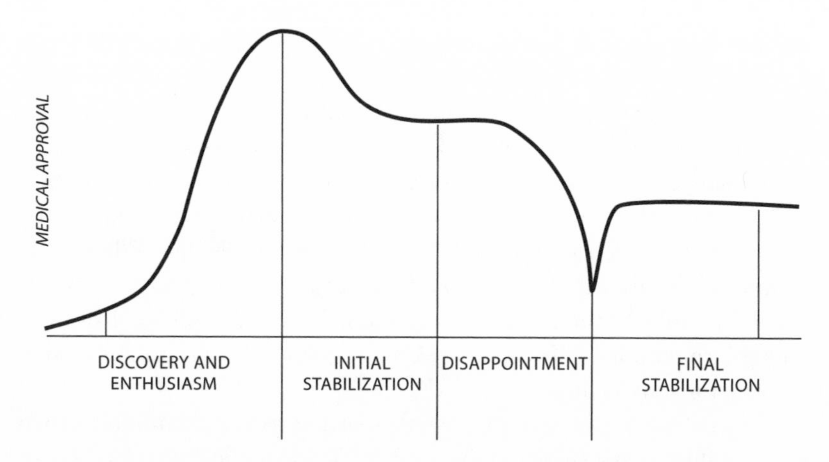

Phases in the Establishment of a New Drug. From Ernest Jawetz, "Infectious Diseases: Problems of Antimicrobial Therapy," *Annual Review of Medicine* 5 (1954): 2. Reproduced with the permission of the *Annual Review of Medicine*.

> We are living in an age in which both physicians and the general public have come to expect miracles of medicine and surgery, leaving little to the powers of Mother Nature. Rare is the physician who has not been urged to prescribe all manner of antibiotics for patients with fever or other equivocal evidence of infection. The pace of living and of medical practice is such that all too often the first question is not "What is the cause?" but rather "Should I take sulfa or penicillin?" There can be little doubt that a considerable part of this attitude is the result of the barrage of publicity concerning the new "wonder drugs" loosed upon the layman and physician by popular "science" writers and pharmaceutical companies. Many a physician is loath to treat a respiratory infection without giving antibiotics lest he be subjected to misinformed relatives of the patient.[158]

Others pragmatically enumerated the pros and cons of "indiscriminate antimicrobial therapy." In the pro column appeared such obvious potential benefits as a more rapid and less eventful recovery if the right bug were fortuitously attacked, the usual lack of toxicity of the medications on an individual usage basis, and the satisfaction of the patients (and of their families) receiving the drugs.[159] However, on the increasingly lengthy con side appeared such concerns as cost, the onset of adverse effects and drug sensitization, the breeding of secondary bacterial and fungal "superinfections" (i.e., occurring after components of the patient's native microbial flora had been wiped out, as most prominently described by Louis Weinstein at the Boston University School of Medicine), the diagnostic sloppiness and potential false sense of security incurred, and the emergence of antibiotic resistance.[160]

The first target of the self-anointed rationalists was the clinical encounter itself, as the reformers invoked a notion of rationality predicated on microbiological and nosological *specificity*: the right drug for the right bug. Here, however, the infectious disease specialists acknowledged the difficulties of the clinical encounter and found themselves divided. While some emphasized the need for antimicrobial therapeutics to be guided by culture data and sensitivity studies, others accused such clinicians of thereby simply "falling into another grievous and expensive error in substituting laboratory tests for sound clinical knowledge and judgment."[161]

Moreover, by the mid-1950s, whether owing to a recognition of their limited ability to assess—let alone alter—individual clinical encounters or to the increasing brazenness of industrial promotion and their own increasing anxiety regarding the regulatory vacuum in which such promotion transpired, the members of the "rationalist" infectious disease vanguard turned their gaze *away* from their fellow practitioners and *toward* the pharmaceutical industry.[162] In parallel, their notion of "rationality" took initial concerns with outcomes as an assumed (even subsumed) starting point and shifted from a focus on nosological and therapeutic *specificity* to a concern with the *sources* of clinicians' therapeutic knowledge and the mechanisms by which antibiotics entered and spread across the marketplace. It was, in effect, a narrowing of their reformist gaze. And thus, despite the fact that the first "Proposed Crusade for the Rational Use of Antibiotics" was declared at the 1954 Antibiotic Symposium in the context of a speech heavily critical of nonspecific telephone prescribing, patient demands, and self-medication with stocked prescriptions, the "Crusade" itself would be coopted by those turning their attention to the marketing efforts of pharmaceutical companies.[163] And there it would remain for more than a decade.

It would be reasonable to expect that Parke-Davis and its Chloromycetin (chloramphenicol)—promoted throughout the decade despite a known, if rare, tendency to induce aplastic anemia—served to catalyze the attack. Harry Dowling (focusing narrowly on the history of the broad-spectrum antibiotics) seemed to imply this connection at the time; James Whorton described the very "therapeutic rationalists" he named as galvanized by chloramphenicol in the 1950s; and Thomas Maeder's thorough and thoughtful biography of chloramphenicol graphically depicts the plight of the victims of chloramphenicol-induced aplastic anemia, as well as the headlines and lawsuits engendered.[164] Yet I would argue that a chloramphenicol-induced 1950s reform effort represents, for the most part, a historical mirage. Instead, the tolerant, even supportive, approach of the rationalists toward chloramphenicol throughout the 1950s further highlights the conception of "rationality" that they would emphasize in the latter half of this decade.[165]

For the infectious disease vanguard of the 1950s, the sin of the effective—if very rarely dangerous—chloramphenicol rested with nonspecifically prescribing clinicians rather than with Parke-Davis, at the very time when such clinicians were generally receding as a site of direct intervention from the purview of the reformers. In April 1953, the University of Washington's James Haviland would remark of the "great 'to do' about chloramphenicol ... [that] every therapeutic agent carries with it a calculated risk," going on to cite a list of additional agents (including the sulfa drugs and streptomycin) surmised to cause aplastic anemia, as well as common side effects associated with other antibiotics.[166] At Henry Welch's annual antibiotic symposium held later that year, Harvard allergist Ethan Allan Brown—perhaps irritated by the degree to which penicillin-induced anaphylaxis had been underreported in the literature—took things further by railing against the negative chloramphenicol publicity, stating that it "caused no one knows how many periods of illness in patients in whom chloramphenicol might have been the antibiotic of choice."[167] Brown had been discussing a talk given by famed Johns Hopkins antimicrobial investigator Perrin Long. And a year later, Long himself would write to Robert Stormant, the secretary of the AMA's Council on Drugs: "It appeared that roughly 20 odd people died as a result of the administration of chloramphenicol. When one considers the total amount of chloramphenicol used, this is an infinitesimal number. More people die each year of anaphylactic reactions of penicillin than they did at the peak of the so-called chloromycetin deaths. More people die of sulfonamide intoxication every year than from chloramphenicol."[168] Über-rationalist Ernest Jawetz summed up such an ethos the following year, arguing that inappropriate chloramphenicol prescribing and inappropriate chloramphenicol avoidance were two sides of the same pill, precisely because they "were based on emotion rather than on sound reason."[169]

Perhaps none of the infectious disease experts would be as influential or consistently supportive of chloramphenicol—when used properly (i.e., specifically)—as Max Finland. Finland was central to the evaluation of chloramphenicol, serving on the National Academy of Sciences–National Research Council chloramphenicol evaluation committees (concerning its relationship to aplastic anemia) in both 1952 and 1960. He would likewise continue to see excesses from both ends of the controversy. In 1952, at the time of the first inquiry, he wrote: "It is quite obvious that part of this furor is induced by the Parke Davis Co. and their representatives who feel that perhaps an analysis of the data or careful collection of the data might reveal that their product is not worse than anyone else's. With respect to some of the chemicals like sulfonamides, I am sure that they would be correct."[170] In 1960, at the time of the second inquiry, he wrote to the National Academy of Sciences–National Research Council's Keith Cannan: "I regret to state that the current reawakening of excitement about this drug is the result of a very personal and emotional interest of one physician who has succeeded in getting a large amount of publicity through an Oakland California newspaper publisher. It now seems that this was all a build-up for a suit against Parke-Davis arising out of a death allegedly caused by chloromycetin."[171]

Finland's course was both representative and directive of the gaze of his infectious disease colleagues.[172] In the early 1950s (as described in chapter 5, in the context of antibiotic resistance), he had indeed focused on individual prescribers. Yet by the mid-1950s, in the setting of a seeming regulatory vacuum, Finland would turn his attention away from concerns with diagnostic specificity and the inappropriate prescribing of *appropriate* drugs, and toward what he perceived as the inappropriate pharmaceutical marketing of *inappropriate* drugs. The contours of over a decade's worth of infectious disease–based therapeutic reform would be set.

The broad-spectrum antibiotics and their enthusiastic investigators and marketers had certainly set Max Finland on edge.[173] An apparent regulatory vacuum, combined with increasing recognition of the dangers of antibiotic use and overuse, had pushed him still further. But the primary focus of Finland and his colleagues' resultant scrutiny would not be chloramphenicol but rather what would come to be called the "fixed-dose combination" antimicrobials, seemingly promoted on the basis of uncontrolled "testimonials." And thus not Parke-Davis and its Chloromycetin but rather Pfizer and its Sigmamycin (introduced in 1956 as a fixed-dose combination of tetracycline and oleandomycin) would bear the brunt of Finland's criticism. Through their efforts, Finland and his colleagues would set in motion a process that would irreversibly alter both antimicrobial therapeutics and the broader course of drug regulation in the United States.

CHAPTER TWO

Antibiotics and the Invocation of the Controlled Clinical Trial

> If this trend is not checked now, the practicing physician will soon be confronted with such a bewildering array of antibiotic combinations supported by multicolored promotional material piling up daily upon his desk that rational chemotherapy will give way to chaos.
>
> —HARRY DOWLING, MAXWELL FINLAND, et al., "The Clinical Use of Antibiotics in Combination" (1957)

THE CHANGING FOCUS of the infectious disease vanguard of the 1950s would have indelible consequences for both the field of antibiotics and the larger domain of pharmacotherapy. With respect to the particular form of "therapeutic rationalism" the antibiotic reformers led by Max Finland would take, the events of the 1950s would lead to more than a decade's worth of focus turned away from individual prescribers and instead on the marketing and development methods of the pharmaceutical industry and the apparent need for measures to counter such influences.

With respect to the larger question of the adjudication and regulation of pharmacotherapeutics, the anxieties and consequent actions of the infectious disease–based reformers and their colleagues in clinical pharmacology would play a pivotal role in the formal regulatory encoding of the controlled clinical trial as the ultimate arbiter of therapeutic efficacy in post–World War II American medicine. Harry Marks has drawn particular attention to the rising prominence of biostatisticians throughout this era.[1] And Daniel Carpenter has traced the normal science process by which the very mandates of the new drug application process led to ever-increasing attention paid to therapeutic efficacy and clinical investigative rigor at the Food and Drug Administration (FDA) throughout this pivotal decade.[2] Clinical pharmacologists such as Max Finland, Harry Dowling, Louis Lasagna, and Louis Goodman certainly show up in their accounts, though in their analyses, conceptual influence derives chiefly from either the biostatisticians (Marks) or the FDA (Carpenter).[3] Yet the anxieties and efforts of the clinical pharmacologists—straddling the domain between practice and the laboratory—would be instrumental in their own right in advancing the essentiality of the controlled clinical trial as a means of taming the therapeutic marketplace and ensuring a "rational" therapeutics in the era of the wonder drug.

Telling the story from this perspective thus not only provides a ground-level view of the first generation of antibiotic reformers led by Max Finland and his colleagues but also helps to further situate the history of the controlled clinical trial itself as a means of ensuring a rational therapeutics. And it would be the fixed-dose combination antibiotics in particular that would serve to catalyze both the anxieties and the efforts of Finland and his colleagues, ultimately contributing to the construction of the Kefauver-Harris amendments in 1962, which mandated FDA proof of drug efficacy via the controlled clinical trial and whose impact extended well beyond the domain of antibiotics alone.

Synergy and Catalysis: The Fixed-Dose Combination Antibiotics

> In the struggle between man and microbes, the latter defend themselves—as we do in war—by developing a growing resistance to the weapons which were lethal at the beginning but to which they may eventually become accustomed. The goal of the researchman, therefore, is to change weapons, to forge a new sword against each new shield used by the bacteria, to create antibiotic agents that, like modern Attilas, will destroy everything they encounter in their paths inside the last strongholds of the microbes.[4]

So spoke Félix Martí-Ibañez, leading off the first Symposium on Antibiotics held in Washington, DC, in October 1953 and urging on the researchers and clinicians assembled to continue to escalate their efforts in the arms race between "man and microbes." Within three years, Henry Welch—Martí-Ibañez's symposium co-sponsor and business partner, as well as director of the FDA's Division of Antibiotics—would echo such arms-race rhetoric in calling for a "third era of antibiotic therapy" marked by the use of "synergistic" fixed-dose combination antimicrobials.[5]

The combination rationale dated back to chemotherapy's origins. Paul Ehrlich, speaking before the Seventeenth International Congress of Medicine in 1913, declared that through combination therapy against infectious microbes, "a simultaneous and varied attack is directed at the parasites, in accordance with the military maxim, march in detachments, fight as a unit."[6] Nearly three decades later, as the sulfa drugs were replacing serum therapy in the treatment of pneumococcal pneumonia in America, Max Finland and Harry Dowling themselves advocated combination serochemotherapy along such lines.[7] Indeed, Colin MacLeod (who would later gain fame for his co-discovery of DNA as the "transforming principle" in bacterial genetics) had shown that combination therapy worked in "synergistic" (i.e., more than additive) fashion in murine tests.[8] And with the widespread use of penicillin, investigators were quick to demonstrate the synergistic effects of combinations of sulfa drugs and penicillin, with clinicians apparently quick to take note.[9]

The introduction of the broad-spectrum antibiotics offered geometrically enhanced opportunities for generating combinations and testing for such synergy, while other theoretical justifications for combination antimicrobial therapy—from increasing the antimicrobial range in mixed infections, through decreasing side effects by lessening the dosages of each agent, to overcoming or delaying the onset of resistance to one agent or another—could be proffered in tandem.[10] And by the early 1950s, the successful combination treatment of tuberculosis, subacute bacterial endocarditis, and brucellosis seemed to validate more broadly the application of such an approach.[11]

But in parallel with the emerging combination ethos appeared concerns regarding its potential dangers. At the annual meeting of the American Medical Association in 1949, a North Carolina clinician warned of the potential threat to diagnostic specificity attendant to empiric combination therapy: "This type of therapy demands a painstaking search for the specific offending organism and in no way justifies 'shotgun' treatment without an etiologic diagnosis."[12] These concerns would be most forcefully advanced by Ernest Jawetz at the University of California–San Francisco, concerned with the apparently widespread flaunting of such a warning.[13] Jawetz, by the early 1950s, was the foremost researcher of combination antibiotic therapy in the country. Yet he would point to two critical stumbling blocks to its easy application, and especially to the application of preassembled fixed-dose combination antibiotics. First, by 1950, it had become apparent that at least in vitro and in murine experimental infections, two or more antibiotics could be *antagonistic* to, rather than synergistic with, one another's activity.[14] By 1951, Harry Dowling and Mark Lepper had found evidence of such antagonism in the *clinical* treatment of pneumococcal mengingitis with a combination of penicillin and Aureomycin.[15] Jawetz, by 1952, had at least offered a tentative framework by which to group the antimicrobials into those generally expected to be synergistic versus those expected to be antagonistic toward one another.[16] What humbled him, however, was his second finding: that the actual outcomes of such antibiotic mixtures were not only bacteria-dependent but even bacterial *strain*-dependent.[17] In other words, one could only apply combination therapy "rationally" when one had isolated the infecting strain at hand and tested the efficacy of the combination in vitro against that strain. Or, as Jawetz warned, "It must be stated with greatest emphasis that fixed 'synergistic' or 'antagonistic' drug pairs do not exist."[18] While developing "simple, practical" laboratory tests to facilitate such application remained an elusive holy grail for the moment, Jawetz and Harry Dowling did not consider that an excuse for what they saw as the "indiscriminate" and "shotgun" application of combinations.[19]

As in the application of antibiotics singly, however, self-styled rationalist emphasis on specificity (as discussed in chapter 1) again yielded to emphasis on pharmaceutical development and overpromotion, this time of fixed-dose combinations of established antibiotics that directly contradicted the rationale of Jawetz's "cumbersome" strain-specific approach.[20] As early as 1951, in his ongoing series of updates on infectious disease, Hobart Reimann had prophetically bridged the concerns, quipping: "It is surprising that, up to now, no commercial interests have exploited the possibility of preparing a combination of antibiotics designed to cure all infectious diseases without the need for bothering with diagnoses. One could even propose a name for it—multimycetin."[21] Reimann would not be aware that Félix Martí-Ibañez, serving at the time as de facto co-host (with FDA Antibiotic Division chief Henry Welch) of the nation's annual antibiotic symposia, would soon suggest something similar to Arthur Sackler at William Douglas McAdams:

> For a house which has a big stake in antibiotics it is particularly important to seek specialties which combine antibiotics with other drugs. Such combinations, if they can be justified medically, are a defense against the current price trend in penicillin and streptomycin. The broad spectrum antibiotics may, eventually suffer the same fate (lack of profit) as the others. (This is the time, therefore, to seek products combining such drugs as Terramycin and Aureomycin with other useful therapeutic agents (with penicillin and/or streptomycin?)[22]

By 1957, Dowling could report at the annual meeting of the AMA on the marketing of *sixty-one* such preparations of fixed-dose combination antibiotics in the United States, with "twenty-nine preparations containing two antibiotics, twenty containing three, eight containing four, and four preparations that contained five antibiotics apiece."[23] It is perhaps fitting that this information came to Dowling by way of Henry Welch, for by 1957, Welch's approach to antibiotic prescribing and promotion—especially as highlighted by his advocacy of fixed-dose combination antimicrobials and as formulated within the conferences and journals he ran with Félix Martí-Ibañez—came to epitomize, for the reformers, the sorry state to which antibiotic development and therapeutics in America had fallen.

Welch's Provocation

The relationship between Henry Welch and prominent members of his journals' editorial boards such as Max Finland and Harry Dowling would prove to be complicated, yet one that grew steadily more acrimonious throughout the mid-1950s. The worst that Finland could speak of Welch in 1954 was that the nation's antibiotic chief ran a dull and diluted symposium in need of better editing.[24] Within a year, Finland and Dowling had downgraded the industry-dependent symposium to "poor" and were engaged in a "running campaign" with Welch to raise its standards.[25] Even in late 1955, though, Finland still remained hopeful that he could influence Welch and was deferential in his direct dealings with him.[26]

Sides began to take firmer shape in early 1956, when Welch informed his editorial board members that Martí-Ibañez and he were transforming *Antibiotic Medicine* into a free-circulation journal (i.e., supported solely by pharmaceutical advertising) to be renamed *Antibiotic Medicine and Clinical Therapy*.[27] For Finland and Dowling, already piqued about rarely being consulted on the papers that actually appeared in the journal, this was a telling step toward moving the journal closer to the domain of throwaway pharmaceutical house organs.[28] An uncontrolled case series appearing that summer in the journal and supporting the admixture of oxytetracycline and "stress formula" vitamins—seemingly "a naked plug for a Pfizer commercial"—removed the last of their restraints, and they planned to create a "united front" with respect to upgrading the level of antibiotic discourse, despite their still worrying it would be "upsetting to Henry."[29]

Welch, however, would beat them to the punch, delivering the opening speech to the participants at the fourth annual symposium on antibiotics in October 1956, with Finland and Dowling in attendance. Praising, rather than condemning, the "cradle to grave" application of antibiotics, Welch pointed out that "the worldwide interest demonstrated in this field can only be appreciated when consideration is given to the tremendous dollar expansion of this young industry during the past 13 years."[30] He acknowledged the subsequent emergence of antibiotic resistance but, rather than highlight such a process as an inducement to restraint, he held it up as a justification for the "trend to rational combined therapy, particularly with synergistic combinations."[31] Having coopted the rationalists' very vocabularly, he remarked of the conference's forthcoming industry-supported papers on fixed-dose combination antibiotics:

> These presentations and others indicate a distinct trend toward combined therapy, not an old fashioned "shotgun" approach, but a calculated rational method of attacking the problem of resistant organisms. It is quite possible that we are now in a third era of antibiotic therapy; the first being the era of the narrow-spectrum antibiotics, penicillin and streptomycin; the second, the era of broad-spectrum therapy; the third being an era of combined therapy where combinations of chemotherapeutic agents, particularly synergistic ones, will be customarily used.[32]

The obvious beneficiary of Welch's invocation was Pfizer, whose Sigmamycin, a fixed-dose combination of tetracycline and oleandomycin, was to be released in November with the obvious imprimatur of the nation's antibiotic tsar.[33] In that same month's issue of *Antibiotic Medicine and Clinical Therapy*, Welch noted that "it is in [general practice] that combined therapy using synergistic combinations of antibiotics will find its greatest usefulness" and continued by remarking that "the combination of oleandomycin and tetracycline [i.e., Sigmamycin] stands out in contrast to the other combinations referred to, because of the synergism demonstrated by this combination."[34] In the same issue, Pfizer included a four-page advertisement extolling the virtues of its wonder drug, bringing "new certainty to antibiotic therapy *particularly* for the 90% of patients treated in home or office."[35] It was to be a world of empiric, nonspecific therapy that flew in the face of findings of antibiotic antagonism and strain specificity.

However, in flaunting the logic and, as it were, the therapeutic morality of the rationalists, Welch and Pfizer had at last provided the "spark" for a full-fledged conflagration of single-minded reforming activity.[36] In the first paper on Sigmamycin offered at the 1956 symposium, not only were no controls used in the study but "pharyngitis, bronchitis, and other upper respiratory infections accounted for the majority of patients treated."[37] Perceiving the fixed-dose combination antibiotics as an exemplar for inappropriate pharmaceutical production, marketing, and usage, Dowling and Finland escalated each other's anxieties and those of their colleagues.

Four months before the symposium, and in the very first pages of the

Maxwell Finland (*far left*) and Henry Welch (*center*) at Welch's fateful 1956 antibiotics symposium, where Welch proclaimed the fixed-dose combination antibiotics as epitomizing "rational" medicine and a "third era" of antibiotic therapy. Finland was less amused than he appears in this photograph; written on the back of the picture is the inscription "Finland [and] Welch and his harem." Box 29, ff 102, Maxwell Finland Papers (H MS c153), Harvard Medical Library in the Francis A. Countway Library of Medicine.

newly christened *Antibiotic Medicine and Clinical Therapy*, Dowling had written ominously of a "new generation" of physicians—of clinicians dulled to diagnostic nuance and clinical investigators failing to ensure proper controls—growing up amid such therapeutic proliferation.[38] And to further situate Dowling's thinking, it's worth noting that it would be in the aftermath of the symposium that he would research and then deliver before the American Medical Association (in June 1957) his famous "Twixt the Cup and the Lip" address concerning the debatable output of the American pharmaceutical industry. Two months before Dowling's talk, Vance Packard's *The Hidden Persuaders* had appeared, examining the degree to which advertisers and those familiar with "motivational research" could use the findings of depth psychology to manipulate consumers of all persuasions (by August the book would be atop the *New York Times* best-seller list).[39] And in his talk, Dowling anxiously noted that "the techniques that had been used so successfully in the advertising of soaps and tooth pastes and of cigarettes, automobiles, and whiskey could be used as successfully to advertise drugs to doctors."[40] As he warned both the pharmaceutical and medical professions: "With the inevitable disillusionment that comes with the failure of each useless modification to make any advance, the pharmaceutical industry will lose its prestige and with this will lose its financial backing. It will fall, and the medical profession will be dragged down with it."[41]

In the wake of the 1956 symposium, Finland and Dowling met and communicated with their network of colleagues, concluding that "the time [was] ripe for a concerted movement," and that "a lobby would have to be formed . . . to compete with the large lobby of the pharmaceutical manufacturers."[42] With the AMA and FDA having apparently failed to provide appropriate leadership, the informal meetings of the infectious disease experts would become a nexus for the construction of manifestoes and the plan to self-consciously flood medical journals with counters to the promiscuous promotion and usage of fixed-dose combination antimicrobials.[43]

In a long, unsigned *New England Journal of Medicine* editorial that November, Finland began with a swipe at Welch, who had lent his "considerable prestige and authority" to "dignifying and encouraging the trend" toward the unproven usage of the fixed-dose combination drugs.[44] And given how high the stakes were, Finland urged physicians, "who readily fall prey to the psychologic warfare embodied in the high-powered technics of modern sales promotion," to be on their guard. In a subsequent joint manifesto, co-authored by nine leading infectious disease specialists (including Dowling, Finland, and Jawetz) and appearing in the *Archives of Internal Medicine*, fixed-dose combination antibiotics were attacked first with the usual concerns: the failure to ensure adequate therapy (through failure to ensure adequate dosing of any single component), the possibility for increased and confusing toxicity, and the potential for engendering resistance. But the final concern of the authors again embodied their underlying rationale: "If this trend is not checked now, the practicing physician will soon be confronted with such a bewildering array of antibiotic combinations supported by multicolored promotional material piling up daily upon his desk that rational chemotherapy will give way to chaos."[45]

Such statements were only preludes to a still more contentious dialogue between Finland and Welch, which appeared (to Welch and Martí-Ibañez's credit) in the pages of *Antibiotic Medicine and Clinical Therapy* in January 1957. Finland's skepticism concerning premature claims of therapeutic effectiveness had deep roots. As noted in chapter 1, he had prominently cautioned physicians in the late 1930s against abandoning antipneumococcal antiserum in favor of the yet incompletely evaluated sulfapyridine in treating pneumonia. Nearly two decades later, with the additional input and imprimatur of eighteen of his infectious disease colleagues in an article entitled "The New Antibiotic Era: For Better or for Worse?" he issued a similar warning with respect to the fixed-dose combination antibiotics: "We would be remiss in our duties as physicians, teachers, and investigators were we to encourage, adopt, and recommend the use of new agents that we cannot consider to be as good as, or no better than, those previously shown to be good, even if they are legally certified."[46] Tellingly, Finland focused on the *nature* of the studies supporting the seemingly imminent expansion of fixed-dose combination antimicrobial usage:

> Much of the clinical information presented [at the 1956 symposium] had the sound of testimonials rather than carefully collected and adequately documented scientific data. To be sure, properly conducted clinical studies may, in the future, support the claims and justify the enthusiasm for these or other combinations of antimicrobial agents, but it is incumbent upon those of us who are intimately concerned with the welfare of our patients to wait until such data are presented before we accept and acclaim any new agents or special formulations and recommend them for general use, particularly in view of their great potential for harm when they are used extensively and indiscriminately.[47]

An emerging hierarchy of medical knowledge was being formulated, with the "testimonial" at one end and the controlled clinical trial at the other. Before examining Welch's and Martí-Ibañez's responses, as well as Finland's subsequent advocacy for the controlled clinical trial, it is useful to situate and unpack the use of the term *testimonial*.

The Invocation of the Testimonial

Indeed, testimonial, serving as the antipode of the controlled clinical trial, was a deeply loaded concept. By alluding to it, Finland invoked several centuries' worth of linked moral and epistemological claims to medical professionalism and knowledge.

The testimonial—the naked and dogmatic claim of therapeutic efficacy for an agent on the strength of the word of either patient or practitioner—had likely served both "regular" and "irregular" (with such distinctions admittedly blurred) medical practitioners alike in the seventeenth and eighteenth centuries.[48] By the early nineteenth century, however, as the medical profession on both sides of the Atlantic attempted to draw a circle around itself and define its norms, the testimonial as a form of promotion was given dubious moral connotations, linked to such professional outliers as oculists and chiropodists, as well as homeopaths and proponents of other "systems" of knowledge. The orthodox medical profession was to avoid lending its name to the claims of such outliers, as when the *Lancet* asked in 1836: "Is it not full time for medical men of *character* and *reputation* to separate themselves from advertising Quacks? Can the profession be respected when the highest names are either lent or sold to the lowest of adventurers?"[49] The situation was considered still worse in the United States, and the fledgling American Medical Association felt compelled to emphasize in its initial Code of Ethics that it was "reprehensible for physicians to give certificates attesting the efficacy of patent or secret medicines, or in any way to promote the use of them."[50]

By the latter half of the nineteenth century, such connotations had been still further displaced onto the purveyors of "patent" (or "proprietary") remedies—composed of oft-secret, and oft-multiple, ingredients—promising broad utility and launched by large-scale advertising in both newspapers and medical journals.[51] As the process was described at the time:

> If by hook or crook a man who can write "M.D." after his name can be cajoled or bribed to sign a testimonial in favor of a patent nostrum, that testimonial will be kept on duty until the paper fades and the ink has lost its blackness. There are various ways in which to get these endorsements, the most business-like being to buy them outright, provided it does not cost too much. But there is another and better scheme than this. It is to get the doctors to use the nostrum for a while; get them to prescribe and endorse it by writing it up in the medical journals and by talking about it in medical societies. To do this takes time, energy and no small amount of tact in advertising.[52]

The problem was that practitioners appeared to believe such testimonials, which pointed to a second connotation of the testimonial, namely, its epistemological underpinnings. It was one thing for the laity or the clergy to render testimonials or to take them on faith, but as consumers of medical knowledge, physicians were supposed to know better.[53] Here, critiques of the testimonial embodied a deeper therapeutic skepticism and conservatism regarding orthodox and unorthodox medicine alike, which Oliver Wendell Holmes most clearly brought forth by drawing attention to both the *vis medicatrix naturae* and the placebo effect in the healing of patients deemed cured by particular remedies or therapeutic systems.[54] As Holmes stated in his address on "Currents and Counter-Currents in Medical Science" in

1860, twenty pages before his more famous quip regarding the sinking of the materia medica to the bottom of the sea:

> Part of the blame of over-medication must, I fear, rest with the profession, for yielding to the tendency to self-delusion, which seems inseparable from the practice of the art of healing. . . . The inveterate logical errors to which physicians have always been subject are chiefly these:—The mode of inference per *enumerationem simplicem*, in scholastic phrase; that is, counting only their favorable cases. . . . The *post hoc ergo propter hoc* error: he got well after taking my medicine; therefore in consequence of taking it.[55]

By 1892, the same year that William Osler (who had taken on Holmes's mantle as the chief therapeutic skeptic in North American medicine) published his *Principles and Practice of Medicine*, a Missouri clinician transferred existing concerns regarding the products of the patent-remedy industry onto those of the increasingly prominent "ethical" pharmaceutical industry.[56] As he intoned, presaging Finland's later warnings, "new remedies, as discovered by chemists, are finding a too prominent place among the therapeutic agents of the average physician, to the exclusion of the older, time-tried remedies."[57] Propelled by free samples and methods "as unscientific as the methods of the ordinary quack medicine vendor," the process of marketing these drugs was further driven by a perversion of the emerging science of clinical investigation, through the incorporation of the testimonial, as "a certain class of physicians are anxious for self-aggrandizement, that they love to see their names appear in print."[58] The outcome placed "the profession in very nearly the same position as the laity in general hold to the advertised patent medicine."[59]

Testimonial thus had become a fighting word by the turn of the century among regulars in their claims to therapeutic efficacy, as well as the antithesis of an envisioned science of clinical investigation.[60] In 1905, Henry Loomis, professor of therapeutics and clinical medicine at Cornell University, declared, "There are so many trained clinicians in the hospitals who have excellent facilities for intelligent tests—so many pharmacologists with elaborate equipment, devoting all their time to laboratory investigations, that there is no excuse for the manufacturer seeking testimonials from those who are not in a position properly to test the effects and efficacy of so-called remedies."[61] By that time, Samuel Hopkins Adams had commenced his muckraking series in *Colliers* against the "Great American Fraud" of the patent medicine industry, inspiring the passage of the 1906 federal Pure Food and Drugs Act and including numerous invocations of the "testimonial," in both its moral and epistemological aspects, in his writings.[62] The AMA had likewise removed patent medicine advertising from *JAMA* and commenced what would be a half-century-long campaign against testimonials in both guises.[63] Its anti-quackery Propaganda Department was formed in 1905 (to be renamed the Bureau of Investigation in 1925), with director Arthur J. Cramp eventually amassing a "Testimonial File" of 13,000 American and 3,000 foreign physicians who had proffered testimonials for proprietary remedies.[64] And the AMA formed its Council on Pharmacy and Chemistry the same year, attempting to inculcate an explicitly "rational therapeutics" amid the proliferation of emerging proprietary and ethical drugs.[65] Both divisions would keep the testimonial before *JAMA*'s audiences over the ensuing decades, with literally hundreds of invocations in their respective "Propaganda for Reform" and "Reports of the Council" articles.[66]

By the time Max Finland was beginning to conduct his own clinical investigations in the late 1920s and early 1930s, the testimonial was being methodologically differentiated from the controlled clinical trial in particular by several established investigators. The University of Minnesota's Harold Diehl, in his study of the treatment of the symptoms of the common cold with opiates (ranging from codeine to morphine), alternated afflicted college students to either active or lactose-tablet placebo groups. Finding the opiates more effective than the lactose tablets, yet finding individual respondents commenting favorably on their apparent response to the lactose tablets, Diehl acknowledged that "some of the comments that were made on the report cards by persons who had received only lactose tablets would serve admirably as testimonials concerning the value of these tablets."[67] Three years later, Finland's former colleague Wheelan Sutliff (then at the University of Chicago), writing more generally on the nature of the therapeutic trial, again contrasted the testimonial with the controlled trial: "Remedies are sometimes used and advocated by lay persons, medical sects, and by physicians, in the treatment of ill-defined groups of diseases, with no rational explanation of the results expected. Data may be presented in support of such claims that have no more value than a testimonial."[68]

In the summer directly before Henry Welch's presentation at his 1956 annual antibiotics symposium, *Science* had opened its June 15th issue with an editorial, "Test by Testimonial," concerning the utility of a marketed battery additive and contrasting the opinions of individual real-world users with the findings of centralized testing by the National Bureau of Standards. The editorial, sticking to methodological concerns, concluded that the Federal Trade Commission's inclusion of real-world opinions in deciding to drop charges of false advertising against the additive's producers "put too much reliance on the testimony of users, which has, it seems to us, doubtful validity as evidence, or, in legal terms, little probative value."[69]

Finland had read the article, and in his telling, testimonials took on not just such methodological connotations but also their more pernicious commercial connotations.[70] Testimonials, as portrayed by Finland and his colleagues, thus exemplified the problematic relationship between shoddy investigators, profit-minded pharmaceutical companies, coopted journals, a complacent (if not complicit) FDA, and duped physicians and patients.[71] For Finland, the state of Martí-Ibañez's publishing empire—encouraging "testimonial papers . . . so as to get bibliographies for the drug companies to quote"—and his model of antibiotic justification and communication represented a dystopic vision of the future of antibiotic affairs.[72] As such, although the FDA still held the legal responsibility to certify novel agents, "clinical investigators and authors of medical and scientific publications [had] the duty to protect the medical profession and the public against the abuse of preliminary scientific information and against the improper and premature exploitation of conclusions based on inadequate data."[73]

The Counter-Invocation of the Controlled Clinical Trial

In their responses published alongside Finland and his colleagues' "New Antibiotic Era," Welch and Martí-Ibañez only antagonized further the coalescing reformers.[74] At a superficial level, not only did Welch (falsely) accuse Finland of raising the concern with the fixed-dose combination antimicrobials "*for the first time* in the present . . . editorial," he also positioned his own announcement at the antibiotic symposium of an imminent third era of antibiotic therapy as "part of the stream of events" that had begun with Finland and Dowling's own delimited advocacy of combination antibiotic therapy (for tuberculosis, subacute bacterial endocarditis, and brucellosis).[75] More fundamentally, Welch attacked the necessary specificity of the rationalists as an ivory tower luxury, as the wide spectrum afforded by combination therapy appeared exactly suited for diagnostically uncertain situations: "In relation to an overwhelming majority of patients being treated in general practice or institutions *without* adequate bacteriologic facilities the question of the preference for variable combinations becomes entirely academic. It would be impossible to impose upon the practicing physician (who administers the greatest amount of antibiotic therapy) the routine restriction of bacteriologic procedures."[76]

Welch's words were surely distasteful to those like Ernest Jawetz who focused on diagnostic specificity. But the chief blow to Finland was to be leveled by Martí-Ibañez, who, in introducing Finland and Welch (in a statement attached to the paper in the guise of a neutral referee), remarked: "The final verdict on the value of a new drug or a new therapy usually comes from one dependable source: the whole body of practicing physicians whose daily clinical experiences extend over many patients treated in actual conditions of practice over considerable periods of time. Medical practice itself provides the sole and ultimate verdict on the true value of a drug, a therapy, or a medical theory."[77] This was therapeutics by vote, a democracy seemingly corrupted by the manipulation of the voters. It was now up to the reformers to set forth their alternative ideal of "rational" therapeutic evaluation and dissemination.

Writing to Selman Waksman after the antibiotics symposium, Finland had already warned, "Something drastic will have to be done to stop this before long."[78] He chose as his "drastic" means of combating the testimonials seemingly encouraged by Welch and Martí-Ibañez a tool he had in fact publicly criticized only a decade and a half before: the controlled clinical trial. The conventional history of the rise of the controlled clinical trial in the twentieth century focused on such methodological advances in the pre–World War II era as the scattered usage of controlled studies (in which the outcome of a group of patients receiving a studied treatment would be compared, in the same time and place, to that of an otherwise presumably similar group of patients not receiving the treatment), followed, in the period after World War II, by the advent of randomization of treatment assignments and the blinding of patients and/or evaluators to such assignments. In such a reading, Austin Bradford Hill's randomized British Medical Research Council study of streptomycin for tuberculosis, published in 1948, served as a self-evident inflection point.[79] Scholarship over the past few decades has deepened our understanding of the evolution and rise to prominence of the controlled clinical trial in the twentieth century. Attention has been drawn, for example, to the increasing post–World War II role of biostatisticians who advanced their commonsense notions of randomization and blinding and established formal relationships with clinicians running therapeutic trials; to the evolving, negotiated lineages of protocols underlying such trials; and to the role of the FDA itself in advancing the conceptual evolution of the controlled trial.[80]

But while the postwar rise of the controlled clinical trial certainly derived from such methodological and institutional evolution, it also represented a reformist response to the flooding of the therapeutic marketplace. Appreciating this response has necessitated, as Harry Marks has described, a "social history of mistrust."[81] And a close examination of Max Finland's own transformation from one of the foremost critics of the application of the controlled clinical trial in the 1940s to one of its foremost proponents by the end of the 1950s provides a key perspective concerning the rise of the controlled clinical trial in American medicine. In this "Marks-ist" reading concerning the taming of the therapeutic marketplace, the clinical trial represented not only methodological, but also social, evolution, and its foil was thus not the case study or series but rather the *testimonial.*

The alternate allocation trial (in which every other patient in a series receives a studied remedy) had gained prominence throughout the first four decades of the twentieth century as a means of judging therapeutic efficacy.[82] The treatment of pneumonia had served as a focus for such studies; and Max Finland himself had conducted one of the several hospital-based, alternate allocation studies of antiserum for the treatment of pneumococcal pneumonia in the 1920s and early 1930s that had been used to justify the use of the costly specific.[83] However, with the advent of the sulfa drugs in the late 1930s and early 1940s, Finland had become a prominent critic of the emerging clinical trial methodology. Concerned with comparisons between the chemotherapeutic and combined serochemotherapeutic approach to pneumonia—especially as represented by a 1941 study of 607 alternated patients at Bellevue Hospital that apparently disproved the utility of combination therapy over sulfa monotherapy—Finland felt that the controlled clinical trial largely remained a heuristic ideal, impeached in practice by the vagaries of chance, the influence of unconscious selection bias in the assignment of patients, and the shallowness of the clinical and laboratory underpinnings of most such studies.[84] Well before the phrase "garbage in, garbage out" entered the lexicon, Finland had reminded his colleagues that statistics could represent mere window dressing in the absence of otherwise convincing data.[85] Instead, for example, in "proving" the efficacy of sulfadiazine for the treatment of pneumonia in 1941, Finland had turned to a case series of its activity in 446 consecutive patients and to a deterministic explanation of its individual treatment failures.[86]

Within twelve years, on the eve of the expansion of the fixed-dose combination antimicrobials, it appears at first glance that Finland had shifted his position in the aftermath of Bradford Hill and streptomycin. He posed a question regarding strep throat to his former fellow John Dingle, then a leading infectious disease authority at Western Reserve University: "I wonder if you know of any studies that have been done on the treatment of exudative pharyngitis with antimicrobial agents in a well controlled manner; that is, were any cases studied in which there was a randomization of therapy and then a classification on the basis of whether or not streptococci were isolated and if so whether antibodies were developed."[87] Yet when Dingle responded with a list of randomized studies and an assertion of the double blinding involved, Finland revealed his actual prioritization: "What I was really interested in when I inquired . . . was whether a careful study had been done with a correlation of bacteriology and the results of antibiotic therapy to determine any possible effect of antibiotics in those cases in which a streptococcal etiology could not be proved."[88] In other words, for Finland, "well-controlled" trials of antimicrobial agents could signify the integrity of the clinical, microbiological, and immunological aspects of such studies as much as it could the formal conduct of randomization and blinding.

Finland's musings on the treatment of "atypical" pneumonia further orient his early 1950s clinical epidemiological coordinates. Typical pneumonia was characterized by the acute onset of shaking chills and high fever, and largely ascribed to the pneumococcus; atypical pneumonia—increasingly prevalent (and certainly increasingly recognized) throughout the 1940s—was clinically characterized by a slow onset and shallower extent of symptoms. It also remained of unclear etiology by the late 1940s. The sulfa drugs and penicillin had proved largely ineffective against such atypical, or "viral" (as it was sometimes and somewhat loosely termed), pneumonia. With the advent of the broad-spectrum antibiotics, however, and their apparent efficacy against rickettsia and such agents at the borderland between viruses and bacteria as those causing psittacosis, lymphogranuloma venereum, and trachoma (what would later be considered Chlamydial infections), their potential for use in the treatment of atypical pneumonia led to intense clinical investigation by the late 1940s and early 1950s.[89] And in the absence of a pathognomonic bacteriological or immunological characterization of atypical pneumonia, and in the presence of a heterogeneity of illnesses typically lumped under the term, an ideal tension was set in place for the various interpretations of the "controlled clinical evaluation" of its treatment.

In the view of Gordon Meiklejohn (then at UCSF), primary atypical pneumonia—given the prevalence of the disease, the low fatality rates entailed (hence rendering it defensible to withhold therapy from controls), and the easily defined use of defervescence as a clinical end point—seemed "exceptionally well suited for controlled evaluation of therapeutic agents."[90] A parallel conversation between Finland and Emanuel Schoenbach of Johns Hopkins, however, revealed deeper uncertainties. Schoenbach, having apparently successfully treated five cases of atypical pneumonia with Aureomycin by October 1948, first hoped to obtain "a much larger [case] series . . . to get away from chance," but soon turned to the ideal of a controlled trial in which patients were alternately allocated to treatment versus control groups so as to answer the question.[91] Here, however, Schoenbach found himself confronted by the heterogeneity of the disease manifestations as presented in the hospital setting. Two options seemed available: to convince John Dingle, then director of the U.S. Army's Commission on Acute Respiratory Diseases (part of the Armed Forces Epidemiological Board), to conduct an alternate allocation trial amid the more homogeneous conditions seen in military camps, or to attempt to validate further his own hospital studies (even if still a case series) by determining the exact microbiological etiology or immunological characteristics of each case.[92] Finland, running his own extensive case series at Boston City Hospital, was far more interested in the latter approach.[93]

This is not to say that Finland denied the epistemic utility of the controlled clinical trial when the conditions were right.[94] But he believed the value of a controlled clinical trial could *only* be determined in the context of the disease being treated, the conditions under which it was being studied, and the degree to which such context was appreciated by the investigators. Harry Marks has noted how in the post–World War II era, "therapeutic reformers increasingly put their faith in methods, not men."[95] For Finland, however, the rubric of the controlled clinical trial could mask all sorts of messiness below; the key to clinical investigative purity had to remain the thoroughness of the men and women conducting the trial, not the formal methodology of the controlled trial itself.

In 1952, Finland wrote a review article for the *New England Journal of Medicine* on the known "controlled" (his quotes) evaluations of the treatment of atypical pneumonia (of then still-unknown etiology) with broad-spectrum agents.[96] Finding no single controlled study conclusive, he would ultimately lump together "observations" (including his and others' own case study findings) with alternate controlled studies by Meikeljohn and others—all of which entailed the attempt at clinical, radiological, and laboratory verification of atypical pneumonia activity—in stating his support for the "strong and acceptable evidence of a favorable action" of Aureomycin in particular.[97] He did report on an alternate allocation study from Johns Hopkins that seemed to disprove the utility of Aureomycin in this setting, but impeaching the study's clinical and laboratory bases, and even invoking the possibility of "unconscious selection" in the assignment of patients, Finland dismissed its results.[98] In other words, amid the nosologically murky setting within which atypical pneumonia proliferated, even clinical "observations," when backed up by laboratory justification, could trump the results of supposedly "controlled" trials.

John Dingle, who had echoed the Hopkins null findings in his own "controlled studies" of the treatment of atypical pneumonia with Aureomycin, begged to differ with Finland. Reviewing a Finland manuscript on pneumonia submitted to the Veterans Affairs technical bulletins, exactly two months after Finland's *NEJM* article was published, Dingle commented: "The one place where Max and I probably differ greatest, is in the interpretation of the data regarding the efficacy of Aureomycin. . . . Max's own study, as well as most of the others, has been uncontrolled. . . . Max and I have argued this frequently in the past and it probably comes down to the degree of confidence that one places in clinical impressions vs. the results of controlled studies."[99] Of all the critiques he received on the paper, Finland appreciated Dingle's the most, yet he remained relatively unmoved, continuing to champion the qualifications and efforts of investigators over the particular methodologies utilized.[100] Five years later, at the same time that he was galvanizing a reform effort against the fixed-dose combination antibiotics, he would still privately declare of the treatment of atypical pneumonia with Aureomycin: "Some of the so-called 'controlled studies' really are not—and have added only confusion by the air of scientific authority contributed to [*sic*] the statistics, when the basic material is at fault. . . . It will not be possible to prove it conclusively until the etiological agents are well-defined and the results correlated with carefully documented etiologic studies."[101]

However, for its potential utility in taming the market, Finland would be willing to choke down his private anxieties and publicly invoke the controlled clinical trial. Once again, the course taken with over-the-counter remedies preceded that taken with prescribed entities. Between 1950 and 1951, Finland had produced a series of unsigned *New England Journal of Medicine* editorials excoriating the publicity given (especially by *Reader's Digest*) to the "cure" of the common cold by over-the-counter antihistamines. Displaying his enduring skepticism and conservatism, Finland critiqued the advocacy of such remedies "before their usefulness has been adequately substantiated and ill effects determined by the usual slow but more thorough and critical methods of clinical investigation."[102] He instead pointed to their need to be studied—and the likelihood of their utility to be impeached—via the admittedly logistically more difficult controlled clinical trial.

Regarding the fixed-dose combination antibiotics, by the conclusion of the 1956 antibiotic symposium, the commercial proliferation of such remedies, buttressed by testimonials, would lead Finland, Dowling, and their reforming colleagues to likewise rally behind the very "controlled clinical trial" over which Finland still privately struggled. As Finland concluded his editorial on "Antibiotic Combinations" in the *New England Journal of Medicine* in November of that year: "Physicians should be particularly careful in accepting drugs purely on the basis of the manufacturer's evidence or on the basis of testimonials provided to the manufacturer. They should demand clear, unbiased, well studied and adequately controlled evidence produced and interpreted by reliable observers. *Caveat emptor*!"[103] And as he and his colleagues more soberly concluded their joint *Archives* manifesto, "It is our firm conviction that the promotion and sale of such combinations should be discouraged until and unless adequate data from controlled clinical investigation justifies this practice, and then only with respect to definite combinations for specific purposes."[104]

Dowling and Jawetz would repeat such proclamations in subsequent editorials, but Finland would carry the banner.[105] Angrily confiding to England's Lawrence Garrod, regarding Sigmamycin, that "the barrage of the manufacturer will have to be counteracted to safeguard the public," Finland unleashed a series of caustic editorials attempting to do just that.[106] In August 1957, he lamented in an unsigned editorial in the pages of the *New England Journal of Medicine*: "The admonition to defer acceptance of the claims of the manufacturers until they are confirmed by reliable and unbiased reports from other laboratories and supported by controlled clinical trials rather than by mere testimonials may have been completely submerged in the deluge of advertising matter that followed."[107] Reflecting the same anxieties that suffused Harry Dowling's "Twixt the Cup and the Lip," he continued:

> The nearly daily mailings to physicians and the repetitive advertisements in medical journals that were willing to carry them, by reiterating the same claims, may have had the intended effect of dulling the senses and perception of the great majority of physicians regarding the underlying truths that they were obscuring. These advertisements have undoubtedly reached more physicians and have been seen and perhaps even read by many more than have read the editorial columns of this journal.[108]

At stake, for Finland, Dowling, and their colleagues, was the very foundation of a rational therapeutics. The nonspecific therapeutics Welch and Pfizer advocated, "far from being a rational approach to antibiotic therapy . . . represent[ed] a recognition of the old authoritarianism and its attendant nostrums as the guiding force in medicine" (note the explicit reference to the days of proprietary combinations).[109] As a counter to such commercial education, Finland proposed "careful scientific work done under controlled conditions [which] usually requires much more time and effort than the writing of advertising copy, which at best exaggerates the truth and all too often only distorts it and renders it meaningless when inferior wares are being peddled."[110]

Finland's depiction of the "controlled" clinical trial may at times have merged components of formal control groups with the expectation of simple laboratory rigor and precision. Yet his rhetoric would complement that emanating from such biostatisticians as Austin Bradford Hill and Donald Mainland—advocating the common-sense merits of the randomized controlled trial in particular—and from such fellow members of the emerging field of clinical pharmacology as Johns Hopkins's Louis Lasagna.[111] Clinical pharmacology, as noted in chapter 1, emerged as an attempt to professionalize and standardize the testing in humans of the drugs pouring out of the pharmaceutical industry and would be led throughout the 1950s and 1960s by such pioneers as Finland, Louis Goodman, Louis Lasagna, and Walter Modell.[112] Lasagna had trained under Henry Beecher at Massachusetts General Hospital in the early 1950s, examining both the nature and the clinical investigative implications of the placebo effect.[113] By 1954, he had been recruited to Johns Hopkins to establish and chair the nation's first division of clinical pharmacology, and he would be a forceful advocate (for a time) for the use of the controlled clinical trial to rationalize therapy. He presented the controlled clinical trial in less moralistic terms than did Finland. As Lasagna opened his influential 1955 presentation on "The Controlled Clinical Trial: Theory and Practice":

> The doctor of today is under constant bombardment with claims as to the efficacy of drugs, old and new. It is difficult, if not impossible, to read a journal, attend a medical meeting, or open the morning mail without encountering a new report on the success or failure of some medication. The clinician who would avoid nihilistic rejection or trusting acceptance of all such claims, or capricious decisions as to their merits, is well advised to adopt a yardstick, a set of criteria, that will improve his chances of making sound evaluations. The investigator is equally well advised to do so.[114]

Yet while avoiding an attack on the pharmaceutical industry itself, Lasagna could insert *some* moralizing into his methodological discussions, directing his aim at this point at both clinicians and clinical investigators. While he derided the "Chi Square Cavalier" and the "Placebo Pusher" who felt that all new interventions required proof of efficacy via the controlled clinical trial, he couldn't help but draw attention to the "Professional Illiterate, or Statistical Hayseed":

> He sneers at the "Ivory Tower Boys" who "just don't realize what the practice of medicine and the problems of clinical research amount to." . . . His prevailing philosophy ranges from absolute nihilism to rapid and complete ingestion (sans mastication) of new claims, since he lacks any standard of reference for evaluation other than an ill-defined, ectoplasmic link with the Unknown referred to as My Clinical Judgment, or My Past Experience.[115]

It was an image that could easily complement Finland's less subtle barbs directed at the pharmaceutical industry, and through the efforts of clinical pharmacologists like Finland, Dowling, and Lasagna, such rhetoric would by the late 1950s begin to reverberate well beyond the halls of academia.

Expanding Audiences

This expansion was no accident. By the late 1950s, such infectious disease–based, self-styled therapeutic rationalists as Max Finland and Harry Dowling were speaking to multiple and often interconnected audiences, as their concerns generalized beyond the specific topic of the fixed-dose combination antibiotics to a general anxiety regarding the state of pharmacotherapeutics in the United States. Finland, while representative of this vanguard, remained unique in his influence, essentially delegated to voice the concerns shared by many of his colleagues.

Finland's primary constituencies remained his fellow clinicians and clinical investigators. He and his Boston City Hospital colleagues served at the forefront of the investigation of fixed-dose combination antibiotics, finding them no more efficacious than their constituent components, and this research lent heft to the therapeutic moralizing of his editorials.[116] However, Finland employed his most moralizing tone in criticizing the general conduct of clinical investigation. As the country's foremost investigator of antibiotics and a mentor to generations of nationally dispersed researchers, Finland expended a good deal of energy in attempting to instill a particular ethics of investigator-pharmaceutical company relations in particular.[117]

This is far from stating that Finland was above smooth-talking pharmaceutical counterparts when he needed supplies of drugs, bugs, equipment, or funding for investigation, and he could offer his own potential "stamp of approval" in exchange to such companies.[118] Warwick Anderson has described the "gift relationship" that clinician-scientist Carleton Gadjusek developed in the course of investigating kuru in Papua New Guinea in the late 1950s and early 1960s, as the controversial future Nobelist exchanged tissue and information among necessary collaborators. During the same years that Gadjusek was navigating the tricky terrain of Papua New Guinea and associated laboratories, Finland was navigating the likewise difficult terrain between the clinic and industry.[119] He was not alone. As esteemed University of Minnesota antibiotic investigator Wesley Spink reflected on the early years of infectious disease–based clinical pharmacology:

> After World War II the financial support of medical science research was severely handicapped by a lack of funds. We were fortunate in the contributions made to the University of Minnesota by pharmaceutical firms to support our work on antibiotics and brucellosis. . . . For a number of years key support for research including salaries of young resident physicians, technicians, supplies and secretarial aid came from this source. . . . The relation was a very fruitful one. Subsequently considerable aid came from the grants of the National Institutes of Health.[120]

And while Spink was quick to insert that "in no way did we compromise our integrity in favor of private industry," his relations with industry, too, throughout the 1950s were characterized by a carefully negotiated exchange of drugs, bugs, funding, expertise, and official support.[121]

But the self-reflective Finland also attempted to provide an explicitly "moral" investigative road map.[122] At Welch and Martí-Ibañez's seventh (and final) annual antibiotic symposium in 1959, Finland spoke on "The Challenge of New Drugs to the Clinical Investigator," attempting to bolster support for clinical pharmacology. One can imagine his frame of mind as he followed Martí-Ibañez and Austin Smith, president of the Pharmaceutical Manufacturers Association (PMA, the predecessor to PhRMA). Martí-Ibañez noted that "at these annual symposia we have listened to the love affairs between antibiotics and germs, with the latter behaving like frivolous girls, now being vulnerable, now resisting the antibiotic onslaught," while Smith remarked of novel pharmaceutical "advertising methods" and "educational techniques" that "some may be considered unnecessary but who is to say what is necessary and what is unnecessary if it means the lessening of suffering or the saving of even one life."[123] Finland, by contrast, described how clinical pharmacologists, working amid a perceived regulatory vacuum regarding therapeutic efficacy, were to play a critical role in parsing signal from noise, wheat from chaff:

> The clinical investigator has an important duty in eliminating the useless and ineffective drugs, in exposing unwarranted claims, in accelerating the obsolescence of the useless or the less active and more toxic agents when

> better ones become available. Here is perhaps the clinical investigator's most thankless and most difficult task, for his is merely a moral duty to his patients, to his colleagues, and to their patients, and this may clash with the more personal vested interests of manufacturers and their associates. . . . Few are the clinicians and investigators with [the] courage and willingness to undertake this task.[124]

And not surprisingly, such trained investigators were to be contrasted with their antithesis, "the tool of the drug house," who could only provide a "'test by testimonial' and yield a type of clinical report that some less critical manufacturers are eager to exploit."[125]

Finland, in essence, envisioned a world of neutral clinical pharmacological experts advising federal agencies and pharmaceutical companies alike on matters of therapeutic efficacy, and he would communicate this vision directly to those audiences.[126] His suggestions anticipated the increased use of extramural advisory committees by the FDA throughout the 1960s (with Finland, Dowling, and Wesley Spink serving in prominent roles on such committees).[127] But in Finland's particular ideal at the time, such clinical experts remained still more independent from the FDA itself. As he wrote to Thomas Bradley of the National Academy of Sciences–National Research Council in early 1957, two options seemed available: either "a revision of the Foods, Drugs and Cosmetics Act" ensuring proper evaluation of drugs prior to their marketing or a "second and from our point of view a more desirable approach . . . [namely, that] some group of disinterested scientists and investigators . . . would arrange for proper and unbiased studies of new drugs and report both to the manufacturers and to the medical profession."[128]

Finland was eager to share his vision with allies in the pharmaceutical industry as well. For better or for worse, he clearly split in his musings concerning industry. He had worked with Lederle Laboratories for over three decades and, as mentioned in chapter 1, had supported and been supported by Wilbur Malcolm (by 1957, the president of Lederle's parent company, American Cyanamid) since Malcolm had been a young advocate of antipneumococcal antiserum in the 1930s. And if Lederle represented for Finland the respectable old guard, then the role of villain was to be played by Pfizer, with its infusion of more aggressive salesmanship into the pharmaceutical industry since the early 1950s.[129] Lederle's leadership—struggling to keep up with Pfizer's aggressive salesmanship—was happy to accentuate such splitting. And at times, it appears they felt a general leveling of the playing field through an industry code of ethics—or even external muckraking—might work to their advantage. As Lyman Duncan, general manager of Lederle Laboratories, wrote to Lederle's director of sales during the broad-spectrum detailing wars: "These are certainly times that try men's souls—particularly those of the Lederle Field force. Not to be pollyannish about it but I wonder if our competitors' scandalous excesses may not begin to help us. . . . If there is an article or even some minor publicity about it, I will bet a number of the boys will run for cover."[130]

Lederle's efforts and stance at the very least complicate a simple narrative of pharmaceutical industry versus academic reformer, and the company's leadership appears to have had a real impact in perpetuating Finland's therapeutic anxieties—and fomenting his demonization of Pfizer—throughout the 1950s. Terramycin, with its initial clinical proof apparently predicated on research conducted by "thousands of 'new clinical investigators' who were showered with the drug" and its promotion embodied by a seemingly biased monograph published by Martí-Ibañez, had certainly agitated Finland.[131] But it was Sigmamycin that had hardened such a Lederle-Pfizer dichotomization, and Finland was concerned that in a competitive race to the bottom, the established firms were being forced to stay abreast of the aggressive newcomers.[132]

In this setting, Finland expressed his concern to Lederle's Malcolm that "when a manufacturer tries to take a temporary advantage over his competitors by promoting products that are not well tested and for which the claims have not already been well substantiated, the physician will be mis[led] and the patient will suffer, directly or indirectly, but in the long run it will also reflect back on the character of the producer and promoter."[133] Finland hoped that instead of spending their money on "massive 'saturation' type of advertising" for poor products, and consequently wasting the time of those who would have to disprove the utility of these drugs, the "ethical pharmaceutical manufacturers would be willing to pool a large proportion of the funds that individual manufacturers now spend on development of new drugs and use them to support activities in a number of clinical centers which are prepared to make reliable observations under properly controlled conditions in the interest of obtaining acceptable appraisals of new products."[134]

By the middle of 1958, however, it had become clear to Finland that despite his considerable reputation and vision of clinical pharmacology, he didn't have much of a stick to back up his recommendations. Even the seemingly "ethical" drug companies, in his view, saw no need for drastic changes to the status quo.[135] Anxiety mounting, Finland reported his criticisms of antibiotic marketing to the vice-president and medical director of Bristol Laboratories:

> I am most unhappy in having been put in the position of the acidulous critic and that some of my good friends happen to be the butt of the criticism. However, I sincerely hope that you and your colleagues appreciate that I would not be doing this were it not for my firm conviction that such criticism is justified and desirable and that if it is taken seriously, the entire pharmaceutical industry as well as its individual members will be better off in the long run. The industry could thus stave off further criticism that would eventually and most certainly lead to stricter and less desirable controls from government.[136]

Finland's conversations regarding the adjudication of therapeutic efficacy, the nature of federal drug regulation, and the role of the FDA in adjudicating therapeutic efficacy in particular, were not taking place in a vacuum.[137] The Federal Trade Commission, motivated by the high and seemingly uniform cost of multiple antibiotics, had been interested in their production and marketing since the early 1950s and launched an inquiry that would by 1958 detail, along with certain patent irregularities, the dramatic expansion in antibiotic marketing over the course of the prior decade.[138] By that time, Rep. John Blatnik of Minnesota had initiated what would ultimately be a four-part series of congressional investigations into "False and Misleading Advertising" within the House's Committee on Government Operations.[139] The hearings, running intermittently from July 1957 through July 1958, provided a forum to explore the relative roles of the Federal Trade Commission, the Post Office Department, and the Food and Drug Administration in regulating (and especially policing) therapeutic claims.[140] They also provided a forum for Harry Dowling to serve as witness and expand on the notions he'd introduced in his "Twixt the Cup and the Lip" talk before the AMA. Echoing Finland's therapeutic skepticism and advocacy for "adequate" controlled clinical studies of the "efficacy and toxicity" of drugs prior to their marketing, Dowling likewise voiced a preference for industry self-regulation over FDA control.[141] The threat of such control clearly hung in the air, however. Dowling specifically spoke during hearings on the "minor tranquilizers," and that same year witnessed the publication of *Pax*, a novel written by two former Pfizer public relations staffers, which featured a Senate inquiry into a new minor tranquilizer ("Pax") produced by Raven, a fictional, Pfizer-resembling "World's Largest Producer of Vermifuges."[142]

Members of the FDA's own Bureau of Medicine increasingly were attempting to insert overt considerations of efficacy into drug evaluations throughout the 1950s.[143] And as the medical director of the Bureau of Medicine, Albert ("Jerry") Holland, stated at the Blatnik hearings: "One cannot make a judgment as to safety . . . without some fair and adequate consideration of efficacy. It is rare that we permit a new drug application to become effective or feel we can't stop it from becoming effective knowing full well that the drug has absolutely no therapeutic value."[144] But Holland did hedge, admitting that "in those instances [in which an inefficacious drug is approved] I think it is both morally and legally incumbent upon us to then legally proceed under another section of the law and attempt to take the manufacturer to task on a charge of misbranding or false and misleading advertising statements."[145]

Leading clinical pharmacologists noted the ambiguity. In 1956, critiquing the apparent wasteland of existing over-the-counter antihistamine sleep aids, Lasagna concluded:

> It is unfortunate that the Food and Drug Administration cannot take regulatory action against a drug whose efficacy for the conditions for which it is offered has not been demonstrated. . . . It appears to be an accepted fact that toxicity and efficacy cannot each be considered separately, *in vacuo* as it were. Is it such a horrendous leap, therefore, from the present position to one where respectable, reliable evidence about a drug's therapeutic potency must be provided by the manufacturer before a compound is released? Must we put the onus of proof on the Food and Drug Administration, insisting that they build up a strong case *against* a worthless drug already on the market, in order to restrain unprincipled advertising?[146]

In April 1959, as the lone academic at the first annual meeting of the Pharmaceutical Manufacturers Association in Boca Raton, Lasagna elaborated further, extending his analysis to all therapeutics covered by the FDA. He offered friendly advice to the pharmaceutical industry regarding its occasionally "sleazy advertising" and promotional claims, revealing that he still felt that he "would much prefer to see reasonable, workable, effective censorship come from within industry than to see unreasonable censorship imposed on it from without."[147] Even while advocating for such intraprofessional self-control, however, he explained the rationale for permitting the FDA to formally pass on drug efficacy:

> Still another inadequacy is the present Food and Drug Act. The foundation of this act—the principle that the government should concern itself with the

safety of drugs and not their efficacy—is nonsensical. . . . Since one cannot make an intelligent decision about safety without knowledge of efficacy, it follows that the FDA *has* to deal with efficacy. It further follows, therefore, that the FDA has to evaluate evidence of efficacy. In point of fact, of course, as you know it does, but in a strange sort of extra-legal way. What I should like to see, therefore, is a change in the law bringing the authority of the FDA legally up to date, and transforming the act from an Alice-in-Wonderland law to a logical one.[148]

The point was not just a semantic technicality. It was that inefficacious drugs—or, at the very least, drugs that could not do what their manufacturers stated they could do—were getting on the market.

Max Finland, for his part, appears to have spoken to one more, increasingly relevant, audience: the press. And this would have lasting consequences. In January 1959, three months before Lasagna addressed the PMA, the *Saturday Review*'s science editor, John Lear, published an article about antibiotics. Lear entitled it "Taking the Miracle out of the Miracle Drugs" and shocked his readers, eliciting the most mail he had ever received for a single piece. Recapitulating the trajectory taken by the infectious disease experts he interviewed, Lear began the article by focusing on physicians and antibiotic overuse but soon focused on "the prevalence of massive advertising pressure" that seemingly fomented such overuse in the first place. As the *F-D-C Reports* (the "Pink Sheet" newsletter for the drug industry) later reported, "Lear warmed over all the cold potatos [*sic*] that had been cooked up in the past three years by . . . long-haired medical educators and researchers."[149] Lear drew attention to Sigmamycin in particular, and to an advertisement (prepared by the William Douglas McAdams agency) in which Pfizer had provided the calling cards of a broad spectrum of specialists to justify the claim that "Every day . . . everywhere . . . more and more physicians find Sigmamycin the antibiotic therapy of choice."[150] It would have been bad enough if such "testimonials" were the sole basis of Pfizer's campaign. But it was worse. As it turned out, none of the physicians actually existed.[151]

As Richard Harris would recount several years later:

One evening, Lear had dinner with an eminent research physician, and afterward the two men visited a laboratory in the hospital where the doctor worked. "He pulled open several drawers that were full of drug samples and advertisements," Lear said later. "'Just take a look at that stuff!' he told me, and then went on to say that a good part of the advertising was misleading—in fact, that some of it was downright fraudulent. Finally, he said, 'Look,

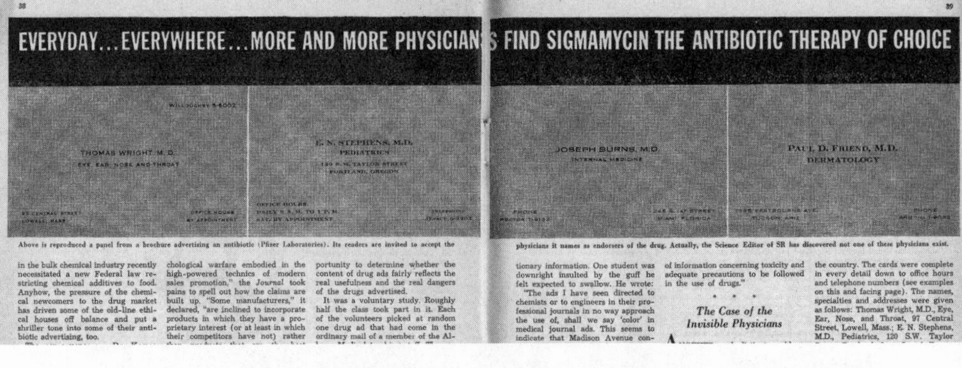
38

39

EVERYDAY...EVERYWHERE...MORE AND MORE PHYSICIANS FIND SIGMAMYCIN THE ANTIBIOTIC THERAPY OF CHOICE

THOMAS WRIGHT, M.D.
EYE, EAR, NOSE AND THROAT

E. N. STEPHENS, M.D.
PEDIATRICS
120 S.W. TAYLOR STREET
PORTLAND, OREGON
OFFICE HOURS
DAILY 9 A.M. TO 1 P.M.
SAT. BY APPOINTMENT

JOSEPH BURNS, M.D.
INTERNAL MEDICINE

PAUL D. FRIEND, M.D.
DERMATOLOGY

Above is reproduced a panel from a brochure advertising an antibiotic (Pfizer Laboratories). Its readers are invited to accept the physicians it names as endorsers of the drug. Actually, the Science Editor of SR has discovered not one of these physicians exist.

in the bulk chemical industry recently necessitated a new Federal law restricting chemical additives to food. Anyhow, the pressure of the chemical newcomers to the drug market has driven some of the old-line ethical houses off balance and put a shriller tone into some of their antibiotic advertising, too.

chological warfare embodied in the high-powered technics of modern sales promotion," the *Journal* took pains to spell out how the claims are built up. "Some manufacturers," it declared, "are inclined to incorporate products in which they have a proprietary interest (or at least in which their competitors have not) rather

portunity to determine whether the content of drug ads fairly reflects the real usefulness and the real dangers of the drugs advertised.

It was a voluntary study. Roughly half the class took part in it. Each of the volunteers picked at random one drug ad that had come in the ordinary mail of a member of the Al-

tionary information. One student was downright insulted by the guff he felt expected to swallow. He wrote:

"The ads I have seen directed to chemists or to engineers in their professional journals in no way approach the use of, shall we say 'color' in medical journal ads. This seems to indicate that Madison Avenue con-

of information concerning toxicity and adequate precautions to be followed in the use of drugs."

* * *

The Case of the Invisible Physicians

A

the country. The cards were complete in every detail down to office hours and telephone numbers (see examples on this and facing page). The names, specialties and addresses were given as follows: Thomas Wright, M.D., Eye, Ear, Nose, and Throat, 97 Central Street, Lowell, Mass.; E. N. Stephens, M.D., Pediatrics, 120 S.W. Taylor

John Lear's exposé in the *Saturday Review* of Pfizer's Sigmamycin advertising. Lear's caption reads, "Above is reproduced a panel from a brochure advertising an antibiotic (Pfizer Laboratories). Its readers are invited to accept the physicians it names as endorsers of the drug. Actually the Science Editor of SR has discovered not one of these physicians exist." "Taking the Miracle out of the Miracle Drugs," *Saturday Review* 42 (January 3, 1959): 38–39.

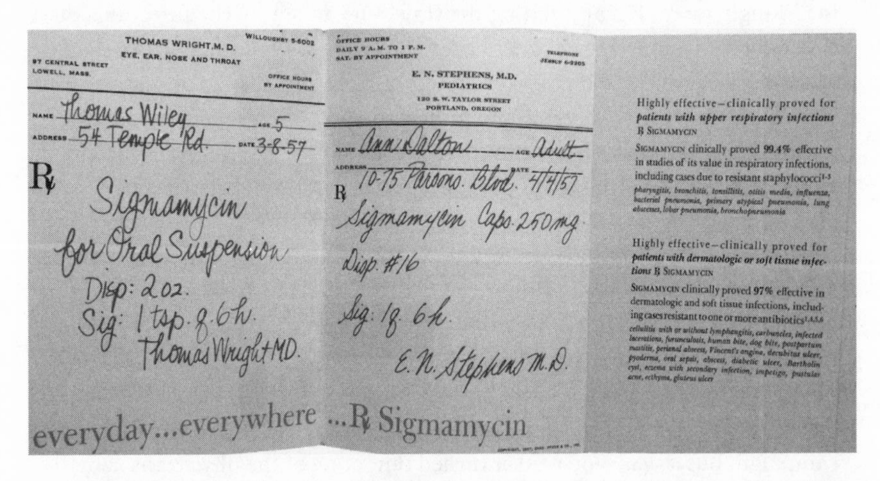

A mailing from the Pfizer campaign for Sigmamycin critiqued by Lear. Note that the text of the mailing recommends Sigmamycin as "Highly effective—clinically proved for ***patients with upper respiratory infections***" [boldface and italics in the original]. FTC Docket 7487, RG 122 (Records of the Federal Trade Commission), National Archives.

> you're walking around a big story. Why don't you step into it?' I said I might if I had enough information."

The "eminent research physician" provided such information.

> "Among other things, he showed me a small folder advertising Sigmamycin, an antibiotic put out by Chas. Pfizer, Inc. Across the top of the folder was a banner of bold type that said, 'Every day, everywhere, more and more physicians find Sigmamycin the antibiotic therapy of choice.' Below that were reproductions of what appeared to be the professional cards of eight doctors around the country, with addresses, telephone numbers, and office hours. The doctor said he had himself conducted some experiments with Sigmamycin, and at one point he had written to the eight doctors to ask the outcome of their use of the drug in clinical tests. As he told me this, he reached into one of the drawers and brought out eight envelopes, all stamped 'Return to Writer—Unclaimed.' I asked him if I might report his experience, and he said that he couldn't get involved in any kind of exposé. He pointed out, however, that there was nothing to prevent me from writing to the doctors myself." Lear did write to them, and all eight letters came back. Then he sent telegrams to the doctors, and was informed that there were no such addresses. Finally, he attempted to telephone them, and learned that there were no such telephone numbers.[152]

While the evidence remains circumstantial, it seems almost certain that Max Finland was Lear's unnamed "eminent research physician," the de facto Strep Throat. Within the article, Lear cited Finland as the "world dean of antibiotic therapists" and quoted him as stating that "any and every combination that comes . . . in a package may be a vicious distortion of the best use of medicine." And Lear's statement that "established ethical drug firms . . . are being jostled and jolted competitively in antibiotic sales by the Madison Avenue 'hard sell' of bulk chemical makers who have invaded the drug field with a great deal of money to spend and no comparable history of restraint in promotion of their wares" came right out of the Finland book of quotations.[153]

The following month, Lear published a follow-up article, "The Certification of Antibiotics," which focused still more on the FDA's evaluation of remedies and exposed the formerly intra-professional debate over the fixed-dose combination antibiotics to a national audience. Lear also drew attention to a novel, and potentially salacious, facet of the antibiotic story: that Henry Welch, while heading the FDA's Division of Antibiotics, derived "significant income" from the antibiotic journals he edited.[154] Lear's exposés would intersect with Congress's increasing attention to the pharmaceutical industry, culminating in Estes Kefauver's hearings into the industry beginning in 1959. Kefauver—the liberal Tennessee senator who had held federal hearings on the mob in the early 1950s, run unsuccessfully as Adlai Stevenson's vice presidential candidate in 1956, and explored the dynamics of the automobile, steel, and bread industries—initially sought to tackle apparently monopolistic tendencies in the pharmaceutical industry, in which drug prices appeared unresponsive to changing times and conditions.[155] Yet as the investigation and proceedings developed, he and his staff (including economist John Blair and counsel and staff director Paul Rand Dixon) would increasingly turn their attention to pharmaceutical marketing, which shone a powerful light on the Food and Drug Administration's inability to explicitly rule on drug efficacy prior to new drug approval.

The hearings would ultimately result in the Kefauver-Harris amendments of 1962, which mandated proof of efficacy via "well-controlled" studies prior to FDA new drug approval. And Kefauver later reflected that Lear's antibiotic articles "helped, as much as anything else, to spur on the investigation and to broaden its range."[156] Both Sigmamycin and Henry Welch would loom large at the hearings, and as the FDA would be forced to confront the definition of "well-controlled" studies, its power to make and unmake the pharmaceutical marketplace, and its role in shaping a "rational" therapeutics in the wake of the passage of the 1962 amendments, it would be another fixed-dose combination antibiotic—Upjohn's Panalba—that would serve as the focus of attention. Chapter 3 describes such events in detail, examining the key and reciprocal relationship between antibiotics and the FDA throughout the 1960s.

CHAPTER THREE

From Sigmamycin to Panalba

Antibiotics and the FDA

> If Dr. Finland does not want to use fixed combinations, he is perfectly free to do so. . . . Taking Panalba, it is used by 20[,000] to 25,000 physicians in this country. Now, can you just say, ex cathedra, that everything these people have done is irrational? . . . Is this the way to run the FDA?
>
> —STANLEY TEMKO (August 13, 1969)

SENATOR ESTES KEFAUVER'S hearings into the pharmaceutical industry—beginning in late 1959, and resulting in the Kefauver-Harris amendments of 1962—were a watershed event in twentieth-century therapeutics. Serving as a model for the power of the investigative hearing in the age of both print and electronic media, Kefauver's highly visible efforts to protect the health and pocketbook of the American consumer were literally daily and nightly news.[1] And if the sulfa drugs and penicillin had ushered in the era of the wonder drug, then the Kefauver hearings, in some ways, marked its end, at least as an unblemished, unproblematic period. Over five decades—and countless investigations into the pharmaceutical industry—later, such congressional hearings have become commonplace, almost clichéd, demonstrations of attempts to protect the public's health. At the time, however, such extended and public excoriation of the pharmaceutical industry was revolutionary.

Kefauver's investigation began with concerns regarding monopolistic pricing, yet as the investigation and proceedings developed, he and his staff increasingly turned their attention to pharmaceutical marketing and the inability of the Food and Drug Administration to rule explicitly on drug efficacy prior to new drug approval. And by the time the bill emerged from months of negotiations, whereas the patent provisions envisioned by Kefauver had been stripped away, the rule that drugs would need to be proven efficacious via "adequate and well-controlled investigations, including clinical investigations, by experts qualified by scientific training and experience to evaluate the effectiveness of the drug involved" would be written and signed into law.

By the end of the decade, on account of the resulting Drug Efficacy Study and Implementation (DESI) process, by which inefficacious existing remedies (or more broadly, remedies that did not do what their manufacturers stated they could do) could be withdrawn from the market, all of the fixed-dose combination antimicrobials would be removed. Exemplified by the withdrawal of Upjohn's Panalba, contested all the way to the Supreme Court, the DESI process represented both a key moment in the strengthening of the FDA, as well as the high-water mark of government regulation of antibiotic prescribing in the United States. It likewise set the stage for a backlash to follow.

Daniel Carpenter, in his magisterial *Reputation and Power*, and in particular in his account of the FDA's increasing insertion of efficacy considerations into regulatory practice throughout the 1950s, has downplayed the significance of the Kefauver-Harris amendments with respect to the incorporation of efficacy considerations into new drug review.[2] He has made an impressive case. Yet the antibiotic story offers additional perspective to Carpenter's analysis. First, as Carpenter relates, the FDA's increasing incorporation of efficacy considerations into its deliberations throughout the 1950s was in fact "ambiguous," both in retrospect and as viewed at the time. By virtue of the 1938 federal Food, Drug, and Cosmetic Act, the FDA could only rule on the *safety* of drugs, though efficacy concerns were always implicit within such safety considerations.[3] At best, such a situation seemed to reformers like Louis Lasagna an "Alice-in-Wonderland" exercise in legalistic and regulatory smoke-and-mirrors; at worst, and more importantly, it seemingly had proved insufficient in keeping *useless* drugs off the market. And as Lasagna and fellow reformers would relate throughout the hearings, useless drugs, even if safe, held their own significant financial and health costs. Second, and drawing on Carpenter's notion of the "conceptual" power (as opposed to the "directive" or "gatekeeping" powers) of the FDA, the Kefauver-Harris amendments and their vague articulation of "adequate and well-controlled investigations" would force the issue of defining just what such well-controlled investigations represented and from what, in particular, they were to be differentiated. By the end of the 1960s, as the FDA, Congress, and the judicial system wrestled with such regulatory notions in the midst of the DESI process, they would elevate the controlled clinical trial—especially the randomized controlled trial—to a position of legal and conceptual dominance it has retained since.

Fixed-dose combination antibiotics would be central to both of these processes. Pfizer's Sigmamycin (tetracycline and oleandomycin)—central to the Kefauver hearings—epitomized the inefficacious drug not prevented from entering the market. Upjohn's Panalba (tetracycline and novobiocin)—central to the DESI process—focused attention on the very definition of the "well-controlled" trial. Fixed-dose combination antibiotics, in this way, would continue to catalyze such tensions throughout the 1960s, at the same time that their resulting removal would affect discourse regarding antibiotic usage itself for the remainder of the century.

Making Drug Efficacy Explicit: The Kefauver Hearings

Senator Kefauver's hearings started with antibiotics, though they originated with a focus on monopolistic pricing and the seemingly inelastic "administered prices" of medications unresponsive to supply and demand. As recounted by Richard Harris, Washington attorney Walter Hamilton noted in 1951 that the prices of Aureomycin, Chloromycetin, and Terramycin for his pharyngitis were identical. This observation triggered the interest of his wife, Federal Trade Commission (FTC) economist Irene Till, who passed it along to her FTC superior John Blair.[4] Over the next seven years, such concerns were tabled for one reason or another, though they would lead to the FTC's 1958 *Economic Report on Antibiotics Manufacture*, which drew attention to marketing, profitability, and patent concerns related especially to the broad-spectrum antibiotics.[5] By late 1958, however, with Kefauver having held hearings on administered prices in the steel and automobile industries, his staff, which now included both Blair and Till, turned their attention to pharmaceuticals.

In the beginning, Kefauver and his staff were indeed chiefly focused on pricing concerns: monopolistic tendencies, profits, patents, and the role of generic versus brand-named drugs.[6] Early activities included a series of "field trips" to small pharmaceutical companies and wholesalers.[7] In John Blair's Valentine's Day 1959 outline of thirty-four "Proposed Points of Inquiry for [the] Drug Hearings," the overwhelming majority of points were related to costs, understandable for a subcommittee on antitrust and monopoly. Even discussion of detail men (pharmaceutical representatives) began with concerns regarding their cost, given that "the cost of supporting a staff of detail men is prohibitive for all except the very largest companies. It is this barrier which, more than anything else, prevents the smaller manufacturer from 'reaching' the practicing physician."[8]

However, by point #30 of Blair's outline, on the "Forced Obsolescence of Drugs" (echoing concerns regarding planned obsolescence, or "style obsolescence," discussed in prior hearings on administered prices in the auto industry), fixed-dose combination drugs came in for special mention:

> Ordinarily, when drugs are first introduced, they are marketed at extremely high prices. In addition to patent protection, they are given intensive advertising promotion, often with extravagant claims made of their therapeutic usefulness. As these claims are subjected to the sad tests of experience, and as other companies enter the field with similar products of a slightly different chemical makeup, there is a tendency for price to decline. To offset this situation, so-called new products are introduced to capture again the high prices fading for the older products. In many cases the new products are merely combinations of older well-known drugs.[9]

Blair's memo was written one week after the appearance of John Lear's *Saturday Review* report on Henry Welch and the relationship between the pharmaceutical industry and the FDA (see chapter 2).[10] And in points #27 and #32, on "Possible Influence by Drug Companies over FDA and Other Government Agencies" and "Fraudulent Practices," Lear's muckraking concerns, and his translation of those of the infectious disease–based reformers, made their way into the interstices of the investigation's own developing structure.[11]

Indeed, a spirit that harked back to the days and achievements of Progressive-era muckraking seems to have increasingly infused the investigation. As former Pfizer employee (turned Pfizer critic and de facto informant) Martin Seidell wrote to John Lear (and passed along to Blair), "You and I both share the hope that, similar to the Flexner revelations, our efforts may do muc[h] to restore the sense of morality and social justice that has been lost in the past decade."[12] And Seidell's very connection to Blair epitomized the network of connectivity developing among Blair and his staff, journalists such as Lear, disgruntled former industry insiders, and academic reformers concerned with the prevalence of extravagant marketing claims rendered in an apparent regulatory vacuum. It was Lear, for instance, who had connected Blair with Seidell, who himself provided a stream of information regarding the interconnected world of Pfizer and Arthur Sackler at William Douglas McAdams.[13] Harold Aaron, the founder of the reformist *Medical Letter*, pointed Blair to academic physicians who "have shown a critical attitude towards promotion methods of pharmaceutical companies"; first on his list was Max Finland, followed by Harry Dowling and Ernest Jawetz among a list of six infectious disease experts.[14] In this way, many of the concerns of the infectious disease–based therapeutic reformers—regarding Pfizer, Sackler, and an increasingly impotent or complicit American Medical Association—became areas of interest for Blair and his staff as well, joining the monopoly, pricing, and patent concerns that continued to occupy them.[15]

It was an organic, inevitably messy investigation, with false leads and dead ends. As Harris points out, less than 20 percent of the investigation had been completed when the hearings commenced in December 1959.[16] But by the time the hearings began, the need for the FDA to be empowered to rule explicitly on efficacy considerations had begun to rank alongside monopoly and price considerations as a leading concern of the hearings.[17] And while the investigation was organic and opportunistic, the public articulation of such efficacy considerations at the hearings seemed to reveal a coherent and progressive logic.

Publicly, the hearings may have been grounded in "the pricing methods of the drug industry" and conducted under the jurisdiction of the Subcommittee on Antitrust and Monopoly, but Kefauver, Blair, and (counsel and staff director, and future FTC chief) Paul Rand Dixon repeatedly attempted to underscore the relevance to their jurisdiction of inquiries into the marketing of drugs.[18] As Dixon stated near the start of the hearings:

> It has been our observation that the representations that have been made for drug products, the promotional campaigns that have been carried on in the sale of the products to and through the doctors, are quite important and should be considered here. It is self-apparent that by this method of promotion and sale there will be found some of the explanations for the high prices that are charged, for the identity of prices that are charged and for some of the exorbitant profits that appear to be made, and also for the inability of small companies to enter and be successful in this field.[19]

And in the context of such marketing concerns, they would continue to insert into the hearings comments regarding the need for the FDA explicitly to pass on drug efficacy.

As a safe, almost symbolic, warm-up act, on the second day of the hearings Kefauver's staff called on Floyd Odlum, Russell Cecil, and E. D. Bransome, from the Arthritis and Rheumatism Foundation, to discuss the outright quackery of proprietary remedies proffered to arthritis sufferers.[20] Odlum and Bransome discussed the worthlessness of the over-the-counter proprietary medicaments advertised directly to patients. Many of them were found "in the old covered wagon and the patent medicine peddler" and capable of doing harm either directly, through monetary loss, or through keeping the patient away from more appropriate remedies. Odlum and Bransome left the clear impression that they felt that such a process occurred in *distinction* to that concerning ethical remedies (marketed only to the physicians who could prescribe them), the manufacturers of which presumably were "required to prove the effectiveness of their drugs on the basis of clinical tests before these drugs can be marketed."[21] When Bransome asked that nonprescription drugs be held to such apparently strict standards, it was Kefauver and his staff's turn to disabuse him of his understanding of prescription new drug laws and to ask whether such efficacy standards should be amended. Bransome, not surprisingly, felt that such "a set of standards . . . could cover the ethical drug just as well as the quack drug."[22]

Having craftily exposed professional ignorance regarding the FDA's inability to rule explicitly on efficacy prior to drug approval, Kefauver and his staff turned to the supposedly ethical drugs themselves. On the third day of the hearings, they called on Louis Lasagna, who echoed Harry Dowling in attacking both the pharmaceutical industry and the ill-informed medical profession for their roles in the precarious path from cup to lip:

> The advertising agencies are being asked to sell to the medical profession a whole bushel basketful of sows' ears for silk purses each year. It is no wonder that there are advertising excesses, and that there are so-called product failures and that obsolescence sets in. This plethora of poor compounds, and of new mixtures of old agents that appears each year confuses physicians. It raises the cost of drugs, I would think, and may harm patients either through keeping them from adequate therapy or by causing them serious side effects. . . . The physician today no longer serves as a satisfactory and adequate shield for the patient against drug toxicity, ineffective drugs, and high costs, and that to rely on the profession to rectify mistakes along these lines is I am afraid unrealistic.[23]

Efforts to demonstrate the relationship between the excessive marketing of inadequately tested drugs and the need for FDA evaluation of drug efficacy continued to gather steam. In late January 1960, Mike Gorman, influential mental health lobbyist and executive director of the National Committee against Mental Illness, spoke on tranquilizing drugs, rendering two linked conclusions:

> First, and fundamentally, the pharmaceutical industry must be annexed to this country. I don't think it has been, yet. In many respects, it is now a pri-

vate feudal enterprise with a dinosaur-laden moat between its kingdom and the rest of the United States. . . . It is given a license to carry on its present activities by various Federal and State statutes. There must be serious consideration of revocation of these licenses when it acts against the public interest. [Second,] the Food and Drug Administration must evaluate the clinical claims now made by the various pharmaceutical houses for individual drug preparations. It is not enough merely to determine that a drug is nontoxic, is therefore nonpoisonous, and therefore safe for human consumption.[24]

Haskell Weinstein, former Pfizer employee and former medical director of the Pfizer-owned J. B. Roerig Company, appeared in February. The preceding December, two days after the commencement of the hearings, Weinstein had written to Kefauver: "There is a very prevalent misconception that the Food and Drug Administration evaluates the efficacy of new drugs. Actually they seem to accept the evidence presented by the manufacturers with very little attention to how this evidence was obtained. As a physician I blush with shame at the quality of some of the 'studies' done by some of my physician brethren."[25] During his testimony, reinforcing the notion that more was at stake than dollars, Weinstein further grounded the need for proof of efficacy in "large-scale cooperative studies, very carefully controlled, with criteria clearly defined before the study is undertaken, and a very objective and unbiased interpretation . . . made of the results."[26]

With such testimony in hand, Kefauver was in a position to test out the idea that "the public would be safeguarded if, in addition to safety or toxicity, the FDA was directed to go into the efficacy of drugs" on University of Iowa professor of medicine William Bean, who agreed on such a "reasonable" course of action.[27] A. Dale Console, former chief medical director of Squibb, voiced his own support (albeit "with hesitation") in April for empowering the FDA to rule on drug efficacy, bashing the "use of the testimonial as scientific evidence of the efficacy of drugs."[28] Console thus evoked the rhetoric of Max Finland's own impeachment of Sigmamycin. And one month later it would be Sigmamycin itself that would garner center-stage attention, providing a crucible in which apparently "shocking" pharmaceutical behavior and FDA inadequacy would yield calls for improved standards, on the very eve of Kefauver's first attempt at constructing a drug bill.

Sigmamycin in the Spotlight

The hearings on May 12, 1960, began with a reference to John Lear's *Saturday Review* articles from the beginning of 1959.[29] And with neither Henry

Welch nor Félix Martí-Ibañez present to act in his own defense (both claimed infirmity), Kefauver's staff started the following day's proceedings by tracing Welch's career, from his hiring at the FDA in 1948, through his acquisition of consent to edit journals (for which he might receive "yearly honorariums" for work done on his own time) to concerns first voiced in 1956 by Merck president John Connors regarding the propriety of having an FDA employee oversee publications ultimately dependent on pharmaceutical advertising for their success.[30] After reporting Welch's assertions at the time that he had "no business connections with MD Publications," Kefauver called forth Francis Engelstad, certified public accountant, to detail Welch's activities over the prior decade. As headlines soon revealed, Welch had received 7.5 percent of all M.D. Publications advertising revenue and, still more lucratively, 50 percent of net reprint sales. Pfizer alone had bought $171,752 worth of reprints from 1953 to 1959, and Welch had made $224,016.70 from M.D. Publications during that span. When combined with his payments from, and interest in, Medical Encyclopedia, his net income from the two publishing operations had amounted to $287,142.40.[31]

Privately, beyond personal financial gain, the very fact of having the chief of the FDA's Division of Antibiotics in charge of journals largely or entirely dependent on advertising revenue (let alone co-published with Martí-Ibañez, an employee of the William Douglas McAdams agency) had placed Welch in a predictably compromised position, as reams of subpoenaed files and hundreds of pages of appendices to the congressional proceedings would demonstrate. For example, on finding out that Pfizer would not be renewing its subscription for 637 copies of *Antibiotics and Chemotherapy* in late 1955, Martí-Ibañez would plaintively implore its advertising director: "To give you an example of the outstanding material that will appear in Antibiotics and Chemotherapy, the February issue of this publication will include an editorial by Dr. Welch reappraising the use of Nystatin in conjunction with broad-spectrum antibiotics. This paper will furnish your people with excellent ammunition with which to counteract the exaggerated claims made for Nystatin."[32] The following year, Welch asked Arthur Sackler to use his influence with the pharmaceutical industry to give the expected controlled-circulation *Antibiotic Medicine and Clinical Therapy* "the extra push that will put it over."[33]

The apparent conflict of interest exhibited was "shocking" to such participants in the hearings as Senator John Carroll (D-Colorado), who stated his disbelief regarding Welch's publications-related income that "anyone could contend that this is just an honorarium at all, and especially with the tremendous sums involved [and] the high responsibility of this type of case by a director of the Antibiotics Division."[34] Welch resigned from the FDA on May 19, 1960.[35] And the participants' attention—and the public's—would soon turn from such dollar amounts back to Sigmamycin itself as the prototypical example of both Welch's own venality and the apparent weakness of the FDA more generally.

On June 1, 1960, after defining antimicrobial *synergy* for his audience, Kefauver again reviewed the relationship between pharmaceutical marketing and his own jurisdiction regarding monopolistic practices, since "the key to obtaining greater market shares in this industry is the ability to persuade practicing physicians that what is being offered is a new and better drug."[36] Paul Rand Dixon then recounted the history of Sigmamycin's introduction, from the new drug application offered by Pfizer on August 26, 1956, to its certification by the FDA on September 28, through its glowing presentation by Welch less than a month later at his annual antibiotic symposium as ushering in the "third era of antibiotic therapy."[37] Dixon next called on Dr. Gideon Nachumi, who had been a medical student working as a Pfizer copywriter in 1956. As Nachumi related, Pfizer had often edited papers submitted to Welch's antibiotic symposia by researchers working with Pfizer products or funds, and he had been advised to edit Welch's introductory address itself. Nachumi's key insertion would be the phrase "Third Era of Antibiotics," which, consequent to Welch's presentation, would be central to all subsequent Sigmamycin marketing.[38] Kefauver next called forth former Pfizer public relations specialists Warren Kiefer (who had co-authored *Pax*, as noted in chapter 2) and Joseph Hixson, who confirmed how Pfizer followed the 1956 symposium by purchasing 238,000 copies of Welch's remarks (from Welch), to the point where "it was a standing joke in the office."[39] Again, Kefauver professed outrage, this time at Pfizer's "shocking" behavior. He then called on Barbara Moulton to indict the FDA itself.

Moulton, trained as a physician, had served for five years in the FDA's Bureau of Medicine, and her characterization (fair or not) of the agency on June 2 would leave a lasting image of FDA capture by the very industry it was supposed to have regulated.[40] She began her statement with the contents of an unsent letter to FDA commissioner George Larrick:

> I believe firmly that a strong militant Food and Drug Administration is necessary for the welfare of the American people, and that a strong militant Bureau of Medicine is necessary for the welfare of the Food and Drug Administration. I believe also that hundreds of people, not merely in this country,

> suffer daily, and many die because the Food and Drug Administration has failed utterly in its solemn task of enforcing those sections of the law dealing with the safety and misbranding of drugs, particularly prescription drugs.[41]

She then reported on the dynamics of new drug approval, yielding an impression of staff pressure from industry, welcomed by industry-sympathetic FDA superiors who were willing to override less industry-sympathetic rulings by their demoralized underlings.[42] As Moulton related:

> When a drug firm submits a new drug application for an important new drug, it is common practice for their representative to call the Chief of the New Drug Branch a few days later and inquire which medical officer is handling the application. The name is promptly supplied. The medical officer then receives a call to the effect that the firm wishes to send representatives in to discuss the application, and he is expected to make such an appointment promptly. If he does not do so, perhaps for the very valid reason that he has not as yet studied the application, he is reprimanded by his Chief as uncooperative, and the appointment is made anyway by a clerk. A day rarely passes without such conferences in the New Drug Branch.

Describing the specific nature of the pressure exerted, Moulton continued:

> If the firm anticipates no difficulties, they send a single representative. If the medical officer has suggested over the phone that the new drug application may not be completely satisfactory, four or five men may appear in his office to argue the case. He may also be invited to attend a medical meeting sponsored by the firm at which the drug in question will be discussed by the clinical investigators. Frequently, at such meetings, the investigators who have been luke-warm or cold about the merits of the drug are not invited to participate. If the medical officer handling the new drug application is still not satisfied with the evidence of safety, the company will frequently make an appointment with the Medical Director, who has not seen the data on the new drug application, to present their side of the story to him. I have known such conferences to be followed by an order to the medical officer to make the new drug application effective, with the statement that the company in question has been evaluating new drugs much longer than the medical officer and should, therefore, be in a much better position to judge their safety.[43]

Finally, Moulton turned to Sigmamycin as the representative case. As she related, her experience as a physician had alerted her to the misuse of antibiotics and "the problem of misleading representations [of drugs] on the part of the pharmaceutical industry," and Moulton appeared to find some cause for the situation once she began working at the FDA, with Dr. Welch in the role of authoritarian, industry-friendly superior.[44] When Moulton reviewed Sigmamycin's new drug application a year after its approval, she found poorly conducted studies by relatively unknown investigators, at times without "an accurate diagnosis having been established, or with a clinical diagnosis of a condition for which antibiotic therapy is not indicated."[45] As she continued:

> In reading this material I found it impossible to believe that anyone with a knowledge of clinical antibiotic medicine could honestly reach the conclusion that it substantiated the claims made for these products. At that period there was a good deal of discussion in the Bureau of Medicine of handling false or exaggerated claims for the efficacy of new drugs under the misbranding sections of the law (which was the province of the branch to which I had been reassigned). Furthermore, there was a hopeful atmosphere that we might soon be able to break the precedent of years and take action against misbranding of prescription drugs, which the Food and Drug Administration has never done.[46]

Moulton never did have the chance to prosecute the Sigmamycin case, yet the episode left her with the belief that a renewed FDA, free of industry ties and pressure, would need to be empowered to rule explicitly on efficacy as well as safety, *prior* to the approval and release of new medications:

> No physician, no one who has ever been responsible for the welfare of individual patients, will accept the idea that safety can be judged in the absence of a decision about efficacy. . . . To attempt to separate the two concepts is completely irrational. . . . For a drug firm to object too strongly to such a change in the law should render it highly suspect. In general, the drug manufacturers claim that they never market a drug . . . until they themselves think they have reasonable proof of its value. If they have such proof, they should not fear its review by the Food and Drug Administration.[47]

This is not the place to gauge the merits of Moulton's claims. Daniel Carpenter has depicted a far more sophisticated contemporary FDA, already following the lead of Bureau of Medicine director Ralph Smith and wrestling with efficacy claims internally.[48] Nevertheless, Moulton's assertions had quite an impact at the time. The following day, Arthur Flemming, secretary of the Department of Health, Education, and Welfare (HEW, of which the FDA was then a component), announced at the hearings that in view of the "serious charges" leveled at the FDA, he had arranged with Detlev Bronk, the president of the National Academy of Sciences, "an outstanding committee of scientists to review the policies, procedures, and decisions of the Division of which Dr. Welch was formerly the chief, as well as those of the New Drug Division of the Bureau of Medicine of the Food and Drug Administration."[49] Flemming would be followed at the hearings by Commissioner Larrick, who expounded at length that "while the new drug provisions of the [existing] act are concerned primarily with safety, the FDA interprets this concept of safety so broadly in the processing of applications for new drugs as to furnish assurance of the efficacy of new drugs for the purposes claimed in the labeling contained in the new drug application." Nevertheless Larrick found himself forced to conclude, like Albert Holland at the Blatnik hearings three years before (as discussed in chapter 2): "The new drug applications are not adequate to insure the efficacy of drugs which are essentially innocuous. We would endorse a proposal that the new drug section of the Food, Drug, and Cosmetic Act require a showing of efficacy as well as a showing of safety."[50] Twelve days later, when Senator Kefauver produced the first version of a bill (S.3667) to amend the Food, Drug, and Cosmetic Act of 1938, the first provision was for the empowering of the FDA to rule on drug efficacy.[51]

By September 1960, the eight-member committee (chaired by University of Chicago professor of medicine C. Phillip Miller and including such infectious disease experts as Max Finland, John Dingle, and Wesley Spink) asked by Flemming and Bronk to review the FDA's Antibiotics Division and New Drug Branch had completed its report.[52] And the committee's first recommendation, following Larrick's lead, was that the "FDA should be given statutory authority to require proof of the efficacy, as well as the safety, of all new drugs."[53] Max Finland and Harry Dowling appeared before Congress during a round of hearings that same month, with Finland describing the prevailing regulatory vacuum and garnering an apparent double-take from the senator for his particular characterization of the AMA's abdication of regulatory responsibility in the wake of the demise of the Seal of Acceptance program (discussed in chapter 1):

SENATOR KEFAUVER: In years past the Council on Drugs [actually its predecessor Council on Pharmacy and Chemistry] of the American Medical Association used to attempt some evaluation, did it not, but they have discontinued that work in recent years?

DR. FINLAND: Well, they have become what I would call sissy.

SENATOR KEFAUVER: They have become what?

DR. FINLAND: Sissy. I do not know whether they use that word down in Tennessee.

SENATOR KEFAUVER: Yes; I hear it very often, but I do not know if it means the same thing in Tennessee as it does in Massachusetts.[54]

It did.

However, by July 1961, when the hearings resumed to discuss Kefauver's revised bill, S.1552 (which included the efficacy provision, along with patent provisions and those geared to encourage prescribing by generic name), the AMA's representatives had the chance not only to defend themselves against such a characterization but also to attack the efficacy provisions of the bill themselves.[55] As Hugh Hussey, chairman of the AMA Board of Trustees, defended the judgment and autonomy of the individual prescribing physician:

> A drug's efficacy varies from patient to patient, sometimes for known reasons such as allergy, and at other times for unknown reasons. Hence, any judgment concerning this factor can only be made by the individual physician who is using the drug to treat an individual patient. A physician can be told many things about a drug, including its chemistry, its mode of action and, to some extent, its toxic properties. But he must judge its efficacy.[56]

Seemingly impeaching the very epistemological tenability of the controlled clinical trial, Hussey continued: "A drug which is, on the average, less efficacious than another, must still be available to every physician since it may be completely efficacious in treating the medical problems of one of his patients. We do not practice medicine on the average—we seek to solve or alleviate the problems of each and every patient."[57] At the individual level, by definition, the findings of the controlled clinical trial could not be generalized to the unique patient at hand. And at the population level, and echoing Martí-Ibañez's response to Max Finland's attack in the pages of *Antibiotic Medicine and Clinical Therapy* from four years prior (see chapter 2), Hussey concluded: "Medical history and experience clearly demonstrate that the only possible final determination as to the efficacy and ultimate use of a drug is the extensive clinical use of that drug by large numbers of the medical profession over a long period of time."[58]

Subsequent hearings that month provided an opportunity for Charles May, Louis Goodman, William Bean, Louis Lasagna, Walter Modell, Dowling, and Finland to refute the AMA's protestations and defend the efficacy provision. Several such would-be reformers, aware of the message the AMA would convey at the hearings, had communicated among themselves prior to the meetings. As Finland wrote to Louis Goodman: "Certainly the complexity of modern medicine, the limitations of individual physicians, particularly those with large active practices, and the need for considerable amounts of detailed, well controlled and specialized studies, entirely preclude evaluation of important drugs by the 'play of the market place' or through 'tests by testimonials.' These must not be permitted to replace careful studies and controlled evaluations by qualified individuals."[59] And at the hearings William Bean (who pronounced himself a good friend of Hugh Hussey) stated:

> Because of its very real and serious apprehensions of the centralization of Federal authority, the American Medical Association in its fear has euchred itself into the astonishing posture of supporting the position that it is better to have nonefficacious drugs or those whose efficacy is as yet unestablished released freely to the American physician and the American public, rather than have those made available only when their usefulness in therapy has been determined or its probability is of so high an order that no one could object.[60]

The proponents of FDA empowerment may have differed regarding how the FDA could accomplish its mission—whether through internal strengthening or through coordination with existing external agencies (like the National Institutes of Health, the National Research Council, or even the AMA's Council on Drugs)—but each supported a more explicit FDA efficacy requirement.[61]

The ensuing and winding legislative course of Kefauver's original S.1552 bill—through the waxing and waning interests of varying components of the executive branch (from the FDA to the president), to the catalytic impact of the thalidomide scandal on ensuring the passage of legislation, before emerging as the Kefauver-Harris amendments—has been recounted before.[62] What is key to this narrative is that while intense disputes ensued over the patent provisions (the initial focus of the inquiry yet eliminated from the final bill) and other economic aspects of the bill, nearly universal accord was given to the inclusion of the efficacy requirements. The efficacy provision mandated that new drugs would need to be proven efficacious via "adequate and well-controlled investigations, including clinical investigations, by experts qualified by scientific training and experience to evaluate the effectiveness of the drug involved."[63] As such, as John Swann and Daniel Carpenter have demonstrated, it echoed procedural workings already afoot within the FDA, but at the same time, it echoed the thoughts, talks, and writings of clinical pharmacologists such as Max Finland, Harry Dowling, and Louis Lasagna.[64]

Such relatively uncontested inclusion supports Daniel Carpenter's interpretation that the efficacy requirement was already gathering steam at the FDA and instead exposed dissent *outside* the FDA, especially from the AMA, which came off poorly in the process. And yet the transformation from what Carpenter has termed the "ambiguous emergence of American pharmaceutical regulation" to an explicit FDA mandate regarding proof of drug efficacy via "well-controlled studies" would set in motion a series of debates, decisions, and regulations regarding the very definition of such terms, as well as regarding their consequences for existing drugs and the autonomy of clinicians. Antibiotics—and especially, fixed-dose combination antibiotics—would remain at the heart of this debate throughout the 1960s.

Interregnum: From Ambiguity to Ambivalence

Upon passage of the Kefauver-Harris amendments, in 1962, the FDA was forced to confront the difficult question of what do to with *existing* drugs approved between 1938 and 1962.[65] This represented both a question of interpretation regarding whether such drugs should be subjected to review and a question concerning the daunting logistics entailed in the actual evaluation of the drugs in the event of such a review. The 1962 amendments empowered the FDA to remove such drugs from the market only after October 10, 1964.[66] However, because the FDA had long been empowered to approve the potency (which, as sometimes construed, implied efficacy) of certain antibiotics, the agency first attempted a "trial run" of its authority to remove existing inefficacious products from the market in 1963.[67] This time, it focused on a different class of fixed-dose combination antibiotics: what were conveniently referred to (and often used) as "cold remedies," combinations of an antibiotic with such substances for symptomatic relief as analgesics, antihistamines, caffeine, and other vasoconstrictors.[68] Yet the FDA would fail to remove such products from the market at this time, the result of ambivalence regarding its authority on its own part and in the eye of academic reformers, setting the stage for much larger battles later in the decade.

The FDA had first certified in 1952 a combination of penicillin with ingredients "for the relief of symptoms and prevention of complications of the common cold and other acute upper respiratory infections."[69] Notions of adherence, convenience, and prophylaxis substituted for those of synergy and broad coverage. And despite their being universally condemned in academic medical circles and publications (with no evidence that antibiotics could either treat the common cold or forestall its complications), sales of such products would grow from 7,000 new prescriptions in 1952 to more than 4 million new prescriptions and $14 million in sales annually by the early 1960s.[70]

In late 1959, just as the Kefauver hearings were commencing, pharmacologist Solomon Garb drew particular attention to an advertisement for Lederle's Achrocidin—a combination of tetracycline with an antihistamine and analgesic, marketed for the prevention of the complications of the common cold—as an example of "objectionable" advertising based on "a minimum of evidence."[71] In private, Lederle's representatives, stating that Achrocidin accounted for 5 percent of *all* antibiotic sales in the nation, countered Garb with a two-part rebuttal that would undergird responses to FDA attempts to restrict existing markets and prescribing autonomy for the remainder of the decade. First, echoing Henry Welch's defense of the fixed-dose combination antibiotics to Max Finland several years before (as described in chapter 2), Lederle's Charles Masur pointed out the difficulty of ascertaining just when a viral cold would be complicated by a bacterial infection, defending the "rational" nature of such empiric therapy: "The use of repeated laboratory tests is possible in teaching hospitals related to medical schools and represents therapy at its best but many physicians practice where all the necessary facilities for scientific medicine are not available. . . . [Achrocidin] is a rational drug for the many physicians who practice in similar areas, but it is not designed for use in medical schools, hospitals, or patients who can afford repeated laboratory work."[72] Second, and extending beyond antibiotics, Masur echoed Félix Martí-Ibañez in defending the authority of the practicing physician and the marketplace over that of ivory tower academicians in the first place, claiming that "medical progress, like all other progress, often consists in going beyond the 'authority,' and I am sure that leaders in medicine recognize that no one group has a monopoly on medical wisdom."[73] Consequently, defending individual physician autonomy over centralized restriction (and anticipating Hugh Hussey's testimony at the Kefauver hearings a year later), Masur wrote to Garb: "Many of the commercially available compounds stem from demands raised by practicing physicians. The prescription of any therapy is the decision of the treating physician, and it is the duty of the pharmaceutical industry to provide the physician with a type of medication he wishes to prescribe, and the 'thinking' physician will have a choice of therapy which he deems prudent for any particular patient."[74]

By March 1962, however, as Senator Kefauver's bill wended its way through the nation's capital, and as the FDA received requests to review three new combination antibiotic plus symptom-relief products, FDA commissioner George Larrick reached out to the National Academy of Science's Detlev Bronk again, requesting "a group of well qualified consultants in

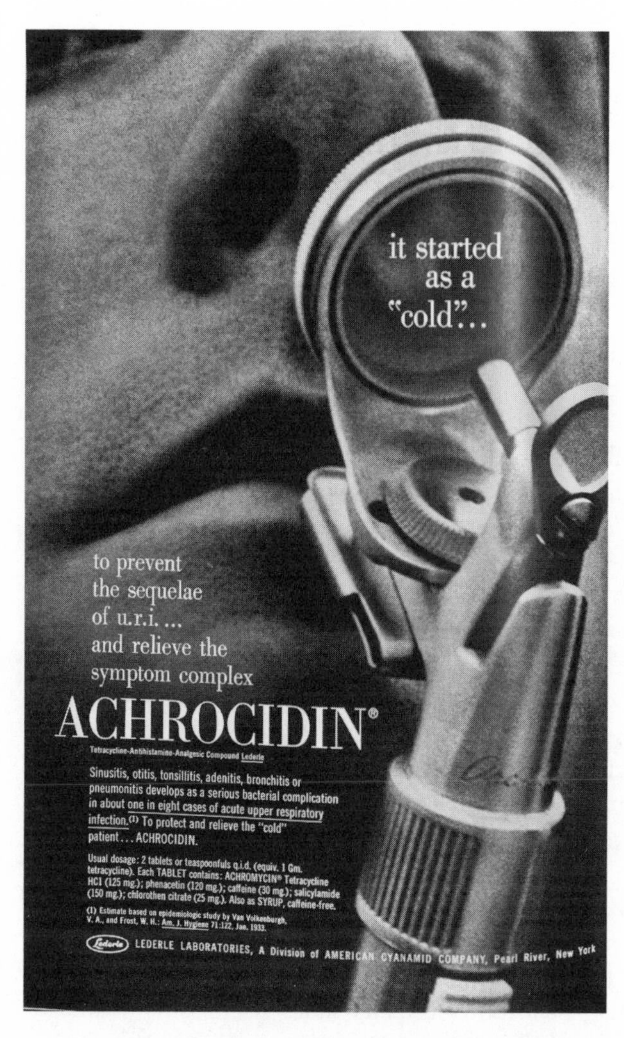

"It Started as a 'Cold.'" Despite decades of study and published articles to the contrary, Lederle cited an article from 1933 in this 1959 advertisement to support its contention that colds could develop into more serious infections and that Achrocidin could therefore help "prevent the sequelae" of upper respiratory tract infections. *Medical Times* (November 1959): 164a.

this area" to weigh in on the remedies more generally.[75] Bronk asked Harry Dowling to lead the panel, and Dowling in turn asked Max Finland, Cornell University's Edwin Kilbourne, Yale University's Paul Beeson, Carl Schmidt of the University of Pennsylvania, and William Jordon of the University of Virginia to complete the panel.[76] By September 1962, while the nation's eyes were riveted on the thalidomide story, the Dowling committee quietly sent its report to Larrick, concluding that "the symptomatic relief provided by the [non-antibiotic] ingredients was questioned, but even if these were effective, in no instance was there justification for any such product to contain an antimicrobial agent."[77]

Larrick sat on the report at the time. In the spring of 1963, in the aftermath of the passage of the Kefauver-Harris amendments, Senator Hubert Humphrey (D-Minnesota) held hearings on "Interagency Coordination in Drug Research and Regulation." There, Hahnemann Medical College's Hobart Reimann—who had first pointed out the possibility for widespread antibiotic "abuse, waste, and commercialization" in 1945—was questioned regarding the federal government's role in protecting the public's health, particularly with respect to the "misuse of antibiotics."[78] Reimann considered that "it is better for physicians to put their own house in order," with government regulation necessitated "only when all other efforts fail."[79] He instead advocated "improved communication" as the best available option, though he admitted: "Even then, it is doubtful how much can be accomplished. Physicians often resent criticism and do not like to be told what not to do, and, of course, the lay public does not like to be disillusioned about these 'wonder drugs.'" As he fatalistically concluded, "Perhaps the misuse of antibiotics will be overcome by time."[80]

By July 1963, though, Ralph Smith, the FDA Bureau of Medicine's acting medical director, felt empowered by both the new drug amendments and the conclusions of the Dowling report to recommend decertification of the existing "systemic oral drugs containing an antibiotic in combination with analgesic substances, antihistaminics, decongestants, or caffeine," in effect attempting to remove them from the market.[81] On August 17, 1963, Larrick published the agency's intention to render such action in the *Federal Register*, offering "interested parties" thirty days to submit their comments to HEW.[82]

Rep. Paul Schenk (R-Ohio) rendered the first expected complaint that "such a determination by the FDA is but one more step toward the socialization of medicine."[83] Ralph Smith, meanwhile, had warned Larrick that the proposal was "likely to be met by substantial industry opposition," and industry did not disappoint.[84] Lederle sent copies of the proposed decertification to 7,500 physicians, stimulating a massive letter-writing campaign to the FDA. While the letter writers bemoaned the "ivory tower," "test tube" pronouncements of the Dowling Committee and the FDA, another commentator noted that "what the protesters have submitted are unscientific testimonials similar to those the AMA itself denounces when offered by sponsors of such [quack] remedies as Krebiozin."[85] At the same time, and perhaps more effectively, Lederle's representatives wrote to such longstanding allies as Harry Dowling and Max Finland, justifying the use of the remedies in certain patients, and Dowling and Finland were at least persuaded to write to Larrick that "there are often two sides to a question and that the party who is ruled against often has some merit to his position."[86]

On September 16, the FDA held a meeting with representatives from Lederle, Pfizer, and the Pharmaceutical Manufacturers Association. But rather than offer studies indicating the efficacy of their remedies, industry representatives again merely produced surveys concerning their extensive use by practicing physicians, on the basis of the (data-free) rationale that "many physicians do not favor a multiprescription approach to the treatment of a specific ailment on the grounds that the patients frequently do not follow the directions prescribed and either fail to use or misuse the various drugs."[87] They thus recommended labeling changes rather than the withdrawal of the medicines.

Subsequent to the meeting, Larrick extended the deadline for filing comments on the proposed decertification for two more months, and the "cold war" that ensued carried forward competing notions of physician autonomy at the same time that it exposed vulnerabilities in the attempt to empower the FDA to unmake existing markets.[88] At one end were those like A. Dale Console, former witness at the Kefauver hearings, who lamented:

> If the drug industry is successful in urging medical leaders to lodge a formal protest against the proposed ban on antibiotic mixtures, . . . the caduceus should be at half-mast. If "thousands of physicians" have found these mixtures useful, it should be easy to collect conclusive data demonstrating their utility. The drug industry can answer the FDA's objections better by collecting and submitting those data than by blowing up an emotional storm over "interference" with physicians' prerogatives. The FDA appears to be interfering with the sale of drugs of questionable merit, not the physician's privileges. A physician has special privileges because he has special training. If he can be goaded into accepting and defending an unscientific attitude by such childishly obvious tactics, he needs reeducation, not privileges. The real need is for data, not protest.[89]

More tersely, a North Carolina practitioner wrote to Max Finland, years before the notion of "choice architecture" was coined, urging him and his panel to "stick to [their] guns" and hold the line in defense of FDA regulation: "The practicing physician will still have the freedom to prescribe an antibiotic for every runny nose he sees if he so chooses. I do not think however that the drug industry should make it convenient for him to be so stupid."[90]

From the other end, exposing ongoing divisions between the AMA's

Council on Drugs and its central administration, and maintaining the position that Hugh Hussey had publicly taken at the Kefauver hearings, came the AMA's response. Its Council on Drugs had in July 1963 taken the stance that the fixed-dose combination antibiotic remedies "tend to make careful clinicians definitely uneasy," and at the end of October the AMA released a statement that it still concurred with the council's assessment.[91] Yet, as the statement continued, "Paradoxical as it may seem . . . the American Medical Association also believes that physicians should oppose the intention announced recently by the FDA to the effect that antibiotics must be removed from the mixtures containing 'cold remedies.'"[92] The implications with respect to the roles of regulation versus education were in full view, and the AMA strove to preserve the autonomy of the individual clinician: "Complete elimination of these mixtures from the prescription market by regulatory fiat is an extreme action that, in our opinion, is not authorized by existing law, and, in any event, attempts to solve by coercion what can be accomplished by voluntary, educational programs and regulation."[93] Or as West Virginia's Eastern Panhandle Medical Society put it, "We deeply resent this proposed usurpation of our prerogative to treat and diagnose our individual patients and our prerogative to err if that be the case."[94]

Along this spectrum, Finland, Dowling, and their clinical pharmacological allies were suprisingly ambivalent. With the passage of the Kefauver-Harris amendments—the result of so much of their own input—the infectious disease–based reformers and their fellow clinical pharmacologists had become slightly wary, collectively, of the power the FDA had apparently assumed.[95] Some appear to have become more wary than others, with Louis Lasagna representing one end of the clinical pharmacological spectrum in particular. Lasagna had been perhaps the most eloquent and quotable of the reformers to have appeared before Kefauver, yet Lasagna, who had studied thalidomide and considered it a promising sedative, felt that in the wake of the "frenzied public response to the thalidomide episode," the Kefauver-Harris amendments had been passed with undue alacrity. In his view, in response to such a "panic," the FDA would be imbued with powers that exceeded its capacity to enforce them, resulting in a potential dramatic brake on pharmaceutical innovation.[96]

And on certain related issues, all the reformers seemed to agree. When, for instance, Rep. Samuel Friedel's (D-Maryland) amendment mandating informed consent in clinical research had been proposed in September 1962, a host of clinical pharmacologists—including Dowling, Finland, Lasagna, Walter Modell, Harry Gold, and even Henry Beecher—wrote to Kefauver protesting that such an inflexible centralized mandate would "cripple" (in

Beecher's words) American leadership in clinical research.[97] Congress pushed forward with an admittedly diluted provision of informed consent in the Kefauver-Harris amendments, but the therapeutic reformers who had united before Kefauver were left with their own private doubts regarding the role of centralized regulation in the conduct of clinical investigation, which seemed to spill over to doubts regarding interference with the practice of medicine as well.[98] As Dowling Committee member Edwin Kilbourne wrote to Commissioner Larrick in October 1963, the day the AMA released its statement defending the role of the autonomous physician:

> I have conceived, perhaps too narrowly, of the panel's function in "the development of sound medical policy" as one of a body of experts qualified in the disciplines of medicine, microbiology, chemotherapy, and pharmacology. Although some of us may wear other hats, I do not yet believe that in the present instance we have been called upon to enter a debate on Federal Paternalism, Individual Freedom, Infringement of Private Enterprise, or the Right of the Physician to Prescribe the Inefficacious (which I hold to be inalienable). Or have we?[99]

Dowling himself wasn't sure at this point. As he had written to Norris Brookens (the former first president of the Illinois Society of Internal Medicine) the preceding month, concerning the centralized withholding of certain antibiotics in reserve so as to preserve their long-term utility, "I suppose we come back to the conclusion that we can only do so much (and only should do so much) by law and that most of our reforms must come by changing the attitudes of the profession."[100]

With prodding coming from neither the Dowling Committee nor Congress, Larrick and the FDA blinked. In justifying to Senator Humphrey his original delay in acting on the Dowling Committee's report, Larrick had explained: "A regulatory agency, whose actions are always subject to court review, should undertake a new control measure when it has the facilities to press forward to a successful termination of the new action. Premature action not only may fail to achieve the desired result, but also may lead to adverse court decisions which, in fact, wipe out regulatory advances that have been achieved with great difficulty over a period of years."[101] By November 1963, the "facilities to press forward" did not seem in place to force the removal of Achrocidin or its ilk.[102] In retrospect, it was felt that the "overwhelming objection of the medical community" was the primary factor in preventing such concerted action.[103] The FDA didn't feel empowered even to change the labeling of such remedies until February 1966.[104]

But by that year, the stage had been set for a much larger conflict over the capacity of the FDA to undo existing markets and delimit physician prescribing authority. Once again, fixed-dose combination antibiotics would catalyze the discussion, though this time Upjohn's Panalba would serve to focus attention. Along the way, the "well-controlled" study and the notion of "substantial evidence" would be transformed from rough notions juxtaposed to testimonials and individual judgment into explicitly articulated definitions and regulations that would indelibly shape both pharmaceutical regulation and the methodology of clinical investigation.

DESI, Panalba, the "Well-Controlled" Study, and the Unmaking of Pharmaceutical Markets

THE EXECUTIVE AND LEGISLATIVE BRANCHES

Concerns regarding the FDA's slow response to addressing the status of existing drugs persisted throughout the mid-1960s. It was indeed a daunting prospect, calling for an imaginative response. As Dominique Tobbell has described, with the FDA's expertise and manpower still deemed unfit to the task at hand, the National Academy of Sciences–National Research Council (NAS-NRC) had in 1964 formed the Drug Research Board (DRB) to serve as a "Supreme Court" for advising on national pharmaceutical policy.[105] Tobbell has further related the degree to which the DRB (in existence from 1964 to 1975) derived from the prior Commission on Drug Safety, formed in 1962 via collaboration between industry and members of academic medicine. The DRB, as Tobbell relates, continued to have heavy industry representation and helped serve the interests of those in both industry and academia concerned about the possibility of undue government influence in the practice of medicine. But the monumental Drug Efficacy Study and Implementation process—initially overseen by the DRB, and by which all existing drugs approved between 1938 and 1962 were to be evaluated for efficacy—would serve as an important exception to this rule, especially with respect to the fixed-dose combination antibiotics.[106]

In March 1966, FDA commissioner James Goddard (who had succeeded George Larrick in January of that year, and who was unabashed in his willingness to confront industry) reached out to Keith Cannan, chairman of the Division of Medical Sciences of the NAS-NRC, to have the DRB oversee the DESI process, reaching an agreement by June.[107] The pharmaceutical industry knew it was taking a "calculated risk" in agreeing to the process.[108] The NAS-NRC eventually oversaw thirty panels, made up of nearly two hundred experts, to review twenty-two drug categories. Led in this effort by Cannan and Duke Trexler, the NAS-NRC formed a Drug Efficacy Study

Policy Advisory Committee, with its own executive committee initially chaired by William Middleton (and later by Alfred Gilman) and including Max Finland as one of its five members.[109]

The panels were initially asked to rely on "factual information" available in the published literature and information available from either the FDA or manufacturers, and to divide the drugs into four categories: "clearly effective," "probably effective," "probably ineffective," or "clearly ineffective."[110] But problems arose almost immediately in both the process and outcome aspects of the program, planting the seeds of the Panalba affair. With respect to the determination of "substantial evidence" from well-controlled trials, Max Finland, at the earliest program planning meeting, drew attention to the thin clinical pharmacological ground on which they actually stood. It would take twenty-five to fifty years to conduct the studies to evaluate all the existing drugs properly, he surmised, meaning that if they wanted categorical answers from the panels, then elements of subjective opinion would have to be introduced into the evaluations. As Finland summarized, unintentionally anticipating later difficulties: "If that is what they want, that can be attained. But if they want something the somewhere nitpickers are going to be able to say: 'Here's exactly the evidence and how well controlled was the evidence that was gotten,' th[e]n I don't think that anybody should volunteer for that sort of a job."[111] And as Cannan summarized toward the end of the study, exposing the Achilles heel of its process to later scrutiny:

> Where evidence was scanty, the panels have not hesitated to supplement it by attempting to seek out a consensus of clinical experience and buttress this, where possible, with deductive reasoning. In brief, the panels have taken a relaxed view of the phrase "substantial evidence" and many of the judgments of the panels should be described as informed opinions that are not solely based on hard and stubborn facts. Such is the state of the art in drug therapy.[112]

Such would also soon be the state of the art in legal contestation.

With respect to outcomes and the four categories initially defined by Commissioner Goddard, fixed-dose combination drugs pointed implicitly to comparative effectiveness considerations, as they at the very least mandated a comparison to their component ingredients. Max Finland would again be at the center of this discussion, recommending that the inclusion of a fifth category—"effective, but with the following qualification"—be added.[113] This additional category was extended beyond the combination drugs themselves and essentially allowed an overt component of comparative effectiveness consideration into the calculus, as certain drugs "were no longer approved forms of treatment because much better, safer, or more conveniently administered drugs were now available."[114]

This was only the beginning of the discussion regarding fixed-dose combination drugs, however. It had been sensed from the start of the program that such combination products would be the most problematic aspect of the program, and while by May 1969, 2,824 reports covering 3,700 drug formulations already had been sent to the FDA, by far the most contentious and broadly consequential concerned the fixed-dose combination antibiotics.[115] Five of the thirty panels were dedicated to antimicrobial agents, which themselves represented 1,037 of the 4,152 products initially targeted for evaluation.[116] Four of the panels, respectively headed by Heinz Eichenwald, William Hewitt, William Kirby, and (former Max Finland fellow) Calvin Kunin, were forced to confront the fixed-dose combination antibiotics.[117] As would become apparent at consequent congressional hearings, "All the [antibiotic] panel chairmen knew each other very well, and had met prior to the initiation of the proceedings of the panels, and . . . staked out their areas and the areas of collaboration."[118] And as opposed to the ambivalence exhibited by those on the Dowling Committee four years previously, panel members appear to have relished the opportunity to adjudicate the merits of the fixed-dose combination antibiotics. Concerns about the impending collapse of clinical pharmacology had not been borne out since the passage of the Kefauver-Harris amendments, and second-generation infectious disease–based reformers such as Kunin felt empowered to act to delimit existing markets. When later asked about a priori bias among the panel members, Kunin would admit that he had felt at the time that "the beauty of this particular opportunity is that we are charged with the efficacy question. And here we have the opportunity to present these opinions fully."[119]

By April 1968, none of the panels was intending to endorse the fixed-dose combination antibiotics. But differences emerged over how to characterize them—as "effective, but" or as "ineffective."[120] While FDA Bureau of Medicine director (and future agency commissioner) Herbert Ley regarded this as a largely semantic distinction (feeling that the FDA could work with either designation), the potential legal implications were obvious to several panel members.[121] As Leighton Cluff, chairman of medicine at the University of Florida, put it: "'Effective, but' or 'Ineffective' resembles Hamlet's 'to be or not to be,' and that is the question. In a sense, one position is to continue suffering an 'outrageous fortune,' while the other position is to arm against a sea of troubles by forthrightly opposing them."[122] A special meeting (including four of the antibiotic panel heads, Gilman, Finland, Ley, Cannan, and others) was thus convened to decide the matter in May 1968, where the fateful sixth category of "ineffective as a fixed combination" was added to the existing categories.[123] The category was later described as developed "to deal rationally with certain combinations of drugs, notably combinations of two or more antibiotics," when the sum of the combination was felt to be no greater than one of its constituent components.[124] And the same month that the meeting took place, James Goddard reasserted his commitment to the "spirit" in which the DESI process was initiated: "to bring about more rational drug therapy by influencing the development of a more rational drug supply."[125] However, such categorizations would serve as the entry point to extensive debate over who was defining such "rational" therapeutics and behavior in the first place.

The relevant panels each ruled the fixed-dose combination antibiotics "ineffective as fixed combinations" and sent their recommendations to FDA commissioner Herbert Ley in July 1968, the same month that Ley replaced James Goddard. It was quite a welcoming present, and the NAS-NRC's Keith Cannan knew that trouble could be brewing. As Cannan admitted to the Council for the Advancement of Science Writing in November of that year: "Combinations are a lucrative segment of the market. They are also a considerable convenience to the busy practitioner. It may not be easy to reconcile the conflicting interests of town, gown and the counting house on this issue."[126] Cannan also knew that it would be difficult for the FDA to back off from the panels' recommendations:

> In the sponsorship of the Drug Efficacy Study, FDA, for the first time invited the biomedical community to participate in the management of drugs in a big way. I say management intentionally because although contractually the Academy is an advisor to FDA, its recommendations will be executive in effect. It is inconceivable that FDA will override the recommendations made to them in any but a very few cases and in these on grounds that reach beyond scientific or medical judgment.[127]

Those grounds, however, would become the very terrain of battle over the methodology and reach of the expert panels and the FDA. And between the time the panels' resulting position piece on combination antibiotics would be rejected for publication in *JAMA* in January 1969 and accepted for publication in the *New England Journal of Medicine* in late May 1969, debate over Upjohn's Panalba in particular would epitomize tensions regarding physician prescribing autonomy and the emerging regulatory authority of the FDA to define the "well-controlled" clinical trial and unmake existing pharmaceutical markets.[128]

Panalba, a combination of tetracycline and novobiocin, had first been marketed by Upjohn in 1957, with 750 million doses prescribed by 1969.[129] Novobiocin had been independently discovered by Merck and Upjohn (and later rejected by Merck) and was considered potentially useful versus staphylococci but known to engender rapid resistance and potentially cause liver toxicity.[130] By 1968, while annual novobiocin sales remained at approximately $200,000, annual Panalba sales approached $20 million, accounting for 12 percent of Upjohn's domestic gross income.[131] Advertisements for Panalba at best reflected the notion of nonspecific "rational" therapy articulated by Henry Welch the prior decade (as discussed in chapter 2). As one Panalba ad, concerning the treatment of "bacterial tracheobronchitis," related: "In the presence of bacterial infection, taking a culture to determine bacterial identity and sensitivity is desirable—but not always practical in terms of the time and facilities available. A rational clinical alternative is to launch therapy at once with Panalba, the antibiotic that provides the best odds for success."[132] At worst, to skeptics like Columbia professor of pediatrics Charles May, the advertisments seemed to demonstrate "a low regard for the intellect of the average doctor" while exemplifying the miseducation of those very physicians.[133]

On Christmas eve of 1968, armed with the NAS-NRC antibiotic panel reports, Commissioner Ley filed an announcement in the *Federal Register* regarding the receipt of the recommendations and his consequent intention "to initiate proceedings to amend the antibiotic regulations where necessary to delete [fixed combinations of tetracycline and novobiocin] from the list of drugs acceptable for certification."[134] This notice set in motion an "explosive" series of backroom negotiations and public debate regarding the process and outcome of such FDA activity.[135] The AMA, so central to prior debates over efficacy and existing remedies, was missing from the discussion, but industry more than made up for the absence.[136] Upjohn, initially given thirty days (soon extended by Ley to 120) to submit "pertinent" data to the FDA, sent in studies, but they also copied Lederle's approach from six years prior and incited the flooding of the FDA with a cascade of letters from presumably aggrieved physicians deprived of their prescribing autonomy and attesting to the utility of the drug in their practices.[137] On April 2, 1969, an unmoved Herbert Ley issued in the *Federal Register* the intended withdrawal of all the remaining fixed-dose combination antibiotics (including Sigmamycin, since renamed Signemycin).[138] Meanwhile, Upjohn and its congressional allies applied pressure on HEW chief Robert Finch to override the FDA process.[139]

In this context—termed a "political holocaust" by the very congressman

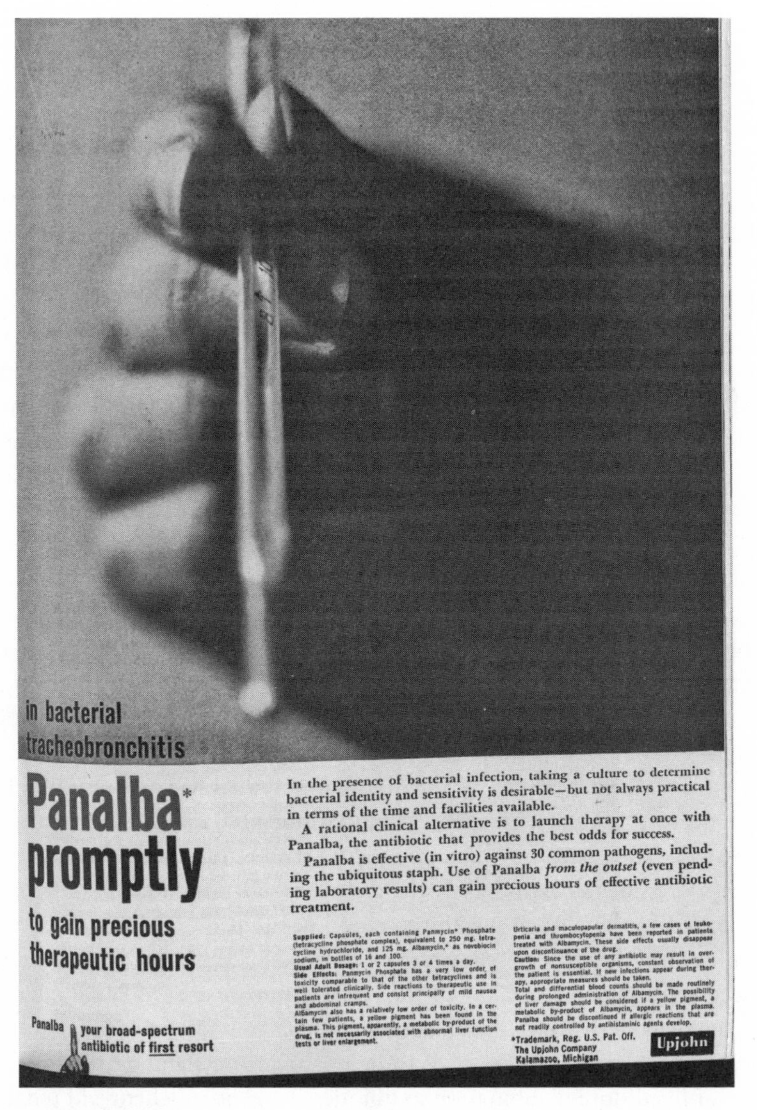

"Panalba Promptly." Upjohn's advertisement for Panalba emphasized the "rational" decision to shoot first and ask questions later in the treatment of "bacterial" tracheobronchitis (which was to be clinically diagnosed even without cultures being taken). *Medical Times* (January 1962): 197a.

who met with Finch—two near-simultaneous congressional hearings were convened in May 1969, though with key differences in their agendas.[140] The first, by Senator Gaylord Nelson (D-Wisconsin), served as the twelfth of an extended set of thirty-four hearings over twelve years on "Competitive Problems in the Drug Industry" and as a final public indictment of the fixed-dose combination antibiotics. The second, on "Drug Efficacy" more broadly and held by Rep. Lawrence H. Fountain (D-North Carolina) and the Intergovernmental Relations Subcommittee, served as a public debate on the very definition of the well-controlled clinical trial.

Nelson's Senate hearing—calling forth NAS-NRC antibiotic panel heads Hewitt and Kunin, among others—served as an orchestrated impeachment of both fixed-dose combination antibiotics and fixed-dose combination drugs more generally. At the same time, it offered an implicit notion of "irrational" medicine as stemming from the dynamics of the unregulated pharmaceutical market. In this telling, Panalba at the very least represented the combination of "a major league with a minor league drug and considering the result a new, superior product."[141] Still more cynically, Kunin invoked a critique of evergreening, years before it would be termed as such, proposing: "If I were in an industry, and I were in danger of losing my patent with which I have reaped my fortune over many many years and wanted to retain that patent, then I would combine that drug [with] something else so that I have a new proprietary agent. This is another way of keeping this within one's own pocket."[142] The litany of potential harms caused by Panalba, deemed both less effective (owing to potential drug antagonism) and more toxic (owing to novobiocin toxicity) than tetracycline alone, was put forth, alongside the announced "theft" of nearly $12 million annually, when compared to the cost of generic tetracycline alone.[143] And an increasingly visible concern, the near-"disaster" of consequent antibiotic resistance, was voiced, exemplified especially by the "indiscriminate use" of Panalba's cousins, the penicillin-streptomycin combinations.[144]

Witnesses were in fact quick to generalize from Panalba to the entire class of fixed-dose combination antibiotics. As Heinz Eichenwald (chair of pediatrics at the University of Texas Southwestern Medical Center) described Dorsey Laboratories' Tain, a combination of triacetyl-oleandomycin with a pain reliever, nasal decongestant, and antihistamine, "One might wonder why the manufacturer did not include vitamins, sex hormones, a contraceptive, and corticosteroids to cover all other eventualities."[145] As compared with the 1963 debate over such combinations of antibiotics and nonantibiotic ingredients, reformers were now clearly prepared both to delimit physician prescribing autonomy and to generalize to fixed-dose combination drugs more broadly. Tulane University and Louisiana State University anesthesiologist and pharmacologist (and unsuccessful nominee, that very summer, to head the FDA's Bureau of Medicine), John Adriani, offered a culinary analogy:

> If you sell salt and pepper, you can mix both together, in a fixed ratio and all you need is one shaker. And this is the argument drug firms use. One capsule or pill. This is ridiculous, because some people like more salt and some more pepper. If you packed salt and pepper together and put it on the grocery shelf it wouldn't sell. Some might call it "piperosal" and say it is delicious, in a big ad. It might then appeal to somebody who might buy it. This is the essence of what you are doing when you mix two drugs together under a brand name.[146]

Adriani, like his fellow witnesses from academia, appeared to relish his role. Four months earlier, he had written to Harry Dowling: "I think we are acting like ostriches. When it comes to a question of evaluation of mixtures and combinations of drugs you recognize that the problem is there, and we hope that if we don't see it that it will go away, but I think it is going to stare us in the face more and more."[147] Still later in his testimony, Adriani felt empowered to summarize, "all of the fixed combinations are irrational."[148]

In this construction, the drugs themselves were depicted as "irrational" or "illogical," the bad medicines at the root of "bad medicine," serving the needs of commerce more than those of patients.[149] Since early in the previous decade, sociologists, pharmaceutical firms, journal editors, and clinician-reformers such as Harry Dowling had taken an increasingly keen interest in the influence of pharmaceutical marketers on physician prescribing habits.[150] Indeed, as mentioned in chapter 1, the most extensive study undertaken, the Fond du Lac study of the multiplicity of prescribing influences on every single physician in a single county in Wisconsin in 1955, had looked specifically at the usage of Lederle's tetracycline antibiotic Achromycin as one of its studied drugs, finding that "three out of five Fond du Lac doctors said that the Lederle detail man was instrumental in getting them to prescribe Achromycin."[151] By the time of Nelson's hearings on the fixed-dose combination antibiotics, in the most extreme reading of this literature, physicians were to be considered innocent, yet "gullible," dupes, "educated" by the pharmaceutical industry.[152] This depiction represented the summary judgment of a decade's worth of infectious disease–based (and eventually, broader) reformist discourse.

Once again, proposed approaches to the ethical pharmaceutical industry followed those formerly used to confront the proprietary drug industry. In the first half of the twentieth century, Arthur J. Cramp, heading the AMA's Propaganda Department against the patent medicine industry, had "sought to emulate his adversaries" and provide counter-educational materials, dragging himself and his lantern slides forth from Chicago to lecture "to any interested party."[153] And in the wake of ethical pharmaceutical wonder-drug marketing, Calvin Kunin used the opportunity to offer, as an offsetting mechanism, the "counter detail man," designed to use the tools of pharmaceutical marketing to "engage in a counterpropaganda program."[154]

Yet while there was room for such counter-education, in the aftermath of the Kefauver-Harris amendments there was room for regulation as well, as worthless drugs could now be removed from the market. The question was on what basis such removal should occur. Kunin made a critical hedge with respect to the fixed-dose combination antibiotics: it might be that in the future, certain fixed-dose combination drugs would in fact be found more effective than their constituent parts. In the case of such drugs, he advised, "we ought to have an open mind and judge each on its merits."[155] And an entire parallel hearing on "Drug Efficacy," led by Lawrence Fountain in the House of Representatives, was to be conducted to determine just how to render such judgments.

Viewed from another angle, the May hearings on "Drug Efficacy" served as a forum on another potential source of "irrational behavior": namely, that of the FDA and its power "arbitrarily to jeopardize the rights of the private sector."[156] Upjohn's supporters maintained that the company had been both caught in the middle of evolving methodological ambiguity and deprived of due process in being given inadequate time to conduct "adequate and well-controlled" studies as mandated by the FDA. It would be their turn to invoke sports metaphors, variously claiming that one ought not "change the rules in the middle of the ball game" and that the FDA was the "referee and [pharmaceutical companies] didn't know the rules."[157] How, in a setting of apparent methodological flux and political pressure, could the FDA apply a fair and consistent "scientific objectivity?"[158]

In response, Commissioner Ley and HEW assistant general counsel William Goodrich stated that companies like (and especially) Upjohn had known since 1962 of the new efficacy requirements and that the regulatory process concerning the approval of *new* drugs had demonstrated the methodological rigor that would be required with respect to the proof of efficacy of *existing* drugs.[159] In *Medicine Makers of Kalamazoo*, a friendly biography of the Upjohn Corporation appearing in the middle of the Kefauver hearings, author Leonard Engel had remarked:

> Several difficulties have arisen to make the testing of new drugs a taxing and often baffling problem. To begin with, conscientious pharmaceutical

> researchers would like results that are completely objective and free from bias. . . . In an effort to avoid such difficulties, many new procedures have been introduced into clinical testing. Most important has been the controlled clinical trial, in which patients are divided into two groups as closely alike as possible, except that one receives the new drug or treatment and the other does not. In a further effort to eliminate researchers' and patients' emotional reactions and so assure trustworthy results, controlled trials are often arranged on the so-called double-blind pattern.[160]

Beginning in 1957, Upjohn itself had supplied the hydrocortisone for Max Finland and his colleagues to perform a multisite, "cooperative," randomized, double-blind study of corticosteroids as a supplement to the treatment of severe infections, predicated on "the hope that the randomized placebo will permit a large enough group of patients to be treated to permit us to determine whether there is indeed a life-saving effect consequent to the use of hydrocortisone as an adjunct to other treatment."[161] The double-blind was rigorously maintained, even when the investigators were "dying to know" whether recovering patients had received the hydrocortisone or not.[162] And such studies were acknowledged by Upjohn medical representatives at the time as a contrast, from a methodological standpoint, to Upjohn's parallel sending out of combination antibiotic-steroid pills "to certain amenable practitioners . . . for case reports only" as a means of achieving market priority.[163]

As Daniel Carpenter has described, such requirements and acknowledgment of a methodological hierarchy increasingly and visibly infused the process of new drug review throughout the 1960s.[164] As early as June 1963, PMA president Austin Smith had admitted as such before the Federal Bar Association: "Clinicians are experiencing serious difficulties in conducting double-blind studies which, ironically, are all the more necessary because of the new regulations."[165] The following month marked the first meeting of the FDA's Advisory Committee on Investigational Drugs, termed the Modell Committee after its chairman, Cornell clinical pharmacologist Walter Modell.[166] Tasked to serve as a bridge between the FDA and the clinical investigator community, one of the committee's first recommendations was for the FDA to publish a manual laying forth expectations for investigators, industrial or otherwise.[167] At the same time, thalidomide heroine and FDA Investigational Drug Branch head Frances Kelsey reported on feedback from psychopharmacological investigators, who requested that the "reasons for rejecting or accepting data should be made available to investigators so that gradually a body of knowledge can be established which will serve as precedents for proper investigation and in essence gradually exclude improper or inadequate data."[168]

Both explicitly and implicitly, the FDA made progress. By October 1963, members of the Division of New Drugs had prepared a Drug Evaluation Manual, stating that Phase III trials entailed "Controlled (e.g., double-blind) studies in specific disease states by clinical experts" (underlined in original), with an extended discussion of "proper controls" that included discussions of randomization, placebo effects, and blinding.[169] Key FDA advisors and representatives would emphasize that too much emphasis could be applied to the "double-blind" nature of individual studies, at the expense of attention to otherwise well-controlled data.[170] But Section 243.4 of the revised manual for the FDA Bureau of Medicine would be presented at the Fountain drug efficacy hearings, stating: "Unless it is clearly not indicated, or not feasible, double or multiple blind placebo controlled studies are required which on medical and statistical evaluation show the test drug to be superior to the placebo. . . . An acceptable study should be confirmed by at least one independent well-designed study on the same formulation."[171] And the same was to hold for combination drugs. As Ralph Smith (director of the Division of New Drugs) and Howard Weinstein (director of the Division of Medical Review) wrote their Bureau of Medicine (and essentially, FDA) medical director Joseph Sadusk in March 1966, just as the DESI process was beginning: "For drug combinations the program of investigation should be designed to show [the] effectiveness of each active ingredient as used in the product. This may be done by comparing the total formulation with the formulation minus one ingredient and with a placebo."[172]

Such emerging notions, however, had received neither formal regulatory codification nor extended public discussion and appeared to be offset by the NAS-NRC panels' being permitted to use "informed judgment and experience" in advising the FDA.[173] Combined with the inherent methodological difficulties in evaluating clinical studies, these vagaries led Rep. Clarence Brown (R-Ohio) to muse at the hearings: "I am beginning to get the impression that medicine is more of an art than a science. And that the determination of efficacy is not quite like the determination of pregnancy, it is more like love."[174] Commissioner Ley replied that "we hope that one leads to the other," and the hearings stood as a public forum regarding the emerging definition of the "well-controlled" study in the years since the passage of the Kefauver-Harris amendments.

To start with, and in a straight line from the public debate between Max Finland, Henry Welch, and Félix Martí-Ibañez over a decade previously, the controlled clinical trial was again contrasted with the testimonial. At one level, *testimonials* referred to the judgments or pronouncements of individual physicians. These could be considered well-intentioned and morally neutral, if methodologically flawed, as when Fountain summarized that "it doesn't matter who a doctor is, or how qualified he is, or how brilliant he is, or how great and good he is; if he expresses an opinion and submits no evidence from adequate and well-controlled studies, you are not in a position to give any weight to it."[175] Or they could be considered the manipulated tools of industry, as when Ley angrily wrote to Upjohn's president in February 1969 about the "testimonial type responses" the FDA had received in support of Panalba and threatened to cancel the very extension Upjohn had been given to submit "substantial evidence."[176] Either way, as William Hewitt intoned at the parallel Nelson hearings, "We emerge into a phase of deterioration, disenchantment, and dismay which I would entitle the evaluation of drugs by the method of popular vote."[177] And at a clinical investigative level, *testimonials* referred to the type of company-sponsored, uncontrolled case report studies criticized over a decade previously, to be explicitly contrasted with "well-controlled studies demonstrating efficacy."[178]

What, then, constituted a well-controlled study? Or as Rep. Guy Vander Jagt (R-Michigan) posited: "In determining what is a well-controlled clinical study, is not that sometimes a complicated matter? . . . Is it not obvious —that the more definite the standards and the more definite the tests and the more adequately [defined] the meaning of a well controlled clinical study, the more definite that is, the greater the chance there will be of arriving at the goal of coming as close to the scientific truth as we possibly can."[179] On the one hand, embodied in the transition from testimonial to controlled study was an elevation of methods over men: "Regardless of who is conducting a study or how well informed they may be, or how much experience they have had, the study must be satisfactorily designed and with proper controls."[180] On the other hand, it continued to be apparent that there were competent and incompetent (or biased) investigators, and the sordid case of the pain reliever Measurin (an extended-release form of aspirin), initially approved on the basis of a single, fraudulent investigation, was brought forth as justification for the need for two well-controlled studies for a given drug.[181] Defining the methodologically rigorous, "well-controlled" study took a bit longer, and an entire afternoon session was devoted to a methodological dissection of Unimed Corp's anti-Meniere's treatment, Serc. In essence, the properly controlled trial was defined by reference to its negative counterpart, as revelations of methodological problems in the Serc trials' seemingly randomized control studies—a poorly defined disease state, insufficient blinding, and inadequate duration of treatment—had led to the

FDA's attempt to withdraw the drug after its initial approval.[182] The properly controlled trial, by inference, was to be contrasted with what was done with Serc, and still more fundamentally and visibly, with Panalba's many trials.

THE JUDICIARY

On May 15, 1969, the last day of the Fountain hearings, Ley republished his order to decertify Panalba in the *Federal Register*.[183] His unaltered stance on the medication was made clear in a speech to be delivered at the PMA's annual meeting five days later: "The best scientific thinking decries the use of antibiotics in this shotgun fashion. Not only are patients exposed to highly toxic drugs unnecessarily, but the indiscriminate use of antibiotic combos can result in the emergence of bacterial strains resistant to the agents used. I can only reiterate what I have said before. The use of two or more active ingredients in the treatment of a patient who can be cured by one is irrational therapy."[184] The burden of proof was clearly to be placed on Upjohn to provide "well-controlled" studies rather than on the FDA to suspend decertification until their completion or to have to justify the need for such studies in the first place. Upjohn, which would in fact submit fifty-four studies to the FDA, filed an objection to the commissioner's order and sued to prevent the FDA from being able to enforce the decertification. And deliberation and contestation over the meaning of "substantial evidence" and "well-controlled" studies would continue in earnest throughout the summer, as the scene shifted from Congress to the courts.

In their formal objection and request for a hearing, Upjohn's attorneys attempted to refute both the Nelson and the Fountain hearings, as well as the NAS-NRC report supporting them.[185] First, once again echoing Welch's support of the fixed-dose combination antibiotics from a decade prior, they noted that the NAS-NRC reports "ignore completely the realities of medical practice. . . . The vast majority of ambulatory infections are treated with antibiotics without preliminary bacteriological studies, which are costly, slow, of limited value in predicting clinical response, and in most cases, unnecessary."[186] With dubious logic, they continued: "Under these circumstances, a broad spectrum antibiotic is a drug of choice, and the panel's contrary statement is in error. Since each of the specified Upjohn products has a spectrum broader than either of its components alone, it is a useful agent in current medical practice."[187] Second, they again questioned the FDA's ability to unilaterally define "adequate and well-controlled investigations."[188] Third, and most convincing to Western Michigan district court justice Wallace Kent, they contested both the initial anonymity of the NAS-NRC panels and the panels' use of "informed opinion," which seemed to contradict the substitution of methods for men in adjudicating therapeutic efficacy.[189]

Justice Kent, citing an attempted imposition of a "rational therapeutics" by the "heavy bureaucratic hand" of the FDA, upheld the injunction.[190] But Ley responded in the *Federal Register* on August 9 with an extended methodological dissection of Upjohn's fifty-four studies. Some were in vitro studies, others simple antibiotic blood-level determinations, others uncontrolled, and still others poorly controlled.[191] To Ley, the nature of the well-controlled study had already been publicly debated at the Fountain hearings. He was willing to hear an oral presentation from Upjohn "in which it can offer its analysis of the reported literature on which it relies and explain its theory as to why it believes the reported clinical experience can and should be accepted as adequate medical support for its claims of effectiveness."[192] But this was to determine which existing evidence fit *Ley's* definitions, not an evidentiary hearing concerning the nature of such definitions themselves.

Upjohn's lawyers apparently missed that half of the memo. At the oral presentation on August 13, Upjohn chief counsel Stanley Temko began by again questioning the very meaning of "adequate and well-controlled investigations," as well as the juxtaposition between the "rational or irrational practice of medicine."[193] Assuming the role of Martí-Ibañez, he defended the value of collective clinical experience: "If Dr. Finland does not want to use fixed combinations, he is perfectly free to do so. . . . Taking Panalba, it is used by 20[,000] to 25,000 physicians in this country. Now, can you just say, ex cathedra, that everything these people have done is irrational? Those people knew Dr. Finland didn't like fixed combinations in 1957."[194] Turning to the reformers' notions of evidence, he advanced: "We don't have just testimonials. We have a very extensive compilation of studies in vitro, animal, human, coupled with very substantial, very substantial clinical experience, and I think, taking that together, it is a factor."[195] As such, in a state of purported methodological flux, the FDA could not determine by fiat the meaning of the well-controlled clinical trial, "say that this is your interpretation of substantial evidence, just say what you will say *a priori* and *ex parte*. This is our definition, we have looked at [Upjohn's] articles, we don't think that measures it, that is the end of the story."[196] Therefore, an evidentiary hearing concerning the definition of the well-controlled study, rather than the "bobtail procedure" of the oral presentation, was necessary.

Commissioner Ley remained unconvinced. As he stated at the presentation, he knew about the challenges of clinical studies, having wrestled with them himself two decades previously when he had been one of the original investigators of chloramphenicol as a member of the Army Medical Department and Graduate School.[197] He had been intimately involved with and invested in the DESI process since his tenure as the director of the FDA's Bureau of Medicine, reporting back in 1967 that the Kefauver-Harris amendments had provided the opportunity for "the establishment of formal legal standards in place of the experience of the market place."[198] The dispute with Upjohn had only stiffened his resolve. In February 1969, in the midst of back-and-forth chidings and negotiations with Upjohn, he had reported to the New York Pharmaceutical Advertising Club:

> The real "gut" issues of the antibiotic combination controversy are exceedingly simple. Are we in this country dedicated to a rational, scientific basis of antibiotics therapy or are we dedicated to the irrational "shotgun" approach to treatment? . . . Where do you stand if you were being treated—or your wife—or child? The Academy has stated its position. We agree with it. And in time, whether we like it or not, I am convinced that even the courts will agree with this position.[199]

By May, amid simultaneous Senate and House hearings on the subject, he had hardened his stance still further, warning the Pharmaceutical Manufacturers Association: "Unless there is a major change in the drug industry's emphasis on sales over safety, the industry as we know it today may well be buried within the next several years in a grave it has helped dig—inch by inch, overpromotion by overpromotion, bad drug by bad drug."[200] More than a decade's debate over the controlled clinical trial was coming "to a head," and the spring and summer of 1969 had given Ley ample opportunity to reflect on the nature of the well-controlled study and its differentiation from bacteriological data, in vitro studies, qualitative impressions, and even historical controls.[201] At this point, his mind was clearly made up and certainly not swayed by Upjohn's protestations.

Within days after Upjohn's oral presentation, Ley would formally define the nature of "adequate and well-controlled investigations." As he published concerning the nature of "control" in the *Federal Register* on September 19:

> Three types of controlled comparisons are possible: (i) Placebo control: The new drug entity may be compared quantitatively with an inactive placebo control. This type of study requires at the minimum that the patient not be able to distinguish between the active product and the placebo. Double blinding, to include the clinical observer, may or may not be desirable, depending on the measurement system used to evaluate the results. (ii) Active

> drug control: The new drug entity may be compared quantitatively with another drug known to be effective in situations where it is not ethical to deprive the subject of therapy. The same considerations to the level of "blinding" apply as with a placebo control study. (iii) Historical control: In some circumstances, involving diseases with high and predictable mortality (acute leukemia of childhood) or with signs and symptoms of predictable duration or severity (fever in certain infections), the results of use of a new drug entity may be compared quantitatively with prior experience historically derived from the adequately documented natural history of the disease in comparable patients with no treatment or with treatment with an established effective therapeutic regimen.[202]

And as applied to the fixed-dose combination antibiotics, as Ley would publish in the adjoining September 19 notice, once again repealing Panalba:

> This means that the experimental factors must be so controlled that the effectiveness of an anti-infective drug on the disease process in patients can be compared with the effect of no treatment or of a recognized effective treatment of patients with the same disease conditions. The Commissioner concludes that with combination drugs purported to have advantages exceeding those of the components, there must be adequate, well-controlled data documenting the claimed advantages. While a controlled investigation does not always require comparisons with placebo or with a known active drug, when that type of double blind clinical study is not done other factors must be controlled for the study to yield meaningful results. . . . The time has come to end the marketing of these combination drugs which fail to meet the legal standards of effectiveness and which involve a significant and unacceptable hazard in the light of the failure of proof of effectiveness.[203]

To industry supporters, the model Ley held up represented "an ideal study, which well fits the platonic notion of the perfect conception which cannot be realized."[204] A series of court cases would ensue, some adding technicalities, others with lasting and generalized consequences. On the day of the repeal, the Pharmaceutical Manufacturers Association filed for an injunction against the FDA's acting on the regulations, citing most persuasively (if technically) the lack of opportunity provided for further invited comment.[205] Between the filing of the case and the court's finding in favor of the association in January 1970, Herbert Ley would resign from the FDA, to be replaced by Nixon appointee Charles Edwards.[206] Edwards would essentially republish the proposed regulations in the *Federal Register* in February 1970, offering opportunity for comment, while reporting to the Drug Research Board of the NAS-NRC (indeed, echoing his predecessor) later that month:

> The reaction to [the NAS-NRC] study by many practicing physicians is disappointing with respect to fixed combinations of antibiotics. I have a rather strong feeling, however, that most practicing physicians value expert opinion as well as scientific evidence; and I believe that the practice of medicine will improve because the majority of our colleagues will be persuaded by the science of medicine rather than the art of medicine, in the final analysis.[207]

By that time, Upjohn's own critical case against the FDA—filed in October (in response to the proposed regulations of September 19) and calling for evidentiary hearings concerning the nature of well-controlled studies—was proceeding through the court system. Lines were clearly drawn in the sand, with the general implications of Panalba for DESI at large—and, in particular, for the FDA to undo existing markets and thereby infringe on prescribing autonomy—clear for all to see. As William Goodrich stated in his supplemental brief, "This case is a test of the FDA's will and its ability to move forward with dispatch in implementing the judgments of the Nation's leading experts in drug therapy expressed in the efficacy review conducted by the NAS-NRC."[208] Competing affadavits were offered by each side, representing a splintering of formerly allied clinical pharmacologists; Louis Lasagna was now on the side of Upjohn, while John Adriani, William Kirby, Heinz Eichenwald, Calvin Kunin, and Louis Weinstein were on the side of the FDA.[209] Lasagna again invoked the role of "personal opinion and experience" in driving the NAS-NRC panels, but in the end, the burden of proof was placed on the pharmaceutical industry, rather than on the panels or the FDA.[210] The Sixth Circuit Court considered that "Congress did not intend that administrative agencies waste their efforts on applications that do not state a valid basis for a hearing," and on March 3, 1970, found in favor of the FDA.[211] When the U.S. Supreme Court declined to hear Upjohn's application for a stay, the FDA was provided with the legal authority to unmake the fixed-dose combination antibiotic market.[212]

It was now time for the denouement. Commissioner Edwards republished for a final time—on May 8, 1970, and with minor modifications—the definition of the "well-controlled investigation," concluding the report with the statement that "the implementation of the drug efficacy study conducted by the NAS-NRC cannot proceed until these regulations are placed into effect."[213] In response, the PMA again filed for declaratory and injunctive relief in court. But this time Judge James Levin Latchum—who had found in favor of the PMA in January 1970 on a technicality—found in favor of the FDA, concluding that "the May regulations are reasonable and valid procedures" and that it was "within the power of the Commissioner to require that a genuine and substantial issue of fact be raised as a condition for granting an evidentiary hearing."[214] A series of subsequent lawsuits—most famously *Weinberger v. Hynson, Westcott, and Dunning*—empowered the FDA to efficiently implement the DESI process more generally, as companies were considered to have a right to an individual hearing only if their studies met the criteria established by the May 1970 rules.[215]

As a final and fitting footnote to the fixed-dose combination antibiotic legal story, Pfizer and American Cyanamid (parent company of Lederle) would each file for injunctive relief for their favored combination antibiotics—Signemycin (Sigmamycin) and Achrocidin, respectively—after the Panalba case had been decided. Pfizer would file that summer, with the courts again denying there was anything "irrational" about the prevailing statutory scheme by which evidentiary hearings could take place.[216] American Cyanamid would file later that fall, and while losing its case as well, it did prompt the FDA to formally define "criteria for rational combination drugs."[217] Within three years, trimethoprim-sulfamethoxazole—branded as Bactrim by Roche and as Septra by Burroughs-Wellcome—would become the first post-DESI fixed-dose combination antibiotic approved in the United States.[218]

Thus, a remarkable era focused on "irrational" drugs had come to a close. From Sigmamycin to Achrocidin to Panalba and back again, the FDA had been forced to define a "rational therapeutics" first grounded in the new drug application and then applied to the withdrawal of an entire class of anti-infectives, the fixed-dose combination antibiotics. This process represented an unprecedented delimitation of existing physician prescribing autonomy. Yet it would represent not only the apotheosis of the controlled clinical trial but also the high-water mark for the ability of federal agencies to impinge on physician prescribing behavior with respect to antimicrobials. Indeed, it would lead to an immediate backlash, with longstanding consequences, as reformers turned from irrational drugs to presumably irrational *prescribers* and a range of regulatory and educational proposals. I turn now to these developments, which continue to influence public policy today.

CHAPTER FOUR

"Rational" Therapeutics and the Limits to Delimitation

"Rational" . . . almost invariably serves as the label for the *opinion* of those who happen to be in authority. —W. CLARKE WESCOE (1971)

THE WITHDRAWAL of the fixed-dose combination antibiotics signified the end of a particular era of antibiotic reform. It also represented the limit to the Food and Drug Administration's (FDA's) willingness to shape antibiotic prescribing in this country. The FDA had been empowered to remove "irrational" drugs, those that did not meet its evolving standard of efficacy, from the marketplace. But no one had been empowered to rein in the seemingly *inappropriate* prescribing of *efficacious* drugs; and when reformers attempted to do so, they met fierce resistance from practicing clinicians, exposing and exacerbating town-gown tensions that had surfaced and festered throughout the debate over the withdrawal of the fixed-dose combination antibiotics.

The decade from the late 1960s through the late 1970s would witness increasing attention to—and quantification of—"irrational" prescribing by individual clinicians. As studied by clinical pharmacologists, pharmacists, epidemiologists, and government agencies, findings of irrational prescribing were promulgated both in the medical literature and at government hearings on the pharmaceutical industry. Such findings appeared all the more relevant in the context of escalating national health care costs after the passage in 1965 of Medicare and Medicaid and in the context of increasing apprehension regarding antibiotic resistance.

Yet the limits to the FDA's restriction of antibiotic prescribing had been reached with the passage of the 1962 Kefauver-Harris amendments and the Drug Efficacy Study and Implementation process. And as academic reformers, government agencies, and national medical organizations debated regulatory versus educational remedies to the problems of apparent antibiotic misuse and overuse, proposed solutions were defined around local and only loosely restrictive programs and, more often, around educational initiatives. As a second generation of would-be antibiotic reformers attempted to inculcate "rational" antibiotic prescribing, they found themselves facing the interrelated challenges of defining such a practice in the first place and of enforcing it among the nation's clinicians. This struggle is a legacy we are left with today, and chapter 5 will carry this story forward with respect to antibiotic resistance in particular.

"Rational" Medicine and "Irrational" Prescribers

The notion of a "rational therapeutics" has long conceptual and rhetorical histories. An explicitly "rational" medicine, as grounded in "reason" or what would come to be considered "science," and as differentiated from healing grounded in magic or religion, dates back to the ancient Greeks.[1] At the same time, and likewise deriving from the Greek and Roman eras, "rationalists" could be differentiated from "empiricists" by their stated emphasis on underlying, a priori mechanical and functional principles over a focus on collections of observed outcomes.[2]

By the first half of the nineteenth century, notions of a rational therapeutics in the United States were multifaceted and ever-evolving: at times the term was used to refer negatively to a rigid system of belief or allegiance to authority, at times to refer positively to a reliance on proper judgment, a grounding in science (as manifested in remedy selection and dosing), a related differentiation from unorthodox medicine (though it could be invoked by the unorthodox as well), or a differentiation from empiricism, which itself could be considered "rational" or not. Complicated indeed, but by the end of the nineteenth century, as physiology and bacteriology increasingly undergirded orthodox medicine, rational medicine in the United States had begun to shed its remaining negative connotations and to stand for a grounding in a self-consciously scientific medicine.[3] By the twentieth century, as a growing stream of therapeutic remedies emerged from the nation's expanding drug industry, another layer had been added. *Rational therapeutics* increasingly came to stand for a self-conscious attempt to adjudicate the merits of such drugs, whether based on mechanistic or empirical grounds. The term had taken on a skeptical, as well as a moral, tone, to be juxtaposed to the influences of commerce, ignorance, or intellectual lassitude.[4]

By the post–World War II wonder drug era, these multiple aspects of rational therapeutics had become amalgamated, and a rational therapeutics could be explicitly contrasted with its counterpart, *irrational therapeutics*. Science and reason, once juxtaposed to magic or religion, could be set against emotion and the unconscious, with physicians seemingly at the mercy of advertisers, the pleading of their patients, wishful thinking, the frenzied pace of medical practice, and fears regarding missed opportunities to treat serious (or soon to be serious) illness.[5] Robert Moser's 1959 volume *Diseases of Progress* (subtitled *A Survey of Diseases and Syndromes Unintentionally Induced as the Result of Properly Indicated, Widely-Accepted Therapeutic Procedures*) related the many potential adverse consequences of *rational* prescribing, the collective medical "side effects" of the wonder drug era; when these effects occurred in the setting of seemingly *irrational* prescribing, they took on still darker moral (and, increasingly, legal) overtones.[6] The Kefauver-Harris amendments, the DESI process, and the regulatory elevation of the controlled clinical trial all represented a particular attempt to promote a rational therapeutics and a particular site of regulatory intervention, but reformers and regulators perceived ample opportunity for the irrational prescribing of effective drugs to persist.

With the passage of Medicare and Medicaid in 1965, rational versus irrational prescribing took on important economic connotations as well. By 1968, a federally convened Task Force on Prescription Drugs—envisioned by Lyndon B. Johnson as "a comprehensive study of the problems of including the cost of prescription drugs under Medicare"—formally employed the terminology *rational* versus *irrational* drug prescribing, with rational drug prescribing defined as the selection of "the right drug for the right patient, in the right amounts at the right times," and irrational prescribing defined as "any significant deviation" from such a norm.[7] Invocations of rational and irrational prescribing would suffuse congressional hearings on pharmaceuticals throughout the late 1960s and early 1970s, while the American Society for Pharmacology and Experimental Therapeutics would change the title of its monthly *Pharmacology for Physicians* (inaugurated in 1967) to *Rational Drug Therapy* in 1971. "Rational therapy" was increasingly prevalent as both an aspiration and rhetorical device, if not, apparently, in actual practice.

Antibiotics, epitomizing the wonder drug era, were at the leading edge of this discussion. The early 1950s had witnessed an absolute growth industry of papers and talks before state and national medical societies regarding antibiotic "abuse" and "misuse."[8] Such a train of talks may be read, as they have been by James Whorton in his invocation of the parallel rise of antibiotics and "therapeutic rationalism," as a site of collective soul-searching and anxiety amid the post–World War II proliferation of the wonder drugs.[9] As an editorialist for the *Rocky Mountain Medical Journal* reflected in 1952, noting the "record attendance" at a meeting of the Denver Medical Society on "The Use and Abuse of the Antibiotics," "We are busy, perhaps over-confident, and do not consistently use the knowledge and laboratory facilities at our command."[10]

Yet no consistent message regarding antibiotic restraint emerged from this uncoordinated litany of presentations. To start with, the talks represented in many respects a consumerist description of the "problems" of choice among a wealth of available and emerging remedies (the challenge facing and to be solved by the "antibioticist," as described in chapter 1), focused as much on the use of the antibiotics in varying clinical situations as on their misuse or abuse. The bridge between use and misuse thus often took the form of an attempt to invoke a "rational" prescribing calculus, "a place for each drug, and each drug in its place," often diluting requests for therapeutic restraint amid directions regarding therapeutic application.[11]

Beyond this, the invocation of "abuse" or "misuse" in many of the titles represented more a catchy, contagious trope than a consistent or collective message. In fact, the messages regarding abuse and misuse were decidedly mixed. At one end were those like the University of Wisconsin's Arthur Lawrie Tatum (noted pharmacologist and father of Edward Tatum, who would share the Nobel Prize with George Beadle and Joshua Lederberg in 1958 for their studies of molecular genetics). After invoking a neo-Darwinian explanation for the emergence of bacterial resistance within the treated patient and coupling this to the emerging evidence of the "free interchange" of bacteria among patients and clinicians, he warned that antibiotics "are to be considered as emergency therapeutic crutches to be used only in seriously threatening conditions."[12] At the other end was Henry Sweany, the chief of research, pathology and allied sciences at the Missouri State Sanatorium. In a 1956 talk on "The Use and Misuse of Antimicrobials," Sweany wrote of flu or severe viral infections that "usually a single dose of penicillin or tetracycline will break this" and that if uncertainty remained regarding the use of a combination of antibiotics, "most drug houses will furnish dosage and drug combinations on request."[13]

In this vein, discussions regarding the misuse of antibiotics could devolve into confessionals, as in the words of a speaker at the Denver Medical Society's above-noted 1952 symposium on "The Use and Abuse of the Antibiotics":

> Another abuse, of which we are all guilty, is that of being trapped into the use of the drugs on advice of our patients. Mr. Jones, an old patient, who

> has no appointment, walks in and says, "Doc, I've got a cold starting and thought you might want to give me a shot of penicillin." You are naturally flattered that he has turned to you in this "crisis." You are also relieved that he is not there to complain of his bill or to show you how small his stream is getting since last month's prostatectomy. You may be interested in how I have solved this antibiotic abuse—I give him a shot of penicillin![14]

At other times, they could take the form of a directly anti-reformist, town-versus-gown sentiment, as when a South Carolina panelist on the "Use and Abuse of Antibiotics" related:

> In panel discussions headed "The Use and Abuse of Antibiotics" what very frequently happens is that the discussion degenerates into something which should be entitled "The Abuse of the Audience." These are the individuals who are supposed to overuse the drugs to such an extent that they are being abused. I, of course, couldn't possibly turn this discussion into anything of that nature because I too am a practicing pediatrician and anything which I said along those lines would be applicable to me as well.[15]

Most representative was a humble middle position that acknowledged the "dilemma" of the practitioner caught between the fear of prescribing potentially dangerous remedies for a "self-limited infection" and the fear of withholding a potentially effective remedy from a patient with "a more serious malady."[16] Such a dilemma could be considered further exacerbated by importuning patients and their families, whipped into a frenzy by over-optimistic media reports.[17]

Rarely a clinician would render a call to "take an oath in blood to protect and assist one another in the battle with the public over the issue of a 'shot of penicillin to knock a cold.' "[18] But this was a fairly isolated call to arms; and (as related in chapters 1 and 2) by the mid-1950s, this decentralized, conflicted series of talks and papers on antibiotic abuse and misuse aimed at the general practitioner would be dramatically overshadowed by the coordinated attack by the vanguard of infectious disease experts against the perceived excesses of antibiotic marketing embodied first by the broad-spectrum antibiotics and then by the fixed-dose combination antibiotics.

Nonetheless, even as the infectious disease vanguard focused first professional and then government attention on pharmaceutical marketers, others began to conduct actual enumerations of antibiotic prescribing habits. The first thorough study took advantage of the records related to the isolated army depot community in Igloo, South Dakota, in which medical care was available to all for the same affordable price and in which medical records were freely available to army hospital researchers. William Nolen and Donald Dille found that from January 1, 1952, through December 31, 1956, of the 763 residents of the community, 703 had received antibiotics, with 52.5 percent of such treatments deemed as "not indicated."[19] And over the ensuing two decades, as researchers continued to study physician prescribing behavior, they could move beyond such natural experiments to prospective studies of individual prescribers. In the process, qualitative notions of abuse and misuse yielded to more quantified enumerations of irrational prescribing.

Foregrounding the outpatient studies of irrational antibiotic prescribing were the evolving twin depictions of the common cold as the nosological litmus test for irrational prescribing and the unwarranted use of chloramphenicol as an even stronger such indicator.[20] With respect to the common cold, in the very foreword to Robert Moser's *Diseases of Medical Progress*, F. Dennette Adams wrote in 1959:

> It is no secret that certain drugs, surgical procedures, and other forms of therapy can, even when properly employed, create unfavorable, often harassing, and sometimes fatal side effects. Unhappily, it is also true that drugs are frequently administered or other procedures performed, apparently without due regard for their disquieting and sometimes dangerous potentialities. One need but mention, for example, the wide-spread use of antibiotics for trivial upper respiratory infections and comparable minor ailments—a practice that seems to continue in spite of the exhortations of many qualified authorities that these agents are, as a rule, ineffective in such cases.[21]

A year later, in September 1960, esteemed sulfa drug and antibiotic pioneer Perrin Long appeared at the Kefauver hearings. There largely to defend the pharmaceutical industry and the role of pharmaceutical marketing as a worthy educational enterprise, Long placed the blame for the overtreatment of the common cold on demanding patients, some of whom would not live long enough to hear his admonition:

> [Sulfa drugs and antibiotics are misused by] the doctors who prescribe them and the patients who threaten to fire the doctor who doesn't give them their pet antibiotic when they have a common cold, a viral sore throat, a viral pneumonia or some other type of infection for which treatment is useless. . . . Not long ago a doctor whom I know was called late one afternoon to see the patient of another doctor who was out of town. She had an acute common cold and also chronic asthma. She imperiously demanded that she be given penicillin, which my friend refused to give her. . . . The patient got very angry, dismissed the doctor, saying, "I'll get a doctor who will do what I say." She

> did. He gave her an injection of penicillin, and in less than five minutes she died from an anaphylactoid reaction produced by the penicillin. This is what doctors all over the country are facing today: patients who want an antibiotic. . . . There seems to be no limit to the ability or the desire of the public to absorb antibiotics.[22]

The following year, writing about "Colds, Drugs, and Doctors" in *Antibiotics and Chemotherapy*, the University of Virginia's William S. Jordan put the blame back on physicians and the pharmaceutical marketers who influenced them, asking, "Is it not time that both the pharmaceutical industry and the medical profession approached the treatment of acute respiratory infections with reason and restraint?"[23] With pressure from the patient on one end, and from the pharmaceutical industry on the other, the harried physician seemed to seldom use such reason or restraint. Paul Stolley and Louis Lasagna could by the late 1960s report that between October 1967 and September 1968, of 1,128 patient visits for the common cold, 60 percent resulted in antibiotic prescriptions.[24]

With respect to the prescribing of chloramphenicol in particular, by the mid-1960s the relationship between Parke-Davis's wonder drug and aplastic anemia had become more convincingly established, and its use for the common cold or for other self-limiting ailments seemed an especially egregious therapeutic sin of commission.[25] And using their own computerized prescription-recording system, "monitoring 85% of the prescriptions dispensed in a defined geographic area," Lasagna's group reported not only that antibiotics remained "by far the most commonly dispensed" class of medications but also that chloramphenicol remained frequently prescribed, "a leading drug in the country" that was mostly prescribed by "general practitioners whom one would not expect to be treating life-threatening infections on an outpatient basis."[26]

Despite such efforts and emerging technology geared toward the evaluation of outpatient therapeutics, it was still easier to capture hospital-based records, perhaps inspired by the nascent effort to promote medical "therapy audits." In 1952, the Kellogg Foundation had funded a Professional Activities Study to conduct therapy audits in hospitals, "in the same manner that the tissue committee evaluates the care of those undergoing operations."[27] Auditing groups were to serve, essentially, as educational oversight committees, providing feedback to providers. As one proponent noted, "We do not consider our audit committee as a Gestapo operation, but rather as a friendly cooperative movement to improve the quality of the medical care in our hospital, as well as the medical ability of our attending physicians."[28] By the late 1950s and early 1960s, the oversight of antibiotic prescribing was already seen by some as the ideal subject for such an audit. As one group suggested, "If the use of antibiotics is to be placed upon a reasonable and scientific basis, it is necessary that the medical staffs of hospitals conduct thorough and continuing evaluations of the use of these drugs in every category of disease treated."[29]

The spread of such audits in the 1960s never matched the dreams of their proponents, who felt that if physicians wanted to maintain their "freedom in the practice of medicine," they would have to conduct such audits before a "third party" chose to do so for them.[30] But those concerned with the public health aspects of antibiotics continued to document the variability and seeming irrationality of antibiotic prescribing in the hospital. Hobart Reimann, who had described both the inpatient and outpatient "misuse" of antibiotics before Senator Humphrey's congressional audience in 1963, would by April 1966 enumerate the wide variability in antibiotic prescription rates among twelve hospitals in the Philadelphia area.[31] In 1967, the Center for Disease Control reported its own study of seven community hospitals, demonstrating not only that "considerable variability in antibiotic usage [existed] . . . within and among the seven hospitals" but also that fewer than 30 percent of the patients studied had evidence of infections when first administered antibiotics.[32] Within three more years, two Ohio State University pharmacists, Andrew Roberts and James Visconti, could title their study of hospital antibiotic use "The Rational and Irrational Use of Systemic Antimicrobial Drugs," judging only 12.9 percent of all antibiotics administered at a 500-bed community hospital as "rational," with 65.6 percent of the administrations deemed "irrational" and 21.5 percent "questionable."[33]

Thus, in both the outpatient and inpatient settings, critiques of "prophylactic" antibiotics—whether those used to forestall bacterial complications of common colds or surgical infections—could be coupled with critiques of apparently insufficient diagnostic testing for offending microbes to produce a general call for a more "rational" use of antibiotics. By the early 1970s, such concerns were likewise fueled by a shift in the risk side of the general antibiotic risk-benefit equation. Reformers were still concerned with individual adverse effects (epitomized by chloramphenicol) and cost (epitomized, in the Medicare and Medicaid era, by the more than doubling in hospital antibiotic costs between 1962 and 1971).[34] But, as discussed at length in chapter 5, reformers were increasingly turning their attention to the systemic "ecological" consequences of antibiotic overuse, a "pollution" manifested by an increasing number of serious gram-negative infections (caused by microbes generally considered normal colonic inhabitants) as well as by the wider emergence of antibiotic resistance.[35] Ironically, despite their being grounded in such "rational" concerns, calls for antibiotic restraint outpaced the production of (what would be admittedly difficult) controlled clinical studies of antibiotic utility in many such "prophylactic" situations.[36] Yet despite such limitations, by 1972 those calling for antibiotic restraint would receive national exposure through another series of government hearings and their fallout.

Remonstration and Resistance

As with so much discourse around rational therapeutics in the late 1960s and early 1970s, a driving force for the crusade after the holy grail of rational antibiotic use would be Gaylord Nelson's protracted hearings on the pharmaceutical industry, lasting from 1967 to 1976. While Nelson's hearings would not provoke the type of sweeping legislative reform brought about by the Kefauver hearings, they would be responsible for smaller-scale changes such as the mandate for patient package inserts for oral contraceptives (the first such patient-directed inserts).[37] Moreover, shining a spotlight on particular antibiotics would lead to dramatic reductions in usage of these antibiotics. Chloramphenicol prescriptions in America, for example, were reduced from 3.3 million in 1965 to 1 million in 1970 consequent to a series of hearings and would never again cross the 1 million mark.[38]

After his examination of the marketing and usage of chloramphenicol and the fixed-dose combination antibiotics in the late 1960s, Senator Nelson turned his attention to "cough and cold remedies" and the "misuse of antibiotics" in early December 1972. The tone of the hearings was set the first day, devoted to over-the-counter remedies, by an exchange between Dr. Sol Katz, of Georgetown University Hospital, and Nelson:

DR. KATZ: I get kind of distressed at industry saying . . . "Well, we are giving the antihistamine to counteract the stimulating effect of the sympathomimetic," when there is no data to show that the oral sympathomimetics are any good in the first place. . . . So if we take out the antihistamine needed to counteract this sympathomimetic and we do not need the sympathomimetic, we do not have a cold preparation. We have no need for one.

SENATOR NELSON: So you need the—

DR. KATZ: You need chicken soup.[39]

On December 7, Nelson called as his first witness the director of the FDA's Bureau of Drugs, Henry Simmons, who related increasing FDA interest in the problem of apparent antibiotic overuse by presenting a letter from FDA commissioner Charles C. Edwards. As Edwards had written of this "national concern":

> We at the FDA are seriously concerned about the increasingly massive use of antibiotics in this country. In the past year 2,400,000 Kilograms, or the equivalent of approximately 10 billion doses of antibiotics, were produced in this country and certified for use. . . . Antibiotics are being massively used when not indicated. . . . One result of such use is iatrogenic disease, in the form of unnecessary adverse reactions and the development of super infections. Another is unjustifiable cost. Perhaps most important, antibiotic misuse carries a threat to the future efficacy of the drugs themselves since bacterial resistance can be and is being caused by such use.[40]

Simmons was followed by veteran reformer and government witness, and by now emeritus professor, Harry Dowling, who incorporated the FDA data to assert that while the agency certified enough antibiotics to provide 50 doses (two full courses) of antibiotics to each American annually, the average citizen likely required an antibiotic only once every five to ten years.[41] The following day, Paul Stolley of Johns Hopkins found himself with a national forum through which to present his previously published data on the usage of antibiotics to treat the common cold and regarding the persistent overuse of chloramphenicol.[42] Simmons and Stolley's research and testimonies formed the basis for their later commentary published in the *Journal of the American Medical Association*, "This Is Medical Progress?" Therein, Simmons and Stolley cited a 30 percent increase in national antibiotic utilization between 1967 and 1971—against a 5 percent increase in the U.S. population over the same period—concluding, "Have we reached the point where the enormous use of antibiotics is producing as much harm as good?"[43]

Dating back to the Kefauver hearings and extending all the way through the FDA's removal of the fixed-dose combination antibiotics on the apparent basis of the input of academic "experts" over prescribing physicians, academicians appearing as government witnesses had gone to great lengths to deny their "ivory tower" removal from the exigencies of daily practice.[44] As William Hewitt, DESI antibiotic panel head and founder and chief of the UCLA division of infectious diseases, reported at the 1969 hearings on the fixed-dose combination antibiotics:

> I would like to emphasize that I am not an ivory tower basic scientist. . . . I have had a practice of my own for 20 years and even to the present rely

> upon this type of activity for one-third of the income I derive from professional activities. . . . [The men on his particular expert sub-panel evaluating the merits of fixed-dose combination antibiotics] were not sitting in libraries writing textbooks and giving lectures to medical students but rather were daily seeing sick patients and caring for their medical and emotional needs. All of us participate liberally in local and national societies, the membership of which consists largely of "general physicians" concerned with both general and specialized medical problems.[45]

Yet by the early 1970s, even as leading reformer Calvin Kunin (then chief of medicine at the Veterans Administration hospital in Madison, Wisconsin) was expressing his "sympathy" for "the man on the firing line of practice who has to face this barrage" of antibiotic requests from patients and their families, he was as apt to be seen as condescending in his analysis of physicians' use of "drugs of fear," with antibiotic misuse a process that could be obviated if the physician was, among other things, "willing and able to perform and properly interpret a few simple tests, such as Gram stains of exudates and appropriate cultures."[46]

Extending in a direct line from previous battles over the fixed-dose combination antibiotics, pharmaceutical companies and physicians fired back when told they were behaving irrationally. Such backlash also reflected more general tensions across several domains. First, Dominique Tobbell has demonstrated the allegiances formed between the pharmaceutical industry and key members of academic and organized medicine throughout the 1960s and 1970s, representing a general effort to forestall further government control of medical practice and research. Such efforts manifested in debates over generic drugs, the formation of a national drug compendium, and the testing of drugs; the resistance of such key figures as Louis Lasagna to government delimitation of antibiotic prescribing can be seen as one manifestation of this larger effort.[47] Second, as Tobbell has described elsewhere, in the aftermath of the passage of Medicare and Medicaid, tensions were all the more heightened as town and gown competed over a seemingly zero-sum supply of both patients and authority.[48]

In this context, the president of Winthrop Laboratories spoke in 1971 of the increasing use of the terms *rational* and *irrational* therapy: "'Rational,' as a term, is vague and indefinite. . . . There are those, I presume, who thought Adolf Hitler rational, and Stalin, too. I did not. In fact, a friend of mine gave me the best definition of rational I know. 'Rational,' he said, 'almost invariably serves as the label for the *opinion* of those who happen to be in authority.' "[49] That same year, the past president of the American Academy of General Practice testified in Congress (at the same set of hearings at which FDA commissioner Charles Edwards attempted to defuse town-gown concerns over the fixed-dose combination antibiotics): "It has become fashionable for nonmedical sources and some academic sources, to castigate us for something called 'irrational prescribing, poor prescribing, overprescribing, et cetera.' Most of us appear to be glibly categorized as little more than a collective band of medical semi-illiterates."[50] And again that same year, the AMA's House of Delegates passed a resolution asking that the term *irrational* be removed from future editions of *AMA Drug Evaluations*.[51]

In January 1974, the *Medical Times* reported on the results of a questionnaire sent to 10,000 family physicians to determine "what the 'accused' had to say about the allegations" of antibiotic overuse; 5,331 physicians took the time to respond. When queried regarding whether they felt "physicians are over-prescribing antibiotics," 55.5 percent agreed, but 44.5 percent denied the charge; when asked whether they "agree that the average person does not require antibiotics more than once in every five or ten years," 89.3 percent responded in the negative.[52]

The responses of the offended physicians clustered around several interrelated themes. Some were simple calls for a reappraisal of the overall cost-benefit analysis of the widespread use of antibiotics. Two years prior, in *The Pills in Your Life*, Michael Halberstam (cardiologist, and brother of historian David Halberstam) had set forth the counter-argument:

> My most heretical belief about antibiotic therapy, one which cannot be proven but which can be inferred from statistics, is that, given the choice between the purist approach and the admitted overprescription of antibiotics, the nation's health is vastly safer with the latter. . . . Another reason why overliberal use of antibiotics is to be preferred to overconservative use (neither is ideal, of course) is that of pure chance. Just as some patients are unlucky enough to be subjected to serious reactions in illnesses they don't really have, so others are fortunate enough to be given antibiotics for completely wrong reasons—and to be cured of an illness the doctor never suspected.[53]

Halberstam would be pilloried by Calvin Kunin both in print and in congressional testimony.[54] But similar sentiments were echoed in the *Medical Times* piece, as when one clinician asked: "Since the massive use of antibiotics for 'fever,' where are the mastoids that you saw—30 or 40 a year, the 40-day hospitalizations for pneumonia, the tubes draining empyema, the rampant fascial dissecting subcutaneous infections? Yes, we *are* 'over-prescribing' antibiotics, and it is good that we are. My God, let's *not* go back to the 'good old days'!"[55]

A more vehement form of backlash released the town-gown frustrations expressed previously with respect to the apparently "ivory tower"–influenced removal of the fixed-dose combination antibiotics:

> How many patients should you watch die from lack of prescribing before you give antibiotics? Those idiots in Washington should try the practice of medicine for awhile, instead of doing it from a test tube.
>
> Are we making more trouble than we are clearing up or preventing? I think *not*, in the absence of some trustworthy studies to the contrary, despite what a few power-hungry physician-bureaucrats, or forgetful, retired professors of medicine may say.
>
> I question whether the honorable emeritus professor of medicine [presumably Harry Dowling] has seen a patient in person from or in his ivory tower for some time. However, in 22 years of vigorous general practice, I haven't had one incidence of super-infection.[56]

And still more deeply, such sentiments could be grounded in a general fear of further government restriction of physician autonomy, as when one physician from Texas declared, "I'll tell you the only thing I think is being over-prescribed and that is a hell of a big over-dose of government being rammed down the esophagus of the medical profession."[57]

Such tensions were heightened over the ensuing three years in the pages of the *Medical Tribune* and *Hospital Tribune*, widely distributed medical newspapers run by Arthur Sackler himself.[58] Sackler and the William Douglas McAdams public relations firm had been promoting antibiotics and the freedom to market and prescribe them since their orchestration of the Terramycin campaign nearly a quarter-century before, and Sackler's newspapers, by the early 1970s, counted René Dubos, Albert Sabin, Bernard Lown, and John Adriani among their editorial board members.[59] Particular attention to antibiotics and therapeutic rationality was stimulated by a letter to Sackler from Louis Lasagna, who had (as described in chapter 3) famously turned from being a leading critic of the excesses of pharmaceutical marketing to serving as perhaps the leading academic critic of the excesses of pharmaceutical regulation. During the Kefauver hearings, Lasagna had been the one reformer to defend Sackler and McAdams.[60] And in March 1974, Lasagna wrote to Sackler to defend physicians against criticisms of the "alleged prescribing at a spinal reflex level of antibiotics for 'the common cold' ": "I suspect . . . that most patients with upper respiratory complaints go to see doctors suffering from a combination of cough, stuffed nose, post-nasal drip, swollen glands in the neck, earache, etc.—in other words, from secondary bacterial complications of the common cold. If this is the case, then the prescribing of an antibiotic is not wrong; rather, the question is only: what antibiotic would be best?"[61] In November 1975—by which time Lasagna was overseeing the "Good That Drugs Do" section of the *Medical Tribune*—President Gerald Ford received antibiotics for a prolonged cold, the first illness of his presidency. Sackler announced that "what's good enough for the President of the United States is good enough for our patients" and printed Lasagna's letter in the same issue. Sackler further lamented: "Attacks on the medical profession by government officials, including top doctors in F.D.A. and H.E.W. continue. Prominent among the charges of professional incompetence is the indictment that practicing physicians are misusing antibiotics for the 'common cold.' "[62]

For Sackler, the antibiotic reformers served (as perhaps they did) as the leading edge of "a total propaganda charging the medical profession with producing 'an overmedicated society.' "[63] In contrast, and consistent with his message for over two decades, Sackler bemoaned an ongoing "undermedicated" society, evidenced by the paradigmatic undertreatment of hypertension in the United States. Lasagna, at one level, took a more measured stance, again invoking a calculus of costs and benefits: "What does the balance sheet look like in regard to cost, toxic reactions, somatic illness, etc., for the 'free' or 'permissive' versus the 'restricted' or 'Spartan' approach to prescribing antibiotics for respiratory infections?"[64] At the same time, for Lasagna—who had championed the need for controlled clinical studies over individual judgment two decades prior (as discussed in chapter 2) but who had clearly switched such allegiances by the time of the Panalba affair—the therapeutic rationalists at the very least ignored the limits of controlled trials in their rush to adjudicate or enforce rational therapy.[65] As he wondered aloud:

> When a physician has a patient who predictably develops an asthmatic attack or prolonged nasal discharge and adenitis if his upper respiratory infections are not treated early but who gets well promptly if they *are* treated, is the doctor wrong to prescribe antibiotics routinely and early in such a case? Does he really need to do a controlled trial to justify such prescribing? Is such a G.P. on any less secure or more tenuous ground than the infectious disease experts who recommend prophylactic antibiotics for patients with valvular disease undergoing dental extraction or genitourinary instrumentation?[66]

To some degree, Lasagna invoked his critique of the "Statistical Snob" from two decades prior (as discussed in chapter 2); only this time, he chose not to criticize the "Statistical Hayseed" as well.[67] More generally, as did Sackler,

The President Catches Cold . . .

. . . and Recovers Uneventfully, **a sequence of clinical events that has a particular moral for practicing physicians. Please see Dr. Sackler's column below and Dr. Lasagna's letter on treatment of colds in the adjacent columns.**

The Common and Not-So-Common Cold

I SHOULD HAVE DONE SOMETHING about this before.

Last year, in fact more than a year ago, I received a letter from an expert whose expertise and judgments have always been, for me, a source of respect and admiration—Louis Lasagna, Professor of Pharmacology, University of Rochester School of Medicine and Dentistry. Of all subjects, it was on the "common cold." I must plead guilty to failing to act sooner on his communication. I do act now, and will try to make up for my default. In exculpation of my guilt, I have asked the editor of MEDICAL TRIBUNE to publish his letter in several issues. You will find Dr. Lasagna's communication in the adjacent columns.

Misuse of Antibiotics?

Why do I bring this up now? It is not just that winter "is-a-comin' in." A number of relevant developments have occurred in the interim. Attacks on the medical profession by government officials, including top doctors in F.D.A. and H.E.W. continue. Prominent among the charges of professional incompetence is the indictment that practicing physicians are misusing antibiotics for the "common cold."

Are the colds for which physicians prescribe really "common colds?" It would appear from Professor Lasagna's letter that one rarely finds a "common cold" in the doctor's office, but rather "uncommon colds."

Now the scene shifts from Mar. 27, 1974, the date of Professor Lasagna's letter, to October of this year, 1975:

President Ford Catches Cold

This was news. Ron Nessen, the White House Press Secretary, issued statements; the wire services buzzed; radio, TV and newspapers throughout the country carried the reports.

The nation was "assured" (that word was used again and again) that it was only a "head cold." According to the *New York Times*, **"Mr. Nessen assured reporters the President was suffering from nothing more serious than the lingering effects of a head cold Mr. Ford has been trying to shake."**

Clearly it was not pneumonia.

The *Times* continued, quoting the President's spokesman, **"It's just not possible to tell how long it will take him to recover from the cold."**

Clearly, it was not pneumonia.

We were informed, **"The illness was**

Continued on page 3

"The President Catches Cold . . . " and receives antibiotics. Arthur Sackler, publisher of the *Medical Tribune*, opined that "one rarely finds a 'common cold' in the doctor's office, but rather 'uncommon colds' " necessitating antibiotics. Sackler used the occasion to print Louis Lasagna's defense of physician antibiotic prescribing, calling forth the *Medical Tribune*'s largest reader response in its history. *Medical Tribune* 16 (November 19, 1975): 1.

Lasagna extended to a critique of top-down regulation and a perceived stifling of therapeutic autonomy and dissent, suggesting: "Full and free debate is proper and helpful; an 'official' line in medicine is clearly not. Lysenkoism is unattractive in any form."[68]

Dissent was in fact not in short supply, as Lasagna's original letter brought forth the *Medical Tribune*'s largest mail response in its history, "from the famous and the unknown, the rural and the urban, the old-timers and the new," with responses initially 10 to 1, and eventually 5 to 1, in favor of Lasagna.[69] Ironically, given that Lasagna's own pioneering efforts to use computerized prescription-recording data to quantify inappropriate chloramphenicol prescribing appear to have been forgotten, several of the retorts were aimed at just such attempted quantification:

> I agree wholeheartedly with Dr. Lasagna! The FDA and HEW doctors should treat people, not computers.
>
> Antibiotic overuse, like adverse drug reaction, is probably a figment of statistical imagination based on the extrapolation of unrealistic or incomplete studies.[70]

More typically, they once again took the form of town speaking to ivory tower–ensconced gown, with Lasagna now lionized as the one gowned professor willing to speak up for general practitioners and their situation amid diagnostic uncertainty, demanding patients, and the fear of untreated serious disease.[71]

Sackler and Lasagna were further supported by a one-two combination from Duke, where legendary internist Eugene Stead argued on behalf of empirically giving antibiotics to his febrile retirement home patients "by protocol without waiting to see the patient," leading a charge for more thoughtful algorithms that could account for the diagnostic uncertainty of the beleaguered general practitioner.[72] Collin Baker, director of undergraduate programs at Duke University Medical Center, went still further from a population standpoint, giving a surprising twist to the 1970s developing world-centered "essential medicines" debate:

> Having lived in a Latin American country in which all drugs (except narcotics) were available without prescription, there is serious doubt in my mind that the present degree of restriction of the use of drugs by the public is in the public interest. It is entirely possible that more harm is done through lack of access to medical care (and hence to effective medications) than would be done by the judicious release of some medications which are now available on prescription only.[73]

Sackler, to his credit, gave space for rebuttal in the pages of his medical newspapers to those who feared Lasagna's stance would "seriously hamper our efforts to obtain some rational approach to antibiotic use by physicians in the United States," including editorial board member Albert Sabin, who had been publicly calling for antibiotic restraint since his tenure as the president of the Infectious Diseases Society of America in 1969.[74] But even the rebuttals revealed an ongoing faith in the judgment of the individual clinician over the rigidity of fixed rules, as when back-to-back would-be restrainers intoned:

> It is the essence of medicine to separate the patient requiring an antibiotic from the one who does not.
>
> It is the ultimate responsibility of the patient's physician to determine the need for antibiotic therapy.[75]

While these were far cries from that of the clinician who declared that the real task of the physician was deciding not whether to use antibiotics but rather "what antibiotic is best" (again harking back to the days of the proposed "antibioticist" described in chapter 1), they were certainly not calls for either restriction or oversight.[76]

The Limits to Regulation

The medical profession's degree of sensitization to the threat of further centralized regulation or oversight was captured most vividly in a letter to the editor of *JAMA* in response to Henry Simmons and Paul Stolley's questioning of "medical progress" in the journal in 1974. Howard Seidenstein, a general physician from New Rochelle, New York, began with a general attack on Orwellian ivory tower pronouncements and perceived amnesia regarding the pre-antibiotic era:

> It was with a feeling of horror that I read the article by Simmons and Stolley and some of the subsequent comments. I hastened to check the date and was only a bit reassured when I saw 1974 and not 1984. . . . It might be that countless thousands of community physicians are as wrong as a $3 bill about antibiotics, but it is just as possible from where I sit that dozens of ivory-tower investigators may be wrong too. My forty years of medical school and general practice span the entire era. . . . I recall full well the 40%–50% mortality of ruptured appendices with generalized peritonitis, the 20% to 25% mortality of pneumococcal pneumonia, and the 100% mortality of streptococcal meningitis. . . . I can still hear the howls of incompletely anesthetized

young patients during myringotomies that barely reduced the terribly high incidence of "mastoids" and conduction deafness and even cavernous sinus thrombosis secondary to "red ears." . . . Following the "indiscriminate" use of sulfas and antibiotics in my practice, I have not had one single case of mastoiditis develop in 35 years. And although I've had a few close calls with anaphylactic shock following parenterally administered penicillin, I've not yet lost a patient.[77]

But it was Seidenstein's concluding paragraph that was most indicative of the prospects that faced would-be antibiotic reformers: "More iniquitous than any possible abuse . . . are the abuses of power suggested by Simmons and Stolley and backed by Kunin." What were those abuses of power? Not the further restriction of offending drugs or prescribers, but only that hospitals should form guidelines, annual reviews, and feedback regarding antibiotic prescribing.[78]

Such proposed antibiotic guideline and review programs should be further understood against the backdrop of the federally mandated, locally conducted professional standards review organizations (PSROs) that were initiated in the early 1970s to evaluate services reimbursed by Medicare; the antagonism to antibiotic review programs may have reflected antagonism to these peer-review efforts to evaluate services more broadly.[79] In any case, the "Gestapo" fears envisioned by proponents of medical audits were indeed widespread. The sensitized resistance to the proposed review measures apparently reflected a deep chord of resentment that a significant proportion of the medical community held against further centralized encroachment on physician prescribing autonomy. Instead, the actual measures brought forth in the early 1970s to improve antibiotic prescribing reveal the limits to delimitation itself, especially as perceived by would-be antibiotic reformers.

More restrictive antibiotic-prescribing measures were in fact publicly proposed and debated, especially by Senator Nelson and particularly with respect to chloramphenicol. At the 1967 hearings on chloramphenicol, Nelson had suggested the possibility of permitting the usage of the drug "only with the written approval of a consultant except in emergencies or special circumstances," and the University of Illinois's Mark Lepper (who had trained under Harry Dowling) had suggested restricting its use to hospitalized patients.[80] Subsequent to the hearings, the FDA convened an Ad Hoc Committee on Chloramphenicol, premised on the view that the potentially dangerous drug "is properly indicated for perhaps 10,000 patients annually but is being prescribed and administered to hundreds of thousands or even a few million patients."[81] Potential remedies mentioned in the charge to the committee—whose members included noted hematologists William B. Castle and William H. Crosby Jr., pharmacologist John Adriani, general practitioner John G. Morrison, and infectious disease specialists Harry Dowling and Wesley Spink—reflected those suggested at the hearing. These included "the imposition of . . . peer judgment in prescribing . . . e.g., a countersigning procedure or . . . drug use committee approval," as well as restriction to the hospital setting, with antibiotic ordering therein conditional upon laboratories studies, stated justification for usage, or committee review.[82] Novel ideas were proffered, including a patient package insert warning about chloramphenicol (two years before such inserts were suggested for oral contraceptives), as were perhaps less-inspired suggestions that included placing "Dear Doctor" FDA editorials in such periodicals as the *Medical World News* and Arthur Sackler's *Medical Tribune*.[83]

At the meeting, held in early 1968, only Crosby and Dowling voted for restricting chloramphenicol to hospitalized patients, with hematologist Crosby also suggesting "a special prescription form for chloramphenicol which would make it cumbersome for the physician to prescribe it."[84] FDA leadership itself prepared a regulation requiring "hospitalization if possible" for those receiving chloramphenicol, wondering aloud if it was "proper for the FDA essentially to direct medical practice by requirements of this nature."[85] However, Crosby's hematologist counterpart William Castle likely spoke for the majority in considering that "the educational process is the only way in the end that will be successful."[86] After a "possible resurgence in use" of chloramphenicol toward the end of 1968, the FDA reconvened the committee, with the AMA's Hugh Hussey now invited as a guest (Castle, Dowling, and Spink were unable to attend). The members were again asked whether to restrict chloramphenicol to hospital or consultant-approved use, or even to have the antibiotic "federally licensed, as with narcotics."[87] However, perceiving that "there would be difficulties in defining a 'hospital,' [that] there are instances where the drug might be used on an outpatient basis, and [that] a countersignature would represent an undue restriction on the practice of medicine," those at the meeting unanimously opposed such measures.[88]

Four years later, Nelson and Philip Lee (at the time, chancellor of the University of California–San Francisco, as well as former assistant secretary for health at HEW, and former chairman of the Task Force on Prescription Drugs) continued to lament the failure of the FDA to regulate chloramphenicol more stringently. Lee again proposed to forbid its use in the outpatient setting and instead to restrict its use (even if the patient in question did not require hospitalization) "to hospital use after consultation" so as to "see this thing brought under rational control."[89] Prescribing general practitioners would therefore have to subordinate their authority to specialists, potential impact on the "physician-patient relationship" be damned.[90] As Lee concluded, what had been thought of "as a problem between the individual physician and the individual patient" needed to be reconceptualized "as a public health problem" so as to overcome the "tacit conspiracy between physician and patient" in the "very act of writing a prescription."[91] In Lee's framework, given the population-wide impact of inappropriate prescribing, physicians had to be further regulated, "just as Congress had to regulate the auto industry."[92] Yet despite Nelson's ability to convince several other witnesses to concede to his and Lee's restrictive position during the 1972 hearings, no broad reformist or political will would be brought forth to qualify the use of any class of antibiotics.[93]

Indeed, the very efforts rendered in the United States in the 1970s to instill a "rational" use of antibiotics reveal the presence of institutional hesitation and the limits of contemporary aspirations. Despite a good deal of activity by such individual members as Calvin Kunin and George Gee Jackson, the Infectious Diseases Society of America (IDSA) would not serve as a primary institutional base from which to approach antibiotic prescribing or resistance in the 1970s (as is described at greater length in chapter 5).[94] Instead, the most important—if muted—institutional push to confront antibiotic prescribing came from the federal government itself, forced to consider costs (after the advent of Medicare and Medicaid), adverse effects, and, increasingly, antibiotic resistance.

With respect to cost, the federal maximum allowable cost (MAC) program, predicated on the rise of generic drugs and the notion of drug bioequivalence, attempted to contain drug prices by enforcing set Medicare and Medicaid price limits on seemingly generic substitutable drugs. Penicillin and ampicillin were the first two drugs for which MAC programs would be established (in 1977), but Daniel Carpenter and Dominique Tobbell have described how such programs were implemented "slowly and sporadically," and the introduction of novel, expensive antibiotics (e.g., cephalosporin derivatives) led to further escalations in overall antibiotic expenditures.[95]

More broadly, when Senator Nelson asked him at the 1972 hearings about the possibility of more restrictive federal control over such potentially dangerous drugs as chloramphenicol, the FDA's Henry Simmons responded that he instead would be "speaking to something . . . that I think is a step in the direction that might finally bring us to some effective means of controlling this and other agents in the public interest."[96] What Simmons actually had in mind was the FDA's convening of a National Task Force on the Clinical Use of Antibiotics, which would include members from both the AMA and the IDSA.[97] As Simmons explained: "The Task Force is to include not just physicians but human behaviorists and others who can examine the question in its broadest sense. . . . This might be a first step, a prototype, of a way the public and the profession can work out problems together."[98] But, by early 1973, control over organizing and convening the meeting shifted to the AMA, which, in addition to inviting representatives from a number of professional medical and pharmacy associations, also invited representatives from private practice and the pharmaceutical industry.[99] And tellingly, during a full day of discussion and debate at the task force meeting in March 1973, while Simmons would ask "if limited approvals [by the FDA for antibiotics] might be justified," one of the first "testable" premises that emerged was that "physicians do not want the major ruling on antibiotic usage to come from Washington."[100]

By the end of the meeting, task force participants, focusing solely on *in-hospital* antibiotic use, put forth a motion for hospital antibiotic control entailing that:

1. Each hospital should form a committee to monitor antibiotic usage.
2. The committee should develop individualized guidelines for appropriate antibiotic usage both for treatment and prophylaxis. These guidelines should be approved by the Executive Committee of the medical staff.
3. The reports of the Antibiotic Committee would be distributed internally to the medical staff and the Executive Committee.
4. There would be an annual review of antibiotic usage by an outside consultant who would submit written recommendations to the Executive Committee.[101]

However, even when this motion came to a vote, the result was a "split decision."[102] And when Simmons and Paul Stolley published the recommendations in their 1974 *JAMA* "Medical Progress?" piece as the output of "an expert committee" that "agreed that there appears to be 'an inappropriate use of antibiotics and a massive overuse,' " six members of the AMA's Department of Drugs (including its chairman, John Ballin) penned a pointed counter-response (while Howard Seidenstein, as noted earlier, penned his own).[103] Ballin and his colleagues qualified the entire tone of the discussion, pointing to the "possible" overuse of antibiotics, while taking pains to "reaffirm that the AMA Department of Drugs is not adopting the position in this commentary that antibiotics are not misused any more than it is accepting the premise that there is massive abuse."[104] Stating that increasing antibiotic use could instead derive from an increasing number of useful antibiotics, from improved laboratory and clinical means for diagnosing infections, and from the increasing (and seemingly appropriate) use of antibiotics for acne, Ballin and his colleagues defended contemporary physician therapeutic identity and proposed that increasing antibiotic use may have reflected increasing, and increasingly powerful, medical care. Instead, pointing to an actual dearth of data on the appropriateness (beyond "limited clinical studies") and consequences of antibiotic usage in the United States, the AMA representatives called for ongoing data collection and the generation of "suggested guidelines," so as to "obviate any need for restrictive regulations for this class of drugs."[105]

In the years that followed, as the AMA declined to take the lead on generating such national guidelines, the American College of Physicians (ACP) formed an Inter-Society Committee on Antimicrobial Drug Usage (ISCAMU), joined in the process by the American College of Surgeons, the American Academy of Pediatrics, the American Academy of Family Practice, and the IDSA.[106] Funded by HEW, the ISCAMU project was formally predicated on the notion, following the AMA meeting, that "PSRO legislation as well as the problem itself demands that further effort must be made to develop some kind of guidelines for the proper use of antibiotics."[107] In the project's initial top-down manifestation, the guidelines were to come first, "to serve not only as educational instruments, but also to provide a basis whereby professional standards review organizations might find a practicable means for determining when excessive or inappropriate use of antimicrobial agents has occurred"; the "development of a data base" regarding antibiotic usage across a wide sample was to come second.[108]

However, Harvard University's Edward Kass, who co-chaired the committee, was far less comfortable having gown dictate to town than his close colleague Max Finland (or fellow Thorndike alumni Harry Dowling and Calvin Kunin) had been. As Kass related at a central planning meeting in November 1974, he had "a mistrust of experts," preferring to reform practice patterns "in a way that doesn't lead to a group of dictators telling everyone how to practice."[109] Kass led the database project's recording of antibiotic prescribing patterns across twenty hospitals in Pennsylvania, and rather than come in with a priori assumptions regarding best practices, he chose to develop the database *first* and "let the data speak," to empirically find "misuse in the most clear and flagrant sense of the word."[110] As he related, "By trying to tap off a 10% to 15% of antibiotic use that everyone agrees is crazy, we move the mean over a little and then we start with the next step."[111] Clearly influenced by the town-gown tensions that surrounded the issue, he continued:

> What I read from government is an immense impatience with the medical profession. Norms have not moved and costs are piling up. How do we get at it? Let's make these trains run on time! I am afraid that may be the approach. Or, is there a way to move the system, with all its inertia, and preserve certain values while achieving a satisfactory result. . . . In my judgment the first step is to lop off manifest excess because there would be no important political disagreement.[112]

Kass and the ISCAMU project thus explicitly eschewed the very generation of broad national guidelines they had been tasked to formulate, instead focusing on the low-hanging fruit of inappropriately excessive surgical antibiotic prophylaxis and the failure of clinicians to monitor renal function in patients taking potentially nephrotoxic (and renally metabolized) aminoglycoside antibiotics.[113] This approach represented a marked de-escalation from their initial task. Yet as Kass justified publicly in early 1976, more extreme regulations would seem to "represent arbitrary judgments passed through debatable areas on the basis of opinions of small groups of experts, . . . [leading to] serious controversy and arbitrary action."[114] Calvin Kay, deputy executive vice president of the ACP, took such sentiments still further at the annual meeting of the AMA in June 1976, deconstructing the very notion of a central authority's capacity to define a "rational" form of antibiotic usage:

> Ideally, the objective of ISCAMU is to develop means by which the right antimicrobial drug will be given to the right person for the right reason and in the right way (to paraphrase a cliché). If attempted, this ideal would be impossible to fulfill. It would require the assumption that the physician will have all of the necessary clinical and laboratory information when he must make his decision. This is rarely ever the case. More importantly, the ideal assumes that there is a right way that is proven and universally agreed upon. This is unrealistic. . . . [Experts] can probably identify a very small number of procedures that are wrong: inappropriate and unacceptable by unanimous agreement.[115]

With the very definition of rational prescribing reduced to cliché, the federally funded group of experts backed off their mandate to broadly reform irrational prescribing.

Instead, it would be left to Kass's onetime colleague at the Thorndike,

Calvin Kunin, to shepherd the most thorough national "guidelines" provided at the time, generated through the Veterans Administration system's Ad Hoc Interdisciplinary Advisory Committee on Antimicrobial Drug Usage and printed as an extensive series of articles in *JAMA* throughout 1977.[116] Even Kunin, though, was careful to frame these as the *beginnings* of conversations, based on contemporary knowledge, to be judged and modified by individual hospitals and their staff: "These are guidelines, not commandments written in stone."[117] They were, in essence, to serve as the very "process audits" that had generated consternation several years earlier but were again a far cry from more restrictive policies at a cross-hospital, let alone national or outpatient, level.[118]

Rather, truly restrictive antibiotic-prescribing regulations would be limited to the inpatient setting, and only in local contexts. As far back as with the spread of staphylococcal resistance in the 1950s, scattered hospital staffs in the United States, appreciative of the apparent selectionist relationship between the use of particular antimicrobials and the development of resistant microbes, had decided to keep one antibiotic or another in "reserve" for otherwise resistant strains.[119] Such measures—taken in venues ranging from small community hospitals to large general ones—were typically dependent on some type of guideline that could be overridden through interaction with a predetermined consultative service.

By the mid-1970s, Boston City Hospital's John McGowan and Max Finland could formally report on the "effects of requiring justification," noting that between 1965 and 1972, the policy resulted in nearly ten-fold reductions in the usage of ampicillin and chloramphenicol during their tenure on the hospital's "restricted" list, with encouraging trends noted with respect to adverse effects, costs, and antibiotic resistance.[120] Others, working closely with clinical pharmacists (discussed below), focused on restricting the use of the expensive cephalosporins and aminoglycosides, and resulting in dramatic cost savings.[121] However, despite the scattered uptake of such measures, they would remain local and limited in nature throughout the 1970s. In the late 1950s, when hospitals were first establishing infection control programs in the setting of increasing staphylococcal disease, such programs had been marked by their heterogeneity.[122] Such a pattern appears to have repeated itself with the uptake of restrictive antibiotic programs, with individual hospitals serving as testing grounds for particular approaches (as described in chapter 5).[123] Indeed, supporters were cognizant that they "would be accepted at very few hospitals," and public grumbling emerged that even the vaunted Boston City Hospital program "was a failure in terms of local acceptance."[124]

The Limits to Education

Instead, throughout the 1970s, other reformers publicly advocated *educational* measures devised to support and enhance the autonomously prescribing physician.[125] As renowned pediatrician C. Henry Kempe had concluded in 1958, tying such autonomy to informed rationality (and "a considerable degree of sales resistance to undue or premature advertising claims"): "Happily, medicine remains a highly individualized vocation. What is more fitting, therefore, than that each of us should work out his own therapeutic philosophy? It is our hope that this philosophy will be at once enlightened and responsible."[126] And by the 1972 hearings on the misuse of antibiotics, while Senator Nelson and Philip Lee may have sought further restriction of existing drugs such as chloramphenicol, the FDA's Henry Simmons's own explicitly articulated role for the agency was not increased regulation but rather "for FDA to improve its communications with the practicing medical community so that practitioners can better utilize available data on the safety and efficacy of drugs."[127]

From late 1973 through 1974, Senator Edward Kennedy (D-Massachusetts), chairman of the Senate Subcommittee on Health, once again placed apparent antibiotic misuse and overuse before Congress and the public. At this set of hearings, James Visconti, by this time the director of the Drug Information Center at the Ohio State University Hospital, could update his prior data to note that in a more recent hospital study, 73.6 percent of all antibiotic use was deemed irrational or questionable, with only 7.2 percent of the prophylactic antibiotic delivery deemed rational. Yet Visconti, too, recommended not restriction but rather "the development of more rational methods by which professionals are provided with and use drug information." Advocating for computerized systems for recording prescribing patterns and providing the basis for enhanced utilization review and feedback, Visconti also advanced the role of the clinical inpatient pharmacist as one who could provide "antibiotic information . . . to assist in the rational prescribing of antibiotics."[128] Visconti's efforts epitomized the goals of the emerging "clinical pharmacy" movement to elevate the role of the pharmacist vis-à-vis the physician in the 1970s.[129] But he was quick to conclude that by bringing in "other professionals who can help him [the physician], not making decisions for him, obviously not prescribing the drugs for him," they intended such assistance to be supportive, not restrictive.[130]

Several groups began to propose and test educational programs. In 1967, the University of Florida's Leighton Cluff (who would serve as president of the IDSA in 1973) had remarked: "Many aspects of drug usage are currently issues of public concern. Up to now the answer to the problems they pose has been to try to put out fires with legislation and regulations. This is like coping with a house with faulty wiring by putting a fire hydrant and a fireman in every room. The answer is to rewire the house. Education is the rewiring."[131] And by the 1970s, while educational initiatives were often grounded in locally derived audit data, they were nevertheless predicated on the notions that "restricting the use of drugs would limit the traditional freedom of a qualified physician to exercise his best judgment in therapeutic decisions involving his patients" and that "any policy that overly restricts the use of drugs runs counter to the tradition of continuous physician education."[132]

One approach was to consider individualized physician (re-)education. In the outpatient setting, Wayne Ray, William Schaffner, and their colleagues at Vanderbilt University utilized the untapped goldmine of Tennessee Medicaid prescription data to determine not only that half of chloramphenicol prescriptions written from July 1973 to June 1974 were for upper respiratory tract infections but also that twenty physicians wrote 55 percent of all the chloramphenicol prescriptions in the state.[133] Similarly, while tetracycline was widely deemed inappropriate for those under the age of eight, 27 percent of the nearly 2,000 physicians analyzed prescribed the medication to such children, with twenty-six physicians writing 54 percent of the prescriptions.[134] As the Vanderbilt team optimistically related, such data seemed to lend themselves to targeted "continuing education" for the identified physicians (a program they would roll out with mixed results by the early 1980s).[135] Another approach was to deliver broader didactic programs, consisting of hospital-based lectures and meetings. But while those who designed the programs remained hopeful, they remained disappointed at the impact of their efforts throughout the 1970s.[136]

As Calvin Kunin had predicted, it appeared that such individually promising educational approaches would have to be comprehensive and coordinated if a "war on antibiotics" were to be effective.[137] And fundamental to this depiction, and bringing reform efforts full circle, it would apparently have to compete with the antibiotic-promoting "educational" efforts of the pharmaceutical industry. It was already clear to the reformers at the 1972 hearings that they were outgunned in this respect.[138] And by the 1974 hearings, when Senator Kennedy called to witness the president of Smith, Kline, and French regarding his company's antibiotic detailing efforts, which recommended, for instance, the push for antibiotics during the presumed summer off-season on the basis that "air conditioning plus the change of season can make upper respiratory infections tough to beat without antibiotic therapy," the magnitude of the challenge became still more apparent.[139] To Kunin, the antibiotic situation was a "symptom" of the larger "default of post-graduate education to the [pharmaceutical] industry," extending all the way to industrial support of formal continuing medical education and necessitating a reconfiguration of the financing and control of post-graduate medical education writ large.[140] And with respect to antibiotics, the situation appeared to mandate a multilateral counter-effort, beginning with educating the public and extending to the teaching of clinical pharmacology in medical school, the provision of suggested guidelines, instruction and feedback to clinicians regarding both their inpatient and outpatient prescribing, and the support of such activities with investment in laboratory facilities.[141]

This would indeed be a fecund era for mobilizing around such educational approaches. Jay Sanford's *Guide to Antimicrobial Therapy*, now in its forty-fourth annual edition, dates to 1970.[142] And this era likewise witnessed the emergence of both academic "counter-detailing"—based on "having medical school faculty members travel around like detail men" and pioneered by such infectious disease–based reformers as Leighton Cluff and Calvin Kunin (and such pharmacoepidemiologists studying the use of antibiotics as Jerry Avorn and Stephen Soumerai)—and wider concern with the apparent industry takeover of formal continuing medical education (CME).[143]

Yet by the end of the decade, those who had hoped to see antibiotic prescribing reformed had to view their project as a failure. As Calvin Kunin—who more than anyone had attempted to inculcate rational prescribing patterns from the late 1960s through the late 1970s—lamented in May 1978:

> A decade ago some of us working in this field believed that we had scored a major victory when the Food and Drug Administration removed fixed-dose combination antibiotics from the market. . . . This was no victory, but abject defeat; these drugs were almost immediately replaced by the more expensive parenteral cephalosporins for surgical prophylaxis and the oral cephalosporins, new tetracylines, and clindamycin in office practice. . . . I do not mean to imply that removing irrational drugs was a mistake in itself, but we expected too much for our efforts.[144]

The irrational use of rational (if expensive) antibiotics had seemingly continued to fill the space left by the use of the irrational fixed-dose combination antibiotics themselves. Practices and problems had been surfaced, but letting the "data speak" had not generated self-correcting measures, nor had any particular restrictive or educational practice been embraced. In part this failure seemed to result from an ongoing lack of data to guide such policy formation. Editorializing regarding yet another account of apparently irrational antibiotic prescribing, *JAMA* editor-in-chief William R. Barclay would write in 1981: "It is not necessary to publish more studies on the pattern of physician prescribing, for the problems have been clearly identified by this and other studies in the literature. What the medical literature now awaits are reports of actions that are successful in correcting poor prescribing habits."[145]

In large part, though, this reflected an explicit rejection of further delimitation or even oversight of antibiotic use in the aftermath of the removal of the fixed-dose combination antibiotics and the perception of ivory tower meddling with the real-world setting of the clinician. It had been a vulnerable period for those who cherished their therapeutic autonomy. It was the era of the Task Force on Prescription Drugs and the PSRO, mandated generic drugs, and the introduction of the Schedule of Controlled Substances that delimited and forced oversight of narcotic and tranquilizer prescribing, to say nothing of second-wave feminism and increasing patient consumerism.[146] However, antibiotic prescribing—so central to the therapeutic identity of both the generalist and specialist physician—had eluded such control.

As Kunin pessimistically surveyed the situation in 1978, he remarked that "we must not expect that some magical surge of money will resolve these problems."[147] They were deeper than any attempt to "declare war" on them.[148] Yet the late 1970s and early 1980s represented, in retrospect, an inflection point with respect to attention to antibiotic use in this country and elsewhere. Such attention was driven by an issue that had come to supersede side effects, superinfections, diagnosic sloppiness, or even cost: namely, the seemingly ever-escalating rise of antibiotic resistance. The "crisis" of antibiotic resistance and the "crusade" that ensued would reframe antibiotic prescribing—as a pragmatic, ecological, moral, and global action—in ways that would indeed generate a surge of money and attention to the issue of antibiotic prescribing.

Nevertheless, while one would no longer find physicians publicly decrying depictions of antibiotic misuse or overuse, one would also not find in this country the widespread uptake of restrictive stewardship programs or even adequate funding to test the utility of restrictive or educational programs in the first place. The impact of the 1960s and 1970s era of remonstration and resistance would persist into the era of the antibiotic resistance "crusade." Chapter 5 examines in detail the history of the framing of antibiotic resistance and the evolution of the "crusade" to combat it, with the first four chapters of this book serving as a critical backdrop to such a reading.

CHAPTER FIVE

Responding to Antibiotic Resistance

> If we accept the seriousness of the problem, the question is: should we continue to deal with this problem by an occasional speech or paper condemning the erroneous use of antibiotics, or, should a more aggressive effort be made to start a crusade for the rational use of antibiotics?
>
> —ALLEN HUSSAR (1954)

IN 1954, Allen Hussar, chief of the medical service at the Franklin Delano Roosevelt Veterans Administration Hospital in New York, called for "a crusade for the rational use of antibiotics" at Henry Welch's annual antibiotics symposium in the nation's capital. Anticipating accusations of overreaction, Hussar summarized the litany of consequences resulting from the "incorrect" use of antibiotics:

> The question arises whether some of us are just overly concerned with the present abuse and misuse of antibiotics, painting a very gloomy picture unnecessarily, or whether there is a real and serious problem with which we have to deal. The answer to this question becomes obvious when one considers that the indiscriminate and incorrect use of antibiotics may cause: (a) unnecessary sensitization to the drug which will prevent its future use in case of real need, (b) increased incidence of serious and fatal reactions, (c) increased incidence of bacterial resistance, (d) increased occurrence of bacterial or fungal superinfection, (e) obscuration of the correct diagnosis, (f) therapeutic failures, (g) loss of faith in therapeutic agents, (h) undue credit given to therapeutic agents, (i) deterioration of diagnostic and therapeutic knowledge, (j) deterioration of professional attitude.[1]

His counterproposals—including such measures as improved antibiotic education at all levels of training and practice, a coordinated "press release

. . . in the major medical journals, condemning the irrational use of antibiotics," education of the "lay public" to complement such efforts, and the formation of a National Registry of fatal reactions—seem quite prescient of measures to come. And challenging his colleagues, he concluded: "If we accept the seriousness of the problem, the question is: should we continue to deal with this problem by an occasional speech or paper condemning the erroneous use of antibiotics, or, should a more aggressive effort be made to start a crusade for the rational use of antibiotics?"[2]

Hussar's proposed crusade, however, would have to wait until the 1990s to fully take off, both in the United States and globally, when Martin Wood could at last proclaim the successful "politicization of antimicrobial resistance."[3] At the time of Hussar's call to arms, antimicrobial resistance was perceived as but one of many reasons for closer attention to rational antibiotic prescribing, but over the ensuing decades, and especially from the 1980s onward, it would become the movement's driving force.

This chapter carries forward from the previous four chapters the history of efforts to promote the rational production and use of antibiotics in both the United States and (increasingly) globally, as well as the debate over the appropriate roles and behaviors of industry, clinicians, and government. But in focusing on the response to antibiotic resistance, against the pharmaceutical, clinical, and regulatory backdrop of the previous four chapters, I will start from the 1940s again, providing another perspective and drawing specific attention to the factors that seem, in retrospect, to have precluded earlier consensus and activity, and that continue to influence today's efforts. This chapter is thus both connected to, and distinct from, the rest of the book, and those who have made it through to this point will be especially primed to appreciate the history described herein.

The "response" to antibiotic resistance, in this telling, refers to (1) the manner by which a topic is surfaced as a "problem" in the first place; (2) the manner by which such a scientific problem gets on the policy agenda once identified; (3) and finally, to the process of *politicization*, as Martin Wood used the term, and as Robert Bud has employed it in describing the British experience, as a process of moving levers of regulation or funding.[4] Such categories are themselves interrelated, and the response to antibiotic resistance ultimately represents the mobilization of—and the mobilization against—antibiotic resistance as issue and process, respectively.

The response in the United States, in this respect, appears to have evolved through four periods: one from the 1940s through 1963, in which, in the context of increasing staphylococcal resistance in particular, calls for restraint in prescribing were both uncoordinated and overshadowed by competing concerns regarding the pharmaceutical marketing of antibiotics, particularly fixed-dose combination antibiotics (as readers of chapters 1–3 will appreciate); a second period, from 1963 through 1981, in which increasing attention to generalized, even "infectious" antibiotic resistance and the public accounting of "irrational" prescribing led to calls for restrictive measures, but ones that would be rejected in the aftermath of the regulatory elimination of the fixed-dose combination antibiotics and amid concerns regarding further government encroachment on physician prescribing autonomy (as readers of chapter 4 will appreciate); a third period, from 1981 through 1992, at which time antibiotic resistance became officially "globalized" and attached to dedicated reform groups like the Alliance for the Prudent Use of Antibiotics, though with disappointing outcomes for such reformers secondary to the prevailing political environment; and a final period, from 1992 up to the present, when antibiotic resistance became fully politicized and globalized (even commodified), though with heterogeneous measures taken under a seemingly standardized banner.

All periodizations are somewhat artificial, running the risk of obscuring important continuities.[5] And the correspondence of such a periodization to larger political and social eras will be obvious to readers. But this four-part periodization has its chief utility in highlighting the degree to which the "crusade" against antibiotic resistance has been contingent on local histories as much as on such larger historical forces. The crusade remains an ever-evolving process, responding to business, science, and politics, and entailing the ongoing development—and at times, collision—of particular therapeutic reform efforts. And it has its own limitations, as described in the book's conclusion. In 1958, Surgeon General Leroy Burney reported on contemporary efforts to confront antibiotic resistance:

> In the heat of a battle, it is not possible to pull together the reports showing just when this squadron took off, or that unit went into action. Nor is it possible to single out the heroes who will be given the medals. All this comes later and a coherent account of the battle is the work of the military historian. So in this invasion of new strains of staphylococcus on an international front, the medical historian of the future will have a difficult, and to him fascinating, task.[6]

The task has indeed been fascinating; my hope is that it will be useful as well.

The Path Not Taken: 1945–1963

To start at the beginning, antimicrobial resistance—as both clinical and laboratory phenomenon—dated to the earliest attempts by Paul Ehrlich and his colleagues in the first decade of the twentieth century to treat infections with chemotherapeutic agents.[7] Moving forward, the sulfa drug experience—marked by large-scale attempts at military prophylaxis during World War II—had demonstrated the possibility of antimicrobial resistance on a widespread clinical scale.[8] And by the penicillin era, Alexander Fleming concluded his own Nobel Prize acceptance speech in 1945 with the following morality tale:

> The time may come when penicillin can be bought by anyone in the shops. Then there is the danger that the ignorant man may easily underdose himself and by exposing his microbes to non-lethal quantities of the drug make them resistant. Here is a hypothetical illustration. Mr. X has a sore throat. He buys some penicillin and gives himself, not enough to kill the streptococci but enough to educate them to resist penicillin. He then infects his wife. Mrs. X gets pneumonia and is treated with penicillin. As the streptococci are now resistant to penicillin the treatment fails. Mrs. X dies. Who is primarily responsible for Mrs. X's death? Why Mr. X whose negligent use of penicillin changed the nature of the microbe. *Moral*: If you use penicillin, use enough.[9]

It was an old morality tale in a new bottle; rather than contract syphilis and infect his wife in the pre-penicillin era, Mr. X contracted a streptococcus and infected his wife with a resistant bug in the post-penicillin era.[10]

Expressing concerns regarding overuse rather than underuse, Roy Fraser, the chair of Mount Allison University's Department of Biology and Bacteriology, wrote that same year to the University of Minnesota's Wesley Spink, a leading researcher of staphylococcal resistance:

> It seems to me to be a tragic thing that self-medication with sulfonamides and the consequent development of resistant strains should partly undo this great advance in chemotherapy. . . . Now we have the perfect commercial build-up for the over-the-counter exploitation of penicillin—and we'll have penicillin tooth-paste and penicillin shoe-polish and penicillin breakfast-foods and penicillin cigarettes and penicillin floor-wax . . . and lots and lots of penicillin-resistant strains of pathogenic bacteria. Perhaps I'm too apprehensive. I hope so. But—"anything that *has* happened, *can* happen!"[11]

And while the blame for engendering resistance would soon be shifted from patients and over-the-counter remedies to clinicians, such moral connotations would continue to proliferate.[12] The following year, Hobart Reimann publicly communicated predictions that "the widespread use of penicillin will gradually induce penicillin-resistant bacteria to arise which after several years will cancel its curative effect."[13]

The clearest instance in which penicillin had been more effective (rather than just less toxic) than the sulfa drugs was in combating staphylococcal infections, so feared in the pre-sulfa and pre-antibiotic eras.[14] Yet staphylococcal resistance to penicillin seemingly emerged in tandem with penicillin's usage, and by the late 1940s it was already assuming alarming proportions.[15] By 1950, the irony that the very staphylococci against whom Alexander Fleming had first noticed the activity of penicillin were likewise the first microbes to exhibit widespread resistance to the medication was painfully obvious to such leading infectious disease experts as Max Finland.[16] The broad-spectrum antibiotics at first appeared to tilt the contest in favor of clinicians.[17] But by the early 1950s, as staphylococcal infections began to predominate over other infections, and as penicillin-resistant staph continued to predominate over penicillin-sensitive strains, staphylococci were apparently becoming increasingly resistant to the novel broad-spectrum agents as well.[18]

By this time, Max Finland would consider staphylococcal resistance the foremost issue before him.[19] In an unsigned *New England Journal of Medicine* editorial in March 1950, he pointed to the probable link between the widespread use of penicillin and the presence of penicillin-resistant staphylococci to conclude of the newer antibiotics that "this should serve as a warning against the indiscriminate administration of such highly important agents, particularly their use over long periods for what is thought to be prophylaxis or whenever there is no real justification for using them."[20] Finland at this stage offered a wide-ranging definition of such problematic "prophylaxis," from antibiotics used to forestall the conversion of upper respiratory tract infections to pneumonias, to those used to forestall surgical complications. Before the New York Academy of Medicine on December 7, 1950, he offered a counterattack to emerging antibiotic overuse:

> One can see here a really universal extension of the use of antibiotics. . . . [Because of] the fact that much of the prolonged morbidity from the common cold and influenza in some individuals is associated with a complicating bacterial infection of the upper respiratory tract or with pneumonia, . . . the "prophylactic" use of such agents in these otherwise mild but very common conditions could be justified. It is only a slight extension of this concept to the idea of treating all persons who might be considered as unusually "susceptible" and from there it is but a short step to include all other people throughout the entire season when respiratory diseases are prevalent,—and perhaps continuously. . . . The surgeons may feel justified in extending the prophylactic use of antibiotics to all preoperative and postoperative condi-

tions in order to prevent the development of wound infections or to prevent the spread of established infections. The fervor for such use will be inversely proportional to the skill of the operator and to the care which he normally takes to avoid infection.[21]

And Finland sardonically extended such prophylactic uses to their logical extremes:

> One may summarize the many ancillary and prophylactic uses of antimicrobial agents in terms used by some of our leading politicians as literally offering protection to the individual from the "womb to the tomb." For, is not the mother, and perhaps the unborn child as well, protected by antibiotics during labor? And is not the first gasp of the infant to be followed almost immediately by the instillation of penicillin in the eye? And is not the infant thereafter to receive some antibiotic each time it sniffles, or looks poorly or has a fever? And are not all of his family to be treated to assure his being freed of sources of infection? And is he not to brush his teeth with an antibiotic-containing dentifrice twice daily as long as he lives or has teeth? And so it goes on until his dying day! And then, as he lies dying with his terminal cerebrovascular accident, or heart failure, or even should he be dying from a severe injury, can he be denied the antibiotics that may prevent a secondary pneumonia, and thus fail to assure him of an adequate antibiotic concentration in his tissues as he is laid to rest?[22]

While Finland in the early 1950s only suggestively linked the worsening staphylococcal situation to increasing antibiotic usage, over the course of the decade a general consensus emerged around the apparent correlation between the quantity of antibiotics consumed in a given hospital and the rate of staphylococcal resistance encountered there.[23] And although staph infections would continue to dominate discussions regarding the ecological consequences of antibiotic administration throughout the decade, by the end of the 1950s Finland's group and others could point to the more general changes in bacterial pathogens encountered in the hospital since the introduction of the miracle drugs.[24] While the pre-antibiotic era had been notable for the devastation brought about by pneumococci and streptococci, the antibiotic era was characterized by the consequences wrought by a shifted collective microbial flora increasingly represented not only by staphylococci but by such gram-negative pathogens as *Escherichia coli*, *Klebsiella*, and *Pseudomonas* that normally resided within the colon of the host.

Leading infectious disease researchers were increasingly likely to draw the ecological and selectionist conclusions from such findings for their audiences.[25] To René Dubos, legendary Rockefeller Institute microbiologist (and discoverer of the antibiotics tyrothricin and gramicidin), such shifting ecology was an inevitable consequence of humans interacting with the world of microbes, leading Dubos to focus ever more on host factors in confronting infections and to deny that mankind would ever be free from ever-evolving disease states.[26] But others were more apt to place a moralizing stance on the process, analogizing from other ecological disasters. In Great Britain in 1954, the same year that Allen Hussar was calling for a crusade for rational antibiotic usage, a British clinician cautioned:

> Those deadly staphylococci, those monilia in permanent possession of the field are not pirates or privateers accidentally encountered, they are detachments of an army. They are also portents. . . . There are parallels in agriculture. We plough the fields and scatter insecticides and selective weed-killers on the land and we find we have killed birds, bees and flowers who minister in various ways to our health and happiness and with whom we have no quarrel. . . . We should study the balance of Nature in field and hedgerow, nose and throat and gut before we seriously disturb it. Again, we may come to the end of antibiotics. We may run clean out of effective ammunition and then how the bacteria and moulds will lord it.[27]

And back in the United States, in 1961, fifteen years after first having warned of penicillin overuse, Hobart Reimann warned:

> Disturbance of natural ecology by similar good intentions or by inadvertency has brought trouble in other fields. One may recall the deliberate importation into new regions of mongooses, rabbits, pigs, mina birds, English sparrows, and even of humans, their overmultiplication and dislocation or destruction of native fauna and flora; the slaughter of carnivores and the resultant over-increase of predatory herbivores in farming areas; the tragic results of measles carried by explorers to a virgin population, and so on. An analogy of antimicrobic disturbance of the microbial flora carried normally by man and its consequences is not too farfetched. Nature seldom can be disturbed without some unwanted consequence.[28]

As Robert Bud has related, some of the anxiety of the 1950s and early 1960s was allayed by hope for a technological fix to the "race" between microbes and mankind, as the initial crop of soil-derived antibiotics was joined by chemically modified descendants (epitomized by Beecher's methicillin, released in 1960) and alternate classes of antibiotics, to say nothing of combinations of antibiotics.[29] Antibiotic resistance reflected a business opportunity for the pharmaceutical industry, eager to market "specialty antibiotics" to

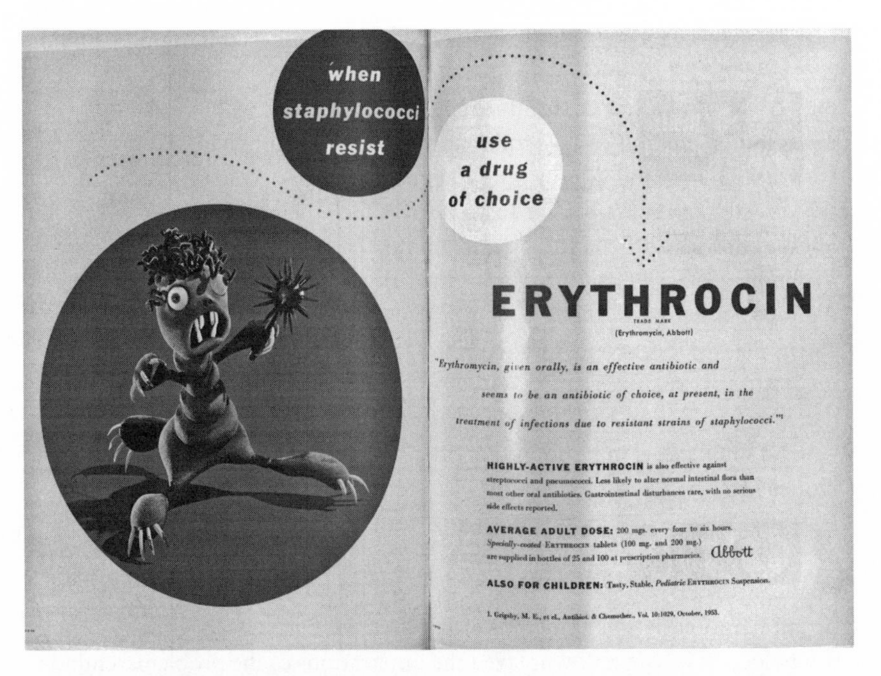

"When Staphylococci Resist." Erythromycin's utility for antibiotic-resistant staphylococci could provide a therapeutic niche, while limiting sales of the drug when it was held in "reserve" for resistant infections. *GP* 9 (January 1954): advertisement between pages 193 and 194.

confront resistant organisms.[30] Such arms-race rhetoric, as described in chapter 1, was likewise the message proclaimed at Henry Welch's antibiotics symposia by co-organizer Félix Martí-Ibañez.[31] It could be heard from the periphery as well, as when pharmacologist John C. Krantz—in a talk on "The Use and Abuse of the Antibiotics" [!]—stated before Pennsylvania's state medical society in late 1954: "It now becomes crystal-clear that many [antibiotics] are available from the metabolic processes of the Streptomyces. Thus the problem of the development of resistant strains becomes ever less important."[32] Or as Selman Waksman put it in a talk on "Man's War against Microbes": "We have not come to the end of 'man's war against microbes.' Numerous problems remain to be solved. The problem of resistance among microbes against the common antibiotics continuously raises its head. Fortunately, we have the tools to proceed further and there is no question as to who the winner will be."[33]

However, the University of London's Lawrence Garrod had already issued a warning in 1951: "So far the supply of new antibiotics has more than matched the capacity of bacteria to resist them, but if this supply should cease—and presumably the number yet to be discovered is limited—the time may come when a few of the more enterprising species will flourish more or less unhindered."[34] By 1956, following reports on staphylococcal resistance by Max Finland and Wesley Spink at the New York Academy of Sciences, the director of Merck's own Institute for Therapeutic Research wrote to Spink that his and Finland's figures "should certainly be a strong memento to the medical profession and pharmaceutical industry alike to approach the topic of drug-resistancy [*sic*] more actively lest the era of miracle drug chemotherapy become a lost paradise!"[35] And by 1963, while Waksman might dismiss such pessimists as "gloomy prophet[s]," it appeared to some that the Earth's soil had been "exhausted," as Ernest Jawetz lamented on Garrod's home turf in London that "the search for truly different new antibiotics has been disappointing."[36]

Where, then, were those who would lead the crusade to confront antibiotic resistance? For starters, neither the federal government nor the American Medical Association chose to focus on prescribers. Food and Drug Administration antibiotic chief Henry Welch, in his own account of drug-resistant staphylococci in 1953, downplayed the implications of the problem, reminding his readers that "it is imperative that these developments be seen in their true perspective so that no patient who really needs penicillin or a broad spectrum antibiotic, and their number is legion, be denied the benefits of such medication whenever clinically indicated."[37] Welch did point, however, to the need not only for the development of new antibiotics but also for improved approaches to reducing in-hospital cross-infections.[38] And both the AMA and the federal government would focus their efforts on such infection control.

The AMA, in November 1957, first held a "Conference on Staphylococcic Infections in the Hospital and Community" attended by representatives from over a dozen national organizations, from the U.S. Public Health Service (USPHS) and the FDA to the American College of Surgeons (ACS) and the American Hospital Association. Yet while a *Journal of the American Medical Association* editorial on the conference concluded that how long the antistaphylococcal antibiotics "will remain so will depend on the wisdom and restraint with which they are used," the conference papers and overall recommendations focused entirely on infection control, or "the establishment and enforcement of adequate directions for asepsis and isolation in the hospital."[39] Even in this respect, the task facing such committees in reining in physicians appeared daunting. As a team from Temple University Hospital reported: "We insist that all infections be reported promptly to the hospital administrator or some similarly central person. . . . Adherence can be obtained from certain reluctant staff members by holding the resident responsible for reporting. Residents usually have sufficient respect for authority to adhere to the rules."[40]

The infection control movement, though, would be off and running in the United States.[41] Nearly one year later, faced with "a real and present threat to national health," the USPHS (through its Communicable Disease Center, the forerunner to the Centers for Disease Control) and the National Academy of Sciences–National Research Council sponsored an even larger conference on hospital-acquired staphylococcal disease, attended by representatives from fifty-nine organizations, ranging from the American Society for Heating and Air-Conditioning Engineers and the Pharmaceutical Manufacturers Association, to the AMA, the ACS, and the Joint Commission on Accreditation of Hospitals.[42] But despite Surgeon General Lee Burney's explicit call to produce something more than the "many well-planned but uncoordinated efforts" cropping up across the country, the conference attendees were again left with little more than a mandate to form locally developed committees on infections. And even in the hospital setting—its focus—the conference yielded few directives regarding antibiotic usage. Despite the "unanimous condemnation of the indiscriminate use of antibiotics and acceptance of the need for plans to govern their judicious usage," the conveners concluded that "no single plan has yet been found to be satisfactory for application in all types of hospitals and therefore specific recommendations cannot be made."[43] Local initiatives were to trump centralized or even collaborative efforts at this point.

In fact, such hesitation regarding more centralized activity likely contributed to the lack of a national response to the 1957–1958 influenza epidemic and its resulting epidemic of staphylococcal pneumonias (especially antibiotic-resistant ones).[44] While the AMA's Council on Drugs did issue a report in September 1957 pointing to the dangers attendant to the routine use of antibiotics for the treatment of influenza or for the prophylaxis of flu-associated bacterial infections, the organization failed to carry such a message forward beyond its isolated warning.[45] Moreover, at Henry Welch's annual antibiotic symposium held a month later, FDA Division of Antibiotics member Howard Weinstein reportedly publicly supported the prevalent treatment of influenza with antibiotics.[46] Instead, the influenza epidemic appears to have provided an opportunity for pharmaceutical companies to up the arms race, epitomized by Panalba itself, introduced in 1957:

> The new combination, christened Panalba, rapidly became one of the most widely prescribed of antibiotic preparations. It was especially in demand dur-

> ing the winter of the great Asian influenza epidemic—1957–1958—because of a marked rise in cases of pneumonia caused by penicillin- and tetracycline-resistant staphylococci. Panalba was prescribed almost routinely in many areas in order to avoid resistant-staph pneumonia as a complication of the swift-moving influenza infection. To meet the demand for Albamycin [novobiocin] and Albamycin-containing preparations, Upjohn at one point had to halt the production of all other antibiotics.[47]

The influenza epidemic may at least have had an impact on the World Health Organization (WHO), but its antibiotics-related efforts remained halting and limited in scope throughout the 1950s and 1960s. The WHO had first set up an Expert Committee on Antibiotics (changed to an Expert Advisory Panel) in the early 1950s, to provide technical assistance to those nations hoping to develop antibiotic production capabilities.[48] However, despite the ambitions of such committee members as Selman Waksman to have the WHO lend its more general support to antibiotic production, testing, and education, the organization had retreated from even its limited initial involvement by the mid-1950s, transferring its technical assistance role to the United Nations Technical Assistance Program and retaining an essentially moribund Expert Advisory Panel.[49] By late 1956, however, certain WHO staff considered that there were still "some needs" for WHO involvement in antibiotics, including regarding antibiotic resistance.[50] And in late 1958, after discussion at the Eleventh World Health Assembly, the WHO prepared again to convene a small meeting of antibiotics experts, with antibiotic resistance as well as antibiotic "use and misuse" to be topics of discussion.[51]

The meeting was held in Geneva in May 1959, with Max Finland and Selman Waksman representing the United States (and Waksman elected honorary chair of the six-person meeting), just as John Blair's investigations leading up to the Kefauver hearings were evolving.[52] "Counter-propaganda" regarding antibiotic "misuse consequent on unscrupulous advertising" was considered, along with proposed efforts to reduce "self-medication" in "countries in which antibiotics are freely obtainable by the public." Members also discussed the need to standardize and coordinate antibiotic resistance testing as a potential prelude to guiding antibiotic usage at both local and global levels.[53]

However, the micro-debate over laboratory standardization would dominate WHO antibiotic involvement over the ensuing decade, narrowing the organization's focus from broad concerns regarding antibiotic usage to the technical aspects of laboratory standardization itself.[54] Initially projected to take two years to complete—at which point it was to be "followed by a world-wide survey of the frequency of resistance and a study of the mechanism by which bacterial resistance to antibiotics develops"—the standardization discussion dragged on through the early 1970s, precluding more ambitious plans.[55] Once again, Selman Waksman in particular articulated plans for a larger agenda (including the establishment of an international antibiotic repository), arguing that the WHO would be the "logical body" to coordinate international efforts. Yet neither funding nor political will would be committed to such an expansion of activities, and the initial WHO expert panel on antibiotics would be dissolved by the early 1970s.[56]

What, then, of the other leaders of the U.S. infectious disease community? Some, such as Cornell University's Walsh McDermott, were surprisingly sanguine about staphylococcal ecology, denying "that the crude equilibrium between the public at large and the ubiquitous staphylococci is shifting for the worse."[57] McDermott, following the lead of his friend René Dubos, instead shifted attention to the improvement of host defense, while at the same time drawing attention to hospital conditions and asepsis.[58] McDermott's and Dubos's interest, in this respect, was directed to the nature of *how* bacteria survived the onslaught of antibiotics—whether through bacterial *persistence* (in which bacteria were sensitive to a particular drug but able to survive owing to particular local "environmental" host factors) or bacterial *resistance*—with obvious implications regarding host versus microbial factors for the effective use of antibiotics.[59] Yet for many other key scientists throughout the 1950s, the nature of antibiotic resistance itself—whether through neo-Darwinian selection of mutated strains or Lamarckian mechanisms of induced resistance—remained the focus of inquiry, generating symposia on "Drug Resistance in Micro-Organisms" that seem remarkably divorced, in retrospect, from their clinical relevance.[60] Indeed, by the time the American Hospital Association presented its bulletin on the "Prevention and Control of Staphylococcus Infections in the Hospital" in 1958, it could accurately report: "It is not known whether resistant strains of [staphylococci] actually acquire resistance after exposure to the antibiotics, or are resistant to begin with and are simply unmasked by the suppression of susceptible strains—although the latter is thought to be the case. From a practical point of view it does not matter which theory is correct."[61]

Others, such as Perrin Long, who had pioneered the use of the sulfa drugs in the United States, remained primarily focused on adverse events.[62] And while debate over the link between chloramphenicol and aplastic anemia drew attention to the overuse of antibiotics in the outpatient setting, the ultimate characterization of chloramphenicol usage as exemplifying "irrational" prescribing would not be rendered (as discussed in chapter 4) until the 1960s. Such discourse in the 1950s only narrowed the discussion to chloramphenicol itself, placed another wedge in the distinction between problems in the outpatient versus inpatient settings, waxed and waned in relation to the debate over the link between chloramphenicol and aplastic anemia, and (as discussed in chapter 1) was considered overblown by such leading infectious disease experts as Max Finland.

This leads to a final question concerning antibiotic resistance in the 1950s, when antibiotic prescribing habits were becoming entrenched: given the emerging network of reformers described in the first half of this book, what about Max Finland? In the early 1950s, Finland had imparted an emphatically moral tone to his implicit linkage of antibiotic overuse and rising staphylococcal resistance. As he began an editorial in the *Journal of Pediatrics* in November 1951, "No honest or self-respecting physician or surgeon, whether his practice be limited to pediatrics, geriatrics, or any other special field of medicine or surgery can help but feel a bit conscience stricken each time he prescribes or administers a sulfa drug or antibiotic after a hurried visit to the bedside of a patient or after a brief interview and examination in his office."[63] He concluded the article by invoking an early notion of antibiotic stewardship: "Are we in medicine, like our counterparts in industry, exhausting our most valuable resources at too rapid a rate? Are we, perhaps, depending too much on the ingenuity of our scientists and industrialists to keep constantly ahead of this decline? Only time will tell. In the meantime, is it not prudent to think more in terms of how these valuable antimicrobial agents can be used most effectively rather than most widely?"[64]

By the mid-1950s, however, we find a striking disappearance of this physician-centered moral stance from Finland's discussions of antibiotic resistance. In part this change reflected prudent backtracking, as he attempted to gather data regarding the details and extent of the evolving resistance issue, while in part it reflected a strategic narrowing of his proscriptive emphasis to decreasing inappropriate surgical prophylaxis.[65] Such data-gathering is reflected in his encyclopedic series of articles on the "Emergence of Antibiotic-Resistant Bacteria" compiled for the *New England Journal of Medicine* in 1955.[66]

More critical, though, was that by the mid-1950s, Finland had shifted his considerable moral and pragmatic emphasis *away* from the harried clinician and onto the pharmaceutical industry as a site of intervention. This transformation was a strategic narrowing of emphasis and, as discussed in chapter 2, was driven by Finland's increasing concern—if not obsession—with the development and marketing of the fixed-dose combination antibiotics.

One finds a useful data point at the time of Finland's talk at the 1953 annual antibiotic symposium in Washington, DC. There, his characterizations of clinician overuse as an attempt "not to miss a trick" and of the dream of the "busy practitioner" to have a truly broad-spectrum "omnibiotic" useful and safe in all instances still superseded discussions of combination therapy.[67] However, by the time he delivered the Charles V. Chapin oration, on "Antibacterial Agents: Uses and Abuses in Treatment and Prophylaxis," before the Rhode Island Medical Society in May 1960—two days before Henry Welch's exploits concerning Sigmamycin would be detailed in Congress—Finland's entire emphasis had shifted to a discussion of fixed-dose combination antibiotics. Echoing Harry Dowling's "Twixt the Cup and the Lip" talk from three years earlier (see chapter 2), Finland evoked a potential dystopia in which clinicians were becoming entirely beholden to the suggestions of pharmaceutical makers. Large-scale ecological concerns had been superseded by those about the very process of medical decision-making, as Finland's notions of rational prescribing focused on the sources of physicians' information:

> Perhaps the greatest objection to the use of fixed combinations of antibiotics, and in fact, in prescribing any of the mixtures of drugs now being marketed by the various pharmaceutical firms, is that they have removed the physician from his important status as an educated and rational individual who acquires his own knowledge, experience, and skill and applies them to the choice of therapy as required for his patient. Instead, these ready-made mixtures put the physician in a position of applying therapy by rote, or because he can more easily remember trade names which are often blasted into his ear by detail men or flashed before his eyes repeatedly in his daily mail or when and if he leafs through his medical journals, very much like the names and theme songs of the popular brands of beers or cigarettes to which he is unwillingly subjected whenever he listens to his radio or watches his television.[68]

With Finland's redirection and relative narrowing of perspective, the major opportunity in this period for broad, unified, and consistent attention to be given to clinician prescribing and the ecology of antibiotic resistance had been vitiated. Indeed, Finland himself appears to have temporarily burned out much of his moral ardor with the passage of the Kefauver-Harris amendments in 1962. That year, the AMA's Council on Drugs, with Harry Dowling on its roster, at last decided to wage a "campaign" against the "abuse of antibiotics." Long-time collaborator Dowling wrote to Finland about the keynote article for the campaign, stating that "we all believe that you are the man to write this article, since you stand for 'Mr. Antibiotics' in America at the present time, and your voice would be heeded more than anyone else's."[69] But Finland ultimately declined, and the campaign itself stalled, bereft of its potential leader.[70]

The Finland-led reform efforts of the 1950s and early 1960s (as described in chapters 2 and 3) had been directed at the pharmaceutical industry, rather than at individual prescribers. With the passage of the Kefauver-Harris amendments, leadership concerning antibiotic resistance would shift to a new generation of reformers, from within both academic and government circles, who would bring antibiotic resistance and individual prescribing to the forefront again. Yet the backlash against the very restrictions to prescribing autonomy brought about in the wake of the Drug Efficacy Study and Implementation (DESI) process would contribute to derailing the movement in this subsequent era.

Resistance: 1963–1981

By early 1963, discourse regarding antibiotic resistance in America would be permanently transformed by the publication of Tsutomu Watanabe's *Bacteriological Reviews* article, "Infective Heredity of Multiple Drug Resistance in Bacteria."[71] Starting in the 1950s, shigellae, the cause of bacillary dysentery, had increasingly been found to be resistant to multiple antibiotics among patients in Japan, and by 1959, Japanese researchers found that this type of drug resistance could be passed *across* bacterial species (e.g., from shigellae to *E. coli*).[72] Over the ensuing several years, Watanabe and others found via bacterial genetic study that the "R factor" (for "resistance") encoding such resistance was in fact a key example of the recently discovered genetic elements that can exist and replicate extra-chromosomally, with such "plasmids" (as they had been termed by Joshua Lederberg in 1952) becoming increasingly central to antibiotic resistance research and discourse throughout the 1960s and 1970s.[73] From a genetic standpoint, the R factor represented a fascinating example of the "infective heredity" posited during the previous decade by such scientists as Norton Zinder (at the Rockefeller Institute) and Joshua and Esther Lederberg (at the University of Wisconsin) in their studies of bacterial genetic transfer.[74] From a medical standpoint, though, Watanabe already communicated to his American audience the potential implications for infective heredity to transform into cross-species "infective drug resistance," which would rhetorically transform further into "infectious drug resistance" by the time it was featured in a *New England Journal of Medicine* editorial in 1966.[75] And while the importance of R fac-

John Osmundsen's use of the term *superbug* in *Look* magazine in 1966 is the first instance I have found of the term being used to refer to antibiotic-resistant microbes. The accompanying cartoon is by noted illustrator James (Jim) Flora (1914–1998). The image originally appeared in John A. Osmundsen, "Are Germs Winning the War against People?" *Look* (October 18, 1966): 140–41. "Superbugs" illustration by James Flora, 1966, © The Heirs of James Flora; courtesy Irwin Chusid/JimFlora.com.

tors in the dissemination of antibiotic resistance would be debated over the next several years, the very possibility of such cross-species transfer of resistance capabilities seemed to have exponentially upped the ante regarding the possible emergence of "superbugs" possessing widespread antibiotic resistance.[76] Certainly the rhetoric concerning antibiotic resistance escalated, such as when the *NEJM* editorialist describing "infectious drug resistance" concluded with the warning "that unless drastic measures are taken very soon, physicians may find themselves back in the preantibiotic Middle Ages in the treatment of infectious diseases."[77]

By the late 1960s, as the decade of Rachel Carson and *Silent Spring* was leading to the formation of the Environmental Protection Agency (with Gaylord Nelson planning the nation's first Earth Day in 1970, even as he was investigating the pharmaceutical industry), such discourse could be still more explicitly linked to larger environmental and ecological concerns.[78] As DDT was transforming from World War II–era panacea for agricultural and economic development into toxic catalyst for the environmental movement, so were antibiotics undergoing their own increasing scrutiny.[79] Some of this scrutiny would be applied to the use of antibiotics as growth promoters and prophylactic medications in animal husbandry.[80] At the same time, a parallel discourse related widespread microbial ecological disturbances to the prescribing habits of clinicians. As Richard Gleckman and Morton Madoff concluded their 1969 *NEJM* editorial on "environmental pollution with resistant microbes":

> Man has succeeded in polluting his environment with an astonishing variety of noxious agents. The development of an antibiotic-resistant microbial milieu might be a logical extension of this self-directed biologic warfare, but it is doubtful if many physicians would wittingly subscribe to such an idea. Perhaps it is time to wonder whether the unwitting accomplishment of the same end, without critical appraisal, is any less serious an offense.[81]

While some researchers warned against "making the mistake of forming an anti-antibiotic crusade," Max Finland once again joined the fray, as antibiotic usage continued to escalate dramatically in the late 1960s and early 1970s.[82] Finland, with a longstanding relationship with Lederle (whose Thomas Jukes had been more responsible than any other scientist for "promoting" antibiotics as livestock growth promoters), remained unconvinced of the dangers of the use of antibiotics in animal husbandry.[83] But by the early 1970s, he was once again publicly decrying antibiotic overuse among clinicians. In a series of lectures and a paper on the "Changing Ecology of Bacterial Infections as Related to Antibacterial Therapy," Finland focused on the increasing incidence of such gram-negative organisms as *E. coli*, *Klebsiella pneumoniae*, *Serratia marcescens*, *Proteus mirabilis*, and *Pseudomonas aeruginosa*—many of them resistant to one or more antibiotics—as causes of bacteremia and death at Boston City Hospital, clearly linking such ecological change to antibiotic selection pressure. Efforts to counter such trends could include developing novel remedies (offering that it would be "prudent to use new and effective agents selectively") or improving host resistance (likely "illusory"). But the key measure was to curtail inappropriate prescribing. As he stated unequivocally, "The unnecessary and improper use of antibacterial agents in treatment, and particularly for prophylaxis, that is not highly specific and sharply circumscribed, will have to be modified or stopped completely."[84] Nevertheless, having published on infectious diseases during each of six consecutive decades, Finland concluded that his article could be taken as "another jeremiad from an old and experienced physician, teacher, and clinical investigator reaching the 'end of his line'" and pointed to "the new generation of 'activists'" who would have to take up the torch and confront the changing world of bacterial infections.[85]

The institutional basis for such activists, in retrospect, looks like it could have been the Infectious Diseases Society of America (IDSA), given its leadership today concerning antibiotic resistance. The IDSA was formed in 1963, in the wake of the discrediting of Henry Welch and the demise of Welch's annual antibiotic symposia in Washington, DC. During the latter years of Welch's meetings, Finland had convened an annual "infectious diseases dinner group" of 50–100 selected members and was asked to form a new organization devoted to the study and teaching of infectious diseases. While Finland had felt that such an organization should evolve under the umbrella of the established American Society for Microbiology, the "new generation" of infectious disease experts—many of them former fellows of his—sought to create an independent organization.[86] Finland, despite his displeasure and subsequent refusal to come to the organizational meeting, was prevailed upon to serve as the first president of the organization, and seven future presidents of the organization were former fellows of his at Boston City Hospital (including Harry Dowling, Finland's first fellow and the IDSA's third president). Initially an exclusive club of 125 members, the IDSA would blossom in its first two decades, taking over responsibility for the *Journal of Infectious Diseases* in 1969 (with former Finland fellow, Edward Kass, its first IDSA-based editor-in-chief), and continually expanding in size, reaching nearly 2,500 members by the time Kass wrote the society's institutional biography in 1986.[87]

Nonetheless, throughout the first two decades of its existence, the IDSA remained an internally focused academic forum, and quite muted with respect to antibiotic resistance. As early as 1964, Harvard's Geoffrey Edsall had "suggested that the IDS[A] may have an important function in the future by becoming a real rallying point for discussion of important problems that may arise at the national level for which specific advice is needed."[88] Yet in 1971, when requests were made for the society "to act in an advisory capacity, or to take a more active role in government or health-related activities," the IDSA's leadership demurred, preferring that "those members of the Society who are so inclined should act as individuals."[89]

In late 1973, Leighton Cluff challenged his colleagues in his outgoing IDSA presidential address:

> We should be actively involved in planning means for monitoring use of antimicrobial drugs, yet many of us seem disinterested in this problem, have given it little or no thought, and frequently accept no responsibility for it. Is it justifiable for experts in infectious diseases, whether in an academic setting, industry, or practice, to be unconcerned about how antimicrobial drugs are used? If this is our area of competence and responsibility, can we reasonably ignore it?[90]

But tensions persisted between the desire to formulate guidelines and guidance for antibiotic control programs, and the ongoing recognition of uncertainty with respect to the efficacy and consequences of proposed restrictive and educational measures (see chapter 4). By the mid-1970s, the expressed town-gown backlash against the removal of the fixed-dose combination antibiotics (as described in chapter 4), felt by such prominent IDSA members as Edward Kass, had led to institutional inertia and a fairly diluted national impact.

However, the 1970s saw continued, and even increasing, attention paid to antibiotic resistance, as the image of microbial "pollution" gave way, in the wake of the Three Mile Island accident in April 1979, to "an image of fallout akin to that from a leaking nuclear reactor."[91] As noted in chapter 4, an ongoing series of investigations and exposures of "irrational" prescribing in both hospital and community settings kept the issue linked to antibiotic resistance on the national radar, while antibiotic audits and restrictive hospital policies were tested in limited fashion.[92] By 1978, the National Institute of Allergy and Infectious Diseases (NIAID), led by Richard Krause, was beginning to reexamine its own mandate, particularly with respect to "its research goals as they relate to the needs of the practicing physician."[93] Concerned that "unless we use antibiotics appropriately, we may not be able to use some of them at all," the NIAID convened a symposium on "The Impact of Infections on Medical Care in the United States—Problems and Priorities for Future Research," chaired by Calvin Kunin.[94] Representatives from academia, government, industry, and the front-line clinical community were brought together "for the first time" to share their findings and opinions over the course of two days.[95] And such nationwide discussion would soon transform into international concern and deliberation.

An Inflection Point

At the level of basic science, the 1970s had seen plasmids increasingly take center stage in biomedical discourse. This attention was due, in large part, to their role in the laboratory, where they would be used as manipulable tools to drive the recombinant DNA technological revolution.[96] Their role in enabling scientists to clone particular genes raised well-publicized concerns regarding the potential for tinkered genes to escape the lab, but their increasingly recognized role in the *natural* spread of resistance genes continued to heighten the threat of antibiotic resistance throughout the decade.[97]

Faced with this threat, several novel potential sites of intervention were articulated, grounded in increasing attention to the role of the patient and the global nature of antibiotic usage, respectively. With respect to patients, as noted in chapter 4 and dating back to the 1950s "use and abuse" literature, clinicians had at times drawn attention to the role of the patient (or parent, in the case of pediatrics) in driving antibiotic overuse.[98] But this emphasis would be sporadic, as attention focused squarely on the other apparent driver of "irrational" prescribing behavior—the pharmaceutical industry—throughout the 1950s and much of the 1960s. However, by the mid-1960s through the 1970s, as sociologists and economists began to examine the "overmedicated society" (and its consequences) more broadly, the patient's role could again be invoked—if not substantiated—as a source of antibiotic overuse.[99] An editorialist in *NEJM* in 1965, describing the "tendency to buy time with antibiotics as sugar pills," understood that the "harried physician" may have little time to explain the rationale for withholding antibiotics from one's patient with an upper respiratory tract infection. Nevertheless, it was the physician's duty to conduct such reeducation on a case-by-case basis: "The most effective area for delivering this message is surely in daily medical practice, and those who would protest lack of time for such education might reflect on whether they are really providing adequate medical care."[100]

Tellingly, though, a different manner of intervention would be proposed from the same starting point a decade later by Robert Moser, then editor-in-chief of *JAMA*, responding in 1974 to Henry Simmons and Paul Stolley's "This Is Medical Progress?" commentary in *JAMA* (described in chapter 4). Moser, who had first written on antibiotics and "diseases of medical progress" over a decade previously, presented "another side of the coin," reflecting contemporary concerns about patient consumerism and the shifting patient-doctor dynamic:

> How did all of this get started? Have we physicians precipitated this crisis in "microbial pollution" through ignorance or slothfulness or failure to heed warnings posted along the way? Perhaps, but I believe the problem has another aspect that penetrates beyond these considerations. It has to do with the changing interface between physician and patient. It has to do with patient pressure—the "operant conditioning" of the general public—the emergence of the "pill culture."[101]

While the AMA intended to "expand its efforts to encourage physicians to resist this pressure" (it didn't), the key was to be the reeducation of patients *before* they entered the doctor's office: "We need a vast program to educate the public. The lay media must assume an obligation to undo much of the damage they have done in dramatizing the 'wonder drugs' and 'wizard cures.' . . . As scientists, we are obliged to practice scientific medicine. In the real world it is not easy, for many times patients are their own worst enemies."[102] No such movement emerged at the time, but attention to the patient's role—and to how antibiotics had been promoted as panaceas to both physicians *and* patients—was soon paralleled by attention to how antibiotics were being distributed and consumed worldwide.

The ecological focus on rapidly spreading plasmids with regard for neither species nor national boundaries played a large role in the globalization of antibiotic discourse. Medical scientists documented large-scale, lethal outbreaks of plasmid-mediated antibiotic-resistant bacillary dysentery in Central America, typhoid fever in Mexico, and salmonella infections around the world.[103] Decrying the advent of chloramphenicol-resistant typhoid, E. S. ("Andy") Anderson, head of the Enteric Reference Laboratory in London, drew the evolutionary implications concerning such spreading plasmids: "Situations such as this, and others analogous to it in many parts of the world, are the result of the long-term indiscriminate use of chloramphenicol and other antibiotics in the affected areas. They can be rectified only by more rational antibiotic usage."[104]

As Robert Bud has related (drawing specific attention to Anderson's advocacy efforts in Great Britain), plasmids would increasingly inform debate and regulation regarding the use of antibiotics in animal feeds.[105] They would also force increased attention to how antibiotics were being marketed and utilized in the developing world (where they were often available over the counter), reflecting broader and increasing efforts by the WHO and others to render international pharmaceutical usage more rational.[106] Characteristic of such efforts, two Swedish researchers reported in early 1981 on the "marketing of obsolete antibiotics in Central America." Whereas in Sweden, 90 antibiotics were marketed, two of them combination drugs, in Mexico 430 antibiotics were marketed, 180 of them fixed-dose combination antibiotics (including Panalba, marketed as Albamycin-T) of the variety eliminated in the United States by the DESI program.[107]

The WHO, continuing its self-conscious efforts to address such broader "global" health concerns, convened a working group in 1973 in Copenhagen on "The Public Health Aspects of Antibiotics in Feedstuffs" and a consultation meeting in 1975 in Brussels on "Antibiotic-Resistant Bacteria in the Environment."[108] At the Brussels meeting, organizers addressed the role of both education and regulation in confronting worldwide antibiotic resistance: "Medical students, physicians and other health personnel, as well as veterinarians, pharmacists, nurses and agricultural specialists should be trained to recognize the dangers of indiscriminate use of antibiotics. . . . It is

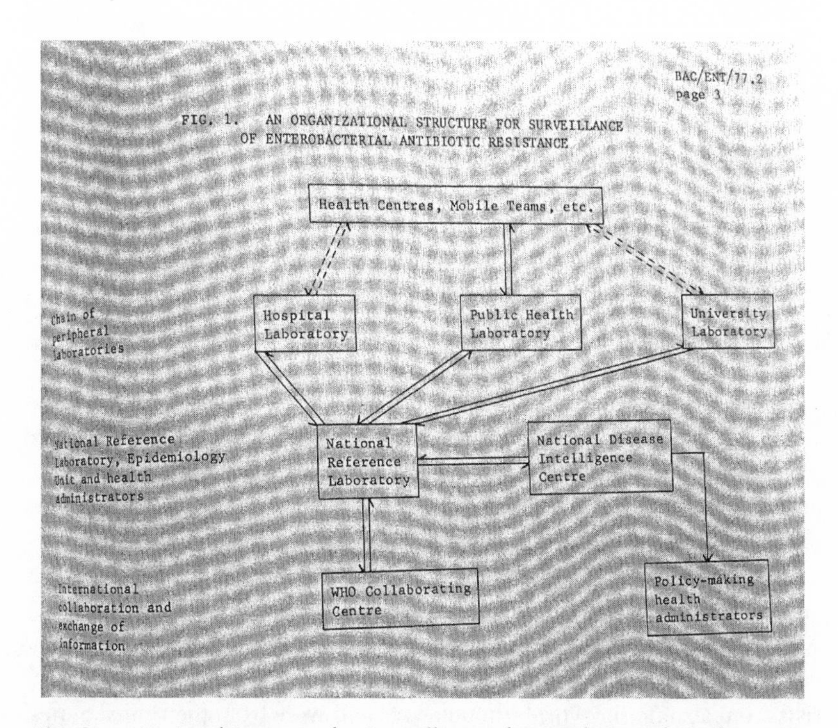

"An Organizational Structure for Surveillance of Enterobacterial Antibiotic Resistance," reflecting the mid-1970s aspirations of certain members of the World Health Organization. Note that enterobacteriacea include *E. coli*, *Shigella*, *Salmonella*, *Klebsiella*, *Yersinia*, *Proteus*, *Serratia*, and *Citrobacter*, which represent some of today's most concerning antibiotic-resistant organisms. D. Barua, Y. Watanabe, and L. Houang, "Organization of Surveillance for Antibiotic Resistance in Enteric Bacteria," in *Meeting on Surveillance for Prevention and Control of Health Hazards due to Antibiotic-Resistant Enterobacteriaceae, Geneva 18–24 October 1977*, BAC/ENT/77.2, World Health Organization Library. Reproduced with permission of the publisher.

desirable that all governments should insist on the supply of therapeutically useful antibiotics to the public only on prescription."[109] And harking back to aspirations from the early 1960s, certain WHO representatives continued to envision (and convened a meeting in Geneva regarding) a central role for the WHO in coordinating a worldwide resistance surveillance network seemingly necessary to undergird rational antibiotic usage.[110]

But the WHO's goals remained largely aspirational throughout the 1970s, and this would provide the backdrop to an international meeting held in Santo Domingo in January 1981 on the "Molecular Biology, Pathogenicity, and Ecology of Bacterial Plasmids." In May 1979, Tufts University clinician-researcher Stuart Levy (who had trained under Tsutomu Watanabe), along with Royston Clowes and Ellen Koenig (who happened to be Levy's sister), had first discussed the possibility of convening an international plasmid conference.[111] And from the earliest planning for the meeting, while many of the topics were to focus on plasmids themselves, "problems of antibiotic resistance" and "antibiotic usage in man and animals [including] the problem of indiscriminate use and over-the-counter sales in developing and developed countries" were to feature prominently, with a workshop on "problems raised by the uses of antibiotics in human health and animal health" (along with the "public health issues of industrial uses of plasmids") to be included.[112] An atmosphere of both scientific exchange and international conviviality pervaded the conference, with a tongue-in-cheek final paper, a survey of participants on whether they had been afflicted by gastroenteritis during the conference, recommending that "plasmid investigators should travel more" (for the sake, of course, of developing immunity).[113]

But the Dominican Republic was also chosen as the site for the conference on account of its being representative of nations in which antibiotics were sold over the counter and life-threatening antibiotic-resistant organisms were prevalent.[114] And at an evening session of the meeting, 147 scientists from around the world (though over half were from the United States, including molecular biologist Walter Gilbert, who had just won the Nobel Prize in Chemistry) signed a joint "Statement Regarding Worldwide Antibiotic Misuse." The collaborative statement pointed to the "worldwide" nature of the "neglected health problem" of antibiotic resistance, ascribing the problem "in large part to the indiscriminate use of antibiotics."[115] Levy and his colleagues presented five particular responsible practices, all sites for educational or regulatory activity: use when not required, promotion of the medications as "wonder drugs," marketing the medications differently in developing and developed countries, use in animal feeds, and dispensing antibiotics without prescription. Back in Boston, Levy would use the statement as a springboard to form the Alliance for the Prudent Use of Antibiotics (APUA), intending for the organization to serve as a unified, explicitly *international* (the thirty-one members of its initial scientific advisory board came from twenty-five different countries) catalyst for efforts to educate the public and the medical profession about antibiotic overuse, as well as to promote uniformity of regulation regarding antibiotic sales and labeling.[116]

With such an international focus, the organization would soon join the WHO in calling for the coordination of worldwide antibiotic resistance surveillance efforts, and it would bring Brigham and Women's Hospital physician Thomas O'Brien aboard as a second vice president by 1986. O'Brien, who

Stuart Levy, as he appeared in *Time* magazine in mid-1981, "speaking at a Boston press conference about the dangers of antibiotic overuse" and in the context of the release of the "Statement Regarding Worldwide Antibiotic Misuse."
"Those Overworked Miracle Drugs," *Time* (August 17, 1981): 63. Reproduced with the permission of photographer Steven Liss.

had pioneered the computerized analysis and display of antibiotic resistance patterns, would serve as an important bridge between APUA and the WHO, establishing the WHONET program to promote standardization and comparison of local resistance data.[117] The WHO also convened a Scientific Working Group on Antimicrobial Resistance (on which Levy and O'Brien, along with Calvin Kunin and Harvard's Roger Nichols, would represent the United States) in November 1981, drawing further attention to the need for improved surveillance and reporting mechanisms, along with a subsequent conference (chaired by O'Brien) on surveillance in November 1982.[118] Surveillance discussions—concerning "tens of thousands of laboratories throughout the world" that remained uncoordinated—continued to point to the relationship between the local and the global, both real (as achieved by microbes) and potential (as yet to be achieved by surveillance structures).[119]

That same year, the University of California–Berkeley's Mark Lappé published the well-researched and highly readable *Germs That Won't Die: Medical Consequences of the Misuse of Antibiotics*. Lappé, while serving in the California Department of Health, had previously warned against the environmental spraying of Malathion to curb fruit fries, and in *Germs That Won't Die*, he argued not only for more restrictive use of existing antibiotics but, à la René Dubos, for more emphasis on strengthening the body's "natural defenses," whether through vaccination or through what would come to be increasingly described as "probiotics."[120] Lappé's book would serve, as Robert Bud has noted, as forerunner to over three decades of public-oriented print and media—some items more sensationalized than others—on the hazards of antibiotic misuse and overuse.[121]

Lappé directly cited the medical community's "violent opposition" to Calvin Kunin's support for antibiotic audits and hospital oversight committees (as described in chapter 4), and in 1982, the same year Lappé's book was published, the IDSA's governing council rejected Kunin's proposal that the IDSA establish and propagate society-sponsored national guidelines on antibiotic usage.[122] The IDSA's membership, however, indicated a desire for the society to take a larger role in public affairs (this largely took the form, at the time, of advocating for increasing funding in infectious disease research and training).[123] In 1984, Kunin spoke to the IDSA membership at their annual meeting on "The Responsibility of the Infectious Disease Community for the Optimal Use of Antimicrobial Agents."[124] In preparation for the talk, he and Stephen Chambers polled members concerning their views on antibiotic usage. They found that 87 percent claimed a "strong" interest concerning antibiotic usage; that the majority felt that antibiotic resistance in the United States stemmed from usage in the hospital (chiefly) and the office (less so), rather than via the developing world or animal feeds; and that an overwhelming majority advocated for educational and regulatory counter-measures.[125]

By 1985, the IDSA's governing council approved the formation of an Antimicrobial Agents committee, with Kunin as its head.[126] And Kunin (who would become IDSA president in 1987) used the committee to publish selected IDSA-sponsored guidelines on antibiotic usage over the next several years.[127] Yet resistance to Kunin's efforts surfaced ongoing deliberation regarding the ability to make definitive recommendations in the setting of insufficient data, a persistent Achilles heel for would-be reformers. Reflecting such caution, Kunin's internally distributed IDSA white paper "Report of the Antibiotic Use Committee" (the committee name change in the title appears to have been an oversight) generated calls "for a softening of language indicating a desire to 'regulate' or 'control' the use of antibiotics or the testing of antimicrobial agents."[128] When push came to shove, advocating for confronting antibiotic resistance was not yet to be a chief mandate of the society.

Indeed, continuing the trajectory he began with the ISCAMU (Inter-Society Committee on Antimicrobial Drug Usage, discussed in chapter 4) project in the 1970s, Edward Kass seems to have remained the IDSA member most opposed to disseminating such guidelines. In hearings on "Antibiotic Resistance" conducted by Rep. Al Gore (D-Tennessee) in December 1984, Kass advised: "I would caution everyone who looks at this problem . . . of trying to develop guidelines that will be helpful. . . . Quick and dirty fixes based on inadequate data are almost certainly going to clutter up the situation [and] give us very little relief from the problem as it exists."[129] The hearings devoted one day to the use of antibiotics in animals and one day to their use in humans. Stuart Levy, Thomas O'Brien, and Public Citizen's Health Research Group director Sydney Wolfe all made appearances, but equal time was given to advocates for the use of antibiotics in animal feed and to the reassuring testimonies of the Pharmaceutical Manufacturers Association's Joseph Stetler and John Jennings.[130] And while Al Gore could point to an inconvenient truth—that while he hoped "that this magnificent pharmaceutical industry we have can continue to come up with ever more potent miracle drugs," the more "sensible course of action" would be "to conserve the . . . susceptibility of the disease organisms to . . . the arsenal of antibiotics we have now"—no course of action appears to have followed the hearings.[131]

Perhaps most representative of the scope, successes, and limitations of antibiotic reform efforts in the 1980s, though, would be the large-scale "International Task Forces on Antibiotic Use" overseen by Levy (O'Brien and Kunin would serve as chairs of two of the six individual task forces) and sponsored by the National Institutes of Health's Fogarty International Center. The path to the task forces had an unlikely origin and a tortuous route. Ernst Freese, the chief of the Laboratory of Molecular Biology at the National Institute of Neurological and Communicative Disorders and Stroke, had spent time in South America and heard from physicians about their inability to treat certain patients sick with resistant organisms.[132] By early 1980, he felt inspired to write to the WHO, warning that "if some bacterial strains should develop resistance to all readily available antibiotics, a major epidemic of the sort reported for earlier centuries cannot be excluded."[133] Freese pointed to worldwide structural factors—from harried doctors in developed countries to doctorless patients in developing countries—promoting antibiotic overuse. After attending Stuart Levy's Santo Domingo conference and signing its "Statement Regarding Worldwide Antibiotic Misuse," he reached out more locally to the Pan American Health Organization and the Fogarty Center in hopes of forming "a task force which will study the current status of the use and abuse of antibiotics in the Western Hemisphere, and in a plenary session involving appropriate officials from countries of the Hemisphere . . . reach a consensus on the activities and legislation needed to overcome the problem."[134] Freese felt that despite the worldwide nature of the problem, "it seems most promising to limit the scope of the task by finding at first a solution for a limited group of countries that are adjacent to each other and whose populations are in frequent contact causing rapid communication of infectious diseases."[135] And the proposed "International Task Force to Analyze the Reasons for, and the Nature of Public Health Problems Caused by Antibiotics and to Recommend Procedures to Prevent Those Problems" soon had a short working title: "Use & Abuse of Antibiotics."[136]

By the spring of 1982, Freese and his colleagues at the NIH invited Levy to advise—and soon, lead—the envisioned task force and subsequent summative conference.[137] And Levy, continuing to focus on the linkages between the local and the global, would expand the reach of the task force beyond the western hemisphere.[138] By 1984, he was overseeing six task forces, respectively concerned with quantitative estimates of worldwide antibiotic usage, similar estimates of antibiotic resistance, international legal frameworks regarding antibiotic administration (focusing chiefly on unregulated, nonprescription use), social and behavioral factors influencing antibiotic usage, potential educational counter-measures, and the economic aspects of antibiotic development, usage, and resistance.[139] The task force findings were ultimately informed by over 100 researchers from around the world, including representatives from academia, industry, foundations, federal agencies, and both national and international public health groups.[140] And while they drew attention to the limitations of the available data and even ongoing difficulties in defining and standardizing such fundamental terms as *usage* and *resistance*, they also pointed to the need for increasing resources, cooperation, education, and (in selected instances) legislation in confronting worldwide antibiotic resistance.

Nevertheless, the Fogarty task forces ultimately failed to catalyze the mobilization of resources or further collaborative activities, reflecting a deeper resistance, during the Reagan era, to the politicization of antibiotic resistance as a "call to action" in the United States. As early as 1983, WHO representatives had apparently suggested changing the title of the proposed conference to "The Appropriate Use of Antibiotics," further cautioning that industry buy-in should be ensured from the beginning.[141] And despite WHO optimism at the time, industry concerns that the task forces served merely as a forum for antibiotic resistance "activists" and as an "outright attack on free enterprise" indeed appear to have contributed to a dramatic NIH scale-back of the planned "major," international, summative conference on antibiotic use and resistance.[142] Sidney Wolfe, of Public Citizen, accused the NIH and the Department of Health and Human Services of "appeasing the drug industry," while pioneering pharmacoepidemiologist Jerry Avorn considered that "the NIH was no longer sacrosanct and appear[ed] to be subject to the same political influences as other [regulatory] agencies."[143] Over a decade later, Levy still lamented that "some influential critics from industry believed that such an activity cast wrongful doubts on the efficacy of their antibiotic products," while stating more broadly that the task forces were "disregarded and even downplayed under the guise that the problem was being overstated."[144]

Equally important, Levy was attempting to mobilize such efforts against a backdrop that saw dramatic public health retrenchment within the United States, and the number of personnel at the National Center for Infectious Diseases at the CDC reduced by 15 percent between 1985 and 1988.[145] For Levy and his like-minded reformers, this represented a disappointing setback. Logic and data had been mobilized, and collaborative networks had been formed, but this was not yet a viable network for leveraging sufficient visibility and resources. However, after having shepherded the antibiotic resistance crusade into a loose, international confederation of reformers throughout the 1980s, Levy would soon find a crucial ally with a longstanding interest in bacterial genetic transfer, an incomparable scientific reputation, and a talent for mobilizing press and government agencies alike to action: Joshua Lederberg.

The Crusade

Lederberg not only had won the 1958 Nobel Prize in Physiology or Medicine (at the age of thirty-three) for his elucidation of the bacterial exchange of genetic information, but also had long maintained an interest in both antibiotic resistance and coming plagues.[146] He also had maintained an interest in (and talent for) communicating about science to broader audiences. With respect to global pandemics, as early as 1968, in a *Washington Post* article drawing attention to the recently contained outbreak of the lethal Marburg virus (and contributing to its naming, in the process), Lederberg had warned:

> The threat of a major virus epidemic—a global pandemic—hangs over the head of the species at any time. We were lucky on this occasion, but it was a near miss. . . . Marburgvirus [*sic*] is but one example of the evils of nature

> that are our real enemies in the living world. It is very unlikely to discriminate between Democrat or Communist or Maoist. And as human society is now organized, our encounters with such threats will not for long be just near misses.[147]

Yet as Robert Bud has identified (and prior to the opening to researchers of the Lederberg papers at the National Library of Medicine, at that), the immediate stimulus to Lederberg's engagement with what would come to be considered "emerging infections" stemmed from his time in New York City as president of the Rockefeller Institute at the height of the AIDS epidemic in the mid-1980s.[148] Lederberg felt that contemporary projections regarding the scope of the epidemic were far too conservative (even fearing that the virus could mutate so as to be able to be spread by air), estimating in 1986 that 10 percent of the U.S. population would be killed by the virus.[149] And as he wrote to the Department of Defense's director of net assessment, Andrew Marshall, in December 1986, "there is no reason to believe that AIDS is the last word in what nature has in store for us."[150]

In October 1987, Lederberg was asked by François Mitterand and Elie Wiesel to present at a conference of Nobelists in Paris on "Facing the 21st Century: Threats and Promises."[151] Lederberg chose to speak on "Medical Science, Infectious Disease, and the Unity of Humankind." Describing an interwoven world of evolving humans and microbes, Lederberg derided "premature complacency" with respect to infectious disease since the advent of the wonder drugs, instead noting with respect to AIDS that "we will face similar catastrophes again."[152] Lederberg came back from the conference and wrote to former labmate (from their time in MacFarlane Burnet's Melbourne immunology lab three decades previously) Sir Gustav Nossal: "It is hard to believe that so many people can remain so complacent about still further plagues, even after the current experience with AIDS. I'm beginning to feel the need to speak out in even higher tones of voice with regard to the importance of looking out for new virus outbreaks."[153]

Lederberg was further stimulated in his thinking by fellow Rockefeller scientist Stephen Morse and spoke in May 1989 at a conference organized by Morse on "Emerging Viruses," recognizing the need to turn scientific interest into government policy.[154] Lederberg would push the Institute of Medicine (IOM), which had recently published a scathing report of the nation's insufficient public health infrastructure, to convene a "multidisciplinary committee" on "emerging microbial threats."[155] Their ensuing landmark report, *Emerging Infections: Microbial Threats to Health in the United States*, broadly defined emerging infectious diseases as "clinically distinct conditions whose incidence in humans has increased," with a focus at the time on those diseases that had "emerged" in the United States within the previous two decades.[156] As Frank Snowden has noted, the committee and report represented a self-conscious public health emergence from an "age of hubris" with respect to infectious diseases.[157] And as Nancy Tomes has described, they likewise emerged against the backdrop of end-of-the-millennium evolutionary pessimism and the increasingly self-conscious process of globalization itself.[158]

While the committee and report focused on viral pathogens, antimicrobial resistance was included as well, with a specific committee recommendation:

> Clinicians, the research and development community, and the U.S. government (Centers for Disease Control, Food and Drug Administration, U.S. Department of Agriculture, and Department of Defense) [should] introduce measures to ensure the availability and usefulness of antimicrobials and to prevent the emergence of resistance. These measures should include the education of health care personnel, veterinarians, and users in the agricultural sector regarding the importance of rational use of antimicrobials (to preclude their unwarranted use), a peer review process to monitor the use of antimicrobials, and surveillance of newly resistant organisms. Where required, there should be a commitment to publicly financed rapid development and expedited approval of new antimicrobials.[159]

Appearing in 1992, the same year as Stuart Levy's *The Antibiotic Paradox: How Miracle Drugs are Destroying the Miracle*, the IOM report set off a host of mutually reinforcing relationship-building, media reports, and resource mobilization regarding antibiotic resistance.[160] At the CDC, the Board of Scientific Counselors (itself strategically formed) of the National Center for Infectious Diseases used the IOM report to mobilize relationships with scientists, the media, and Congress around the issue of emerging diseases more broadly.[161] And Robert Bud has described well Lederberg's skill in communicating with sympathetic science writers, especially former student Laurie Garrett, whose Pulitzer Prize–winning *The Coming Plague: Newly Emerging Diseases in a World out of Balance* would enunciate the themes of the IOM report to a global audience.[162] And whether in *Science*'s "Post-Antimicrobial Era," Mike Toner's Pulitzer-winning series in the *Atlanta Journal-Constitution* on "When Bugs Fight Back," *Newsweek*'s "The End of Antibiotics," *Time*'s "Revenge of the Killer Microbes," or Garrett's chapter on "The Revenge of the Germs, or Just Keep Inventing New Drugs," the general scientific morality play regarding antibiotic resistance in particular depicted mutating bacteria responding to Darwinian selection pressures, a host of blameworthy actors (from demanding patients to defensive physicians to pharmaceutical marketing departments) generating such selection pressures, and an increasingly pessimistic view of the "arms race" between microbes and mankind, with the potential, in an era of increasing globalization, to culminate in "medical disaster" or a "post-antibiotic era."[163] Antibiotic resistance was increasingly viewed by the media as a "hot" story, setting in motion a continuous wave of media and popular reiterations over the ensuing two decades.[164]

At the same time, the IOM report led to a cascade of more soberly worded government and expert policy reports on antibiotic resistance, including those by the Congressional Office of Technology Assessment (OTA) and the American Society for Microbiology (ASM).[165] With respect to emerging diseases more broadly, the CDC's National Center for Infectious Diseases (led by James Hughes) produced its own landmark report, "Addressing Emerging Infectious Disease Threats: A Prevention Strategy for the United States," relying on the feedback of more than two hundred individuals both internal and external to the agency.[166] Such fruitful work was followed by an increase in CDC funding for emerging infections from $1 million in 1994 to $7.7 million in 1995 to $59.1 million in 1998.[167] The agency launched the journal *Emerging Infectious Diseases* in 1995, and with respect to antibiotic resistance in particular, partnered with the FDA and the U.S. Department of Agriculture to establish the National Antibiotic Resistance Monitoring System (NARMS) in 1996.

Just as the experts were advocating "networks" of surveillance and reporting, such activities themselves were driven and enhanced by ongoing collaborations and interlocking activities among key players and organizations. Levy introduced himself by letter to Lederberg in August 1993, sending along a copy of *The Antibiotic Paradox*; within several years they would serve together on both the OTA and ASM panels and were strategizing regarding ongoing activities in their mutually constituted "campaign."[168] Gail Cassell, who had chaired the NCID's Board of Scientific Counselors and served as president of the ASM in 1994, chaired both the OTA advisory panel and the ASM task force that produced their respective reports.[169]

Several key recommendations pervaded nearly all of the antibiotic resistance reports that appeared throughout the decade (most robustly explored in the OTA report), as well as the congressional hearings on "the threat and risk of certain old and new infectious diseases on the nation's health" held in October 1995.[170] The first was the continuing call for increasing—and increasingly coordinated—surveillance, grounded in improved and standardized laboratory infrastructure. The second was for decreasing antibiotic usage, mandating studies and interventions at the levels of physicians, patients, and the animal husbandry industry. The third was a request for increased research funding from NIAID, the CDC, or elsewhere regarding antibiotic resistance. And the fourth was a call for FDA regulatory changes to incentivize industry to develop novel antibiotics while refraining from overmarketing such uniquely self-limited commodities.

Two morally charged and tightly intertwined themes underlay the reports. The first, dating back to the 1988 IOM public health report and Lederberg's talk in Paris the same year, was the implication that the present emergency emanated from systemic *complacency* and the collapse of public health infrastructure. While antibiotic resistance could still serve as a moral barometer of physician prescribing activity, it had become equally linked to this larger correspondence between emerging infections and the need for reconstituting the public health system and improving physician-public health system cooperation.[171] Emerging infections would be grounded, in the wake of AIDS, in the notion of a depleted system, deficient in responding to both present and coming plagues.

The second was the multifaceted dynamic—pragmatic and moral—forged between the global and the local. Lederberg had initially linked these concerns in Paris, concluding his Nobel panelist talk: "As one species, we share a common vulnerability to these scourges. No matter how selfish our motives, we can no longer be indifferent to the suffering of others. The microbe that felled one child in a distant continent yesterday can reach yours today and seed a global pandemic tomorrow. 'Never send to know for whom the bell tolls; it tolls for thee.'"[172] And as he would state later that year in a symposium on modern plagues: "Our neglect of infectious disease in the poor majority of the world is not just a humanitarian disgrace; it leaves unchecked the seeds of our parochial infection."[173] The vector (and vectors) pointed in both directions. Pragmatically, while globalization in human transportation and microbial genetic exchange rendered seemingly distant events locally relevant, warranting calls for global cooperation in surveillance and standardization of regulatory practice, René Dubos's decades-old maxim to "think globally, act locally" was being applied to calls for individual hospital infection control and antibiotic stewardship programs. Morally, not only were global health concerns increasingly (if haltingly) reconfigured as mandating the gaze of the world's medical and public health communities in their own right, but also the most local of actions (i.e., inappropriate antibiotic prescribing) increasingly were reconfigured by reformers as having global public health consequences.[174] The "collapse" of public health was thus being paralleled by a collapsing of the very space between the global and the local.

Nicholas King has illustrated how such a process reflected a self-conscious employment of "scale politics" to expand both the scope and significance of the emerging infections discourse.[175] And the emerging infections-antibiotic resistance linkage would have implications for policy makers around the globe. The WHO, which had continued to develop its WHONET program but which had again become relatively quiet with respect to antibiotic resistance since the mid-1980s, convened a series of working groups and meetings on antibiotic resistance throughout the 1990s.[176] And European countries, amid the formation of the European Union (though some, as described in the conclusion, had been invested in antibiotic stewardship since well before the 1990s), became individually and collectively highly attuned to antibiotic resistance. In Great Britain, a cascading series of expert panels culminated in a 1998 report by the House of Lords. The report borrowed the notion of "prudence" from Levy while seeming to borrow that of complacency from Lederberg; either way, it would have a major impact in its own right, leading to the establishment of the European Antimicrobial Resistance Surveillance System (EARSS) and a 100 million-euro infusion of research funding.[177] Countries across Europe established individual antibiotic resistance policies, while countries outside Europe, though slower to make such changes, could soon draw upon the WHO's "Global Strategy for Containment of Antimicrobial Resistance" (in addition to APUA) in national and international planning efforts.[178]

In the United States, political momentum likewise continued to develop. Antibiotic resistance was included as the second of four workshops conducted by the IOM as part of its Forum on Emerging Diseases, which concluded its 1998 policy report with the bold premise that "whatever frictions might ensue from shaping and implementing such policies would be more than offset by the medical and hospital costs and, most importantly, by the deaths and disability avoided."[179] By February of the following year, Senator Bill Frist (R-Tennessee) held hearings on "Antibiotic Resistance: Solutions for this Growing Public Health Threat." There, Senators Frist and Ted Kennedy, FDA commissioner David Satcher, NIAID director Anthony Fauci, and CDC National Center for Infectious Diseases director James Hughes continued to enunciate themes of complacency and the need for a revived public health–private practice alliance to offset the financial and human costs of antibiotic resistance. In an ironic twist from Félix Martí-Ibañez's call in the 1950s for chairs of "antibiotic medicine" at every university in the nation (see chapter 1), Stuart Levy now called for "people teaching about antibiotic or drug resistance in ever[y] major university."[180] Senator Barbara Mikulski (D-Maryland) offered up a new set of rationales, and metaphors, relating the proposals to those to enhance national defense:

> These little microbes are "bioterrorists" every day. I worry about these little bugs. They are out there, outsmarting us and outwitting us. They are fast, they are flexible, they adapt. They are like little terrorist organizations, and they lurk in hospitals and other areas. Although we can say it in a humorous way, we need to really see what it is we're dealing with and develop our public health infrastructure for research, early detection, and public education to help both practitioners and those out in the communities.[181]

Within two years, of course, such reasoning would be still more formally linked to the theme of bioterrorism in seeking funding, especially for the CDC and by those attempting to improve microbiological surveillance.[182]

From the mid-1990s onward, antibiotic resistance would thus become a focus of research funding, journalistic attention, and organizational rebranding and strengthening.[183] Two transformations were most emblematic of the shape of the response to antibiotic resistance in the United States at this time. The first was the federal response, embodied in the formation, in the wake of the 1999 Senate hearings, of an Interagency Task Force on Antimicrobial Resistance co-chaired by representatives from the CDC, the NIAID, and the FDA. By 2001, the task force members had produced a "Public Health Action Plan to Combat Antimicrobial Resistance" comprising eighty-four action items and thirteen "Top Priority" actions that included, once again, increasing surveillance efforts and research funding, promoting public education and "appropriate" physician prescribing habits, and developing a more supportive environment for "product development."[184] Within five years, the CDC was spending $17.2 million on surveillance and antimicrobial resistance, having launched its public-facing "Get Smart: Know When Antibiotics Work" national media campaign in 2003. The NIAID was allocating nearly $200 million annually for antimicrobial resistance by that time, increasing to $292 million by 2009.[185]

Equally telling was the transformation of the IDSA. While the AMA would reposition itself to promote not only antibiotic development but the "judicious use" of such resources, it would be the transformation of the IDSA that would have the greatest impact at the national policy level.[186] This transformation reflected centralized planning, internal committee dynamics, and the surveyed opinions of members who reflected on an era that had seen AIDS converted into a treatable disease while antibiotic resistance and "emerging infections" had been transformed into public health crises.

As a starting point, despite Calvin Kunin's efforts in the 1980s to have the IDSA shift from an internally facing academic society to a more externally facing voice of the infectious disease expert community—and despite the IDSA's tentative foray into the production of guidelines, as well as IDSA-FDA cooperation (led by Kunin and David Gilbert) concerning the FDA's evaluation of antimicrobials—a 1993 retreat had still found that while the society's journals were considered its primary strength, its role in contributing to public policy remained a key weakness in the eyes of its membership.[187] By 1999, however, a self-consciously revitalizing IDSA approved a new mission statement, in which advocating for policy positions immediately followed the promotion of excellence in clinical care and advancing research (contributing to public awareness now ranked fifth).[188] And among policy needs, "there was consensus that IDSA needs to become the source of expertise for developing a plan for antimicrobial resistance."[189] In March 1998, the IDSA created an Emerging Infections Committee (EIC) to help inform Congress and federal agencies concerning such issues as biodefense, food safety, and antimicrobial resistance.[190] And while by 2001 the EIC had disbanded and its functions had been redistributed (some to a "re-energized" Antibiotics Use and Clinical Trials Committee, others to the Public Policy Committee), a 2002 member-needs survey now ranked advocacy of policy positions as the single most important function of the society (with patient care and research tied for second) and "combating antimicrobial resistance" the single most important policy issue.[191]

By this time, leaders at the IDSA had become alarmed by changing FDA policy that would seem to increase dramatically the number of patients needed for the testing of novel antimicrobials.[192] Such concerns would in the short term lead to a productive series of meetings between IDSA and FDA leadership; in the long term, they would indelibly shape IDSA policy.[193] In March 2003, IDSA leadership formed a Task Force on Ensuring the Future Availability of Anti-Infective Therapies under the leadership of John Bartlett.[194] It was to have two areas of focus: "one on developing methods to prevent overuse and the second on developing strategies to encourage the creation of new anti-infectives."[195] Members communicated with and researched the major pharmaceutical companies working on antibiotic development, coming away with the sense that industry was withdrawing from the antibiotic market in favor of more lucrative treatments for chronic illness.[196] And while the IDSA continued, for example, to work with the Society for Healthcare Epidemiology of America to inform antibiotic stewardship programs (concerning the conservation of existing antibiotics), its focus on the drying pharmaceutical pipeline for new antibiotics began to dominate its public discourse.[197]

In July 2004, the IDSA released its own landmark report, *Bad Bugs, No Drugs: As Antibiotic Discovery Stagnates . . . A Public Health Crisis Brews*. IDSA leadership was well aware of both the supply and demand aspects of antimicrobial resistance yet agreed that the focus of *Bad Bugs, No Drugs* should be "to highlight the need for incentives to spur antibiotic research and development."[198] The report advocated consideration of a host of potential incentives—including tax credits for research and development, wild-card patent extensions (in which a company's discovery of a novel antibiotic would be rewarded by the extension of a patent period for another existing drug of the company's choice), and liability protection—for industry to once again enter the age-old arms race with microbes.[199] In a historical irony not lost on the likes of Calvin Kunin, a society founded by such industry-wary infectious disease experts as Max Finland and Harry Dowling now found itself recommending: "Industry must take the lead to ensure success. Industry decision-making is not perfect from a public health perspective, but the focus on financial incentives has made industry successful in the past, and new incentives can lead to future successes."[200] By 2010, the IDSA's recommendations had been crystallized into a call for "10 by '20" (the development of ten new, safe and effective systemically administered antibiotics by 2020). And if the WHO's ambitious "3 by 5" program (the treatment of 3 million patients afflicted with HIV in low- and middle-income countries with antiretroviral therapy by 2005) had called for loosening the implementation logjam in getting effective HIV medicines to those who needed them, the IDSA advocated for incentivizing industry to develop novel antibiotics in the first place.[201]

The IDSA ran the risk at times of constructing a one-sided public argument, focusing on the supply side (novel drugs) of the equation while unintentionally downplaying, in defense of this approach, efforts to influence the demand side (reduced or more "appropriate" usage of existing antibiotics). As the members of the IDSA's Antimicrobial Availability Task Force (AATF) responded to Calvin Kunin's criticisms in 2008, they echoed the very sentiments expressed against Kunin and his colleagues with respect to antibiotics and physician autonomy from decades earlier:

> The IDSA is an organization of ~9000 professionals and currently has no regulatory authority for "antibiotic stewardship" for the >1,000,000 licensed physicians and health care workers who are entitled to use antibiotics in any

> way they feel is appropriate. The complex problem that these unregulated physicians and health care workers face is that of being asked to place a subset of their patients at risk for mortality due to curable infectious diseases by either withholding antibiotics or shortening the course of therapy. In addition, unlike in many other countries, they are asked to take this potentially high-risk, conservative medical approach in a nation in which medical lawsuits are frequent and risk avoidance strategies are highly valued by the general population.[202]

However, by this time, the IDSA was contributing to an overall more integrated crusade. As of 2007, its antibiotic resistance efforts were divided among three committees: the AATF, the Research in Resistance Working Group (advancing the need for increasing NIAID funding for antimicrobial resistance efforts), and an Antibiotic Resistance Working Group (working on surveillance, data collection, and stewardship efforts).[203] These efforts would be translated into the IDSA's advocacy for more balanced support of both the supply and demand sides of the equation in its lobbying efforts and in congressional testimony.[204] The IDSA has pointed out in repeated government testimony that federal efforts remain uncoordinated, that surveillance remains porous (especially as compared to European efforts), and that antibiotic resistance remains underfunded relative to both the overall NIAID budget and the impact of resistance.[205] Among the IDSA's recommendations, enunciated in several versions of the Strategies to Address Antimicrobial Resistance (STAAR) Act and other legislative proposals in Congress, has been the formation of an Antimicrobial Resistance Office within the Department of Health and Human Services (serving to coordinate the efforts of the Interagency Task Force) and the establishment of a nongovernment board of experts to advise the Interagency Task Force, along with the increased funding deemed necessary to support such efforts.[206]

Such legislation has been put forth annually the past several years. Components of it have passed, including extended market exclusivity for "qualified infectious disease products."[207] But the STAAR Act and more comprehensive legislation have not. Perhaps indicative of the anxieties and tensions attendant to both ongoing antibiotic resistance and perceived congressional inertia, the rhetoric of "crisis" has itself become further heightened, as phrases like "superbug" and "post-antibiotic era" have become endemic in academic journals, government publications and congressional testimony, newspapers and popular media, and the blogosphere alike.[208] In this setting of recurring "doomsday" prophecies, congressional supporters of antibiotic resistance–based legislation have at times found themselves having to deny the possibility of the public health community's "crying wolf."[209]

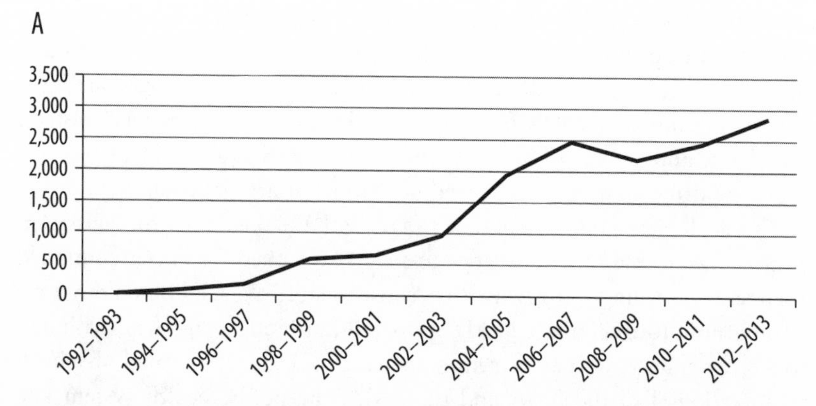

Use of the term *superbug* over time, charted as citations per two-year intervals and displaying a series of growth phases. Derived from a LexisNexis Academic search for "superbug," restricted by "search within results" for "antibiotic resistance," and filtered by de-duplication of articles of "moderate similarity."

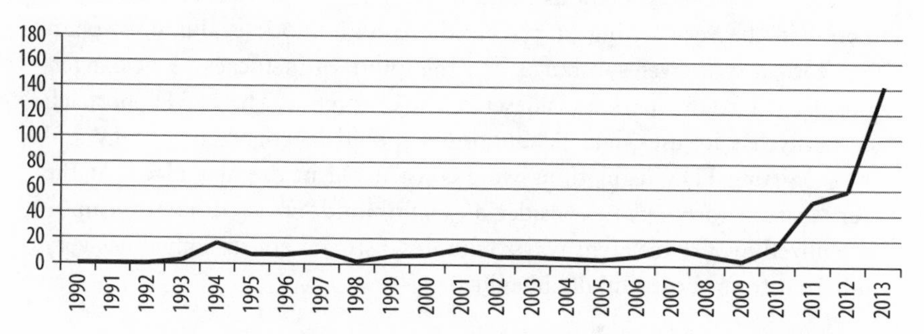

Use of the term *post-antibiotic* over time, charted as citations per yearly intervals and showing more recent growth. Derived from a LexisNexis Academic search for "post-antibiotic" within five words of "era," "age," "future," or "apocalypse," and filtered by de-duplication of articles of "moderate similarity."

Yet such advocacy and legislative efforts have certainly had an impact, including the apparent recent redirection of FDA leadership with respect to incentivizing industry to invest in antibiotic development, especially concerning the treatment of resistant pathogens.[210] This has manifested most visibly in recent years in relation to the Limited Population Antibacterial Drug (LPAD) Approval Mechanism, by which antibiotics geared toward highly resistant organisms "would be studied in substantially smaller, more rapid, and less expensive clinical trials."[211] The downside would be less precise estimates of both efficacy and safety, and an expectation that through labeling (indicating that the drugs have been tested in less rigorous fashion and are intended solely for the populations in which they were tested), education, and stewardship programs, clinicians would avoid using such approved drugs in more generalized or inappropriate situations.

This is all happening in real time today (with the possibility for legislation to be passed while this book goes to press), and there has been continued discussion regarding the downstream consequences of such limited trials, regarding how to disconnect LPAD-approved antibiotic usage from volume-driven revenue, as well as regarding the relative roles of national physician organizations like the IDSA and the AMA, the public health system, the pharmaceutical industry, the FDA, and individual stewardship programs in regulating versus educating physicians regarding potentially emerging antibiotics.[212] In these discussions we see the ongoing invocation of antibiotic "crises" and now "apocalypse." But we also see the collision of competing reformist impulses. The irony, after all, is that FDA antibiotic regulations—indeed, new drug regulations more broadly—were first articulated in the 1960s in the very setting of perceived loose antibiotic evaluations, pharmaceutical style over substance, and the influx of inefficacious or dangerous drugs into the market (as described in chapter 3). The LPAD approach certainly has its uses, but its supporters should be cognizant of why and how existing FDA regulations were constructed in the first place. At the very least, it forces us to consider how antibiotic "crisis" discourse can be mobilized in very different ways, with downstream effects sometimes very different from those initially imagined.

Coda

The response to antibiotic resistance has thus entailed a complex process of surfacing and marshaling data; of promoting academic, government, and global interest in the topic; and of translating such efforts into the mobilization of funds or the enactment of regulations. It remains to be seen whether the existing network of medical organizations, government and public health agencies, the pharmaceutical industry, and congressional supporters will be able to agree on a coordinated and internally consistent plan of confronting antibiotic resistance moving forward. But as the various players at the table continue to bring forward approaches to curtailing such resistance and to defining a "rational" therapeutics for the 21st century, they would do well to reflect on historical forces from both the supply and demand sides of the equation.

From the supply side, concerns regarding the role of industry in confronting antibiotic resistance reflect neither paranoia nor knee-jerk anti-market screeds, but rather date back at least to the invocation of the fixed-dose combination antibiotics in the 1950s as the supposed cure for staphylococcal resistance. The very development of the "well-controlled" trial, as defined by therapeutic reformers and the FDA in the 1950s and 1960s, derives in significant part from discussions regarding the development and promotion of antibiotics. Such history challenges us to remain ever mindful of the tension between possible therapeutic innovation and commercial opportunism.[213] Suggestions, especially those from Kevin Outterson and Aaron Kesselheim, to balance such "production" components of combating antibiotic resistance with linked measures for industry reimbursement to be conditional on appropriate antibiotic usage and ongoing utility ("conservation" measures) must be read in this context.[214]

From the demand side, attempts in the United States to support antibiotic stewardship programs, or to change or delimit physician prescribing practices, will need to confront a legacy of physician autonomy carefully articulated against central oversight, mandating an unpacking of this history and a thoughtful exploration of potential supports versus delimitations of such autonomy itself. Tellingly, despite the apparently widespread belief among both academic medical center and community hospital leaders that "good antibiotic stewardship programs are essential," only 29 percent of the community hospitals in the United States surveyed between 2001 and 2003 had instituted an antibiotic approval process (whereby permission to use certain antibiotics required approval from someone with oversight).[215] Decades after the heralded proposal of such measures, stated obstacles to their implementation continued to include both physician resentment of "limitations on their autonomy to prescribe" and displeasure "in policing the prescribing of fellow physicians."[216]

Finally, those working on both sides of the equation, when brought to a global level, will need to confront not just cultures of therapeutic autonomy and perceived panaceas but also deeper structural issues shaping the provision and receipt of antibiotics worldwide. The conclusion explores such themes more fully while attempting to place the "Antibiotic Era" in a global context.

Conclusion

> What is the future of antibiotics? Antibiotics are here to stay. Others, less toxic and more active . . . will be found. Costs will be reduced. Side effects will be either eliminated or controlled. Physicians will have well equipped laboratories at their disposal to evaluate each antibiotic and determine at once the particular antibiotic required for the treatment of a specific disease. Self medication will be reduced to a minimum. Government control of antibiotics will be tightened, so as to render antibiotics safer, more useful, and less expensive. The antibiotic era will accomplish what nature has intended it to be: Man's control over infectious diseases and epidemics that have plagued mankind since prehistoric times. —SELMAN A. WAKSMAN (1963)

> Antimicrobial resistance is a ticking time bomb not only for the UK but also for the world. We need to work with everyone to ensure the apocalyptic scenario of widespread antimicrobial resistance does not become a reality. This is a threat arguably as important as climate change for the world. —DAME SALLY DAVIES (2013)

THE TITLE PAGE of soil microbiologist Selman Waksman's own book *The Antibiotic Era* (1975) bears a biblical-sounding quote: "Out of the Earth Shall Come thy Salvation."[1] Antibiotics were indeed miracle drugs, but they represented the fruits of rational medical science, of the combined efforts of a research-intensive pharmaceutical industry and an increasingly powerful medical profession. Selman Waksman, discoverer of streptomycin, was a Nobelist in science, not a biblical miracle-worker (it appears he was not a prophet, either). His above-cited quote from a half-century ago encapsulates the dream of a rational therapeutics—the right drug for the right bug at the right cost, guided by the right information at the right time, with just the right balance between provider autonomy and government oversight to support the whole system.

But from the beginning, antibiotics have served to focus our collective therapeutic anxieties—that clinicians have been misusing this bounty, practicing irrational medicine, guided by incomplete or even biased information, goaded by pressures all about them. This could result in missed diagnoses, increased costs, adverse effects, superinfections, and now . . . the New Apocalypse. As one pharmaceutical researcher stated before Congress in 2000: "The worst case scenario for the New Apocalypse. . . . Well, I would have to say it would be pandemonium. There would be large loss of life. There would be flooding of hospitals and it would be a crucial time in the health history of the United States."[2] However, the fear of the New Apocalypse, as we have seen, is not new at all. The "end" of antibiotics was envisioned almost from the beginning.

Medicine continually rewrites its pasts, its presents, and its futures (be they utopias or dystopias), especially in the name of particular reform efforts. Sometimes these take the forms of halcyon pasts, to contrast with more complicated or corrupted modernity. As former IDSA president Merle Sande reflected in 2000:

> I really do not know how we got into this mess. You know, 50 years ago, we were told by astute microbiologists and infectious disease clinicians that this would happen. They knew then that if we misused and abused antibiotics, their life expectancy would be limited. And in those days when a physician treated an infection, he made a diagnosis. He isolated the organism and he treated very specifically and very effectively. These antibiotics were treated like gold and something happened over the last 30 or 40 years that I really do not understand. Somehow, it has become standard of care to treat runny noses and fevers and coughs with antibiotics.[3]

More often, they appear as frightening dystopias. In the 1950s jeremiads of Max Finland and Harry Dowling, this took the form of a medical-pharmaceutical edifice threatening to collapse, its foundation eroded by redundant drugs, marketing ballyhoo, and clinicians who had abdicated their roles as thoughtful educators (of their patients and of each other). From the 1990s warnings of Stuart Levy and Joshua Lederberg onward, they have taken the form of a post-antibiotic era, in which the benefits of modern medicine—transplantation, chemotherapy, intensive care—are threatened and even lost.[4]

The reformist impulses deriving from these dystopias may collide, as chapter 5 shows with respect to the incentivizing of new drug development. But this collision is far from inevitable, and both the dystopias described above have been countered by visions of a rational therapeutics and attempts to ensure its application through regulation or education. This is a process very much in motion, and as we simultaneously look back and look ahead at our attempts to ensure rational antibiotic therapy, a few concluding thoughts may help us to frame directions taken and still to be considered.

From Sackler to Sackett

To begin with, of course, the dream of a rational therapeutics is not new, and the contours of the reforms taken in its name have been shaped by a multitude of actors who have left their legacies not only in the policies they have helped shape and the levers of funding they have helped to mobilize but in the networks of colleagues and organizations they have formed or influenced, and in their direct teaching and writings.[5] Twentieth-century attempts to inculcate a "rational therapeutics" were simultaneously an attempt (social, epistemic) to ground medicine in science and perhaps a less conscious application to medicine of all the calculating, bureaucratizing tendencies of twentieth-century society. But they also derived from a lineage of therapeutic skepticism dating back centuries.[6]

In the mid-twentieth century, against the backdrop of the wonder-drug revolution—and the wonder-drug era of marketing, epitomized by the promotion of antibiotics by Arthur Sackler's William Douglas McAdams advertising agency—this strand of therapeutic skepticism took shape around a cohort of like-minded infectious disease experts, with indelible consequences. The initial targets of such reformers as Max Finland and Harry Dowling were seemingly irrational *drugs*. And the reforms they helped set in motion—the passage of the 1962 Kefauver-Harris drug amendments, the regulatory encoding of the controlled clinical trial, the conduct of the Drug Efficacy Study and Implementation process, and the removal from the market of the fixed-dose combination antibiotics—focused on the drug industry as the point of intervention.

Subsequent efforts in the United States, aimed at irrational *prescribers*, have been far less successful, whether stymied at the regulatory level or relatively ineffective at the educational one. These efforts, however, have drawn further attention to the forces—especially pharmaceutical marketing and patient pressure—promoting such irrational prescribing. Such efforts, moreover, cannot help but draw attention to the very historical momentum of the process in which such forces are entangled. As longtime pharmaceutical critic and activist Sydney Wolfe described at Al Gore's hearings on antibiotic resistance three decades ago: "Patients are not born into this world with the view that antibiotics are required for common colds. It is learned from their friends who learned it from their doctors when they went, and so forth, and from the indiscriminate promotion of a lot of antibiotics for upper respiratory tract infections, more in the past than now."[7] There are ironies to this story. Joshua Lederberg served as a Sackler Foundation Scholar at the Rockefeller University, while Stuart Levy works adjacent to the Sackler building at the Tufts University School of Medicine. I point out these ironies not to be snarky but rather to draw attention to our ability to collectively forget prior historical forces and initiatives. We are now awash in educational countermeasures (from counter-detailing to the CDC's patient-centered Get Smart campaign) to help offset such pressures, as well as ongoing calls for better surveillance and diagnostics to help ensure a rational therapeutics. But as recent data indicate, while we are improving, we still have our work cut out for us.[8]

We are also in the midst of the era of evidence-based medicine (EBM). And while its first stirrings may have developed in the 1970s in the context of advances in clinical epidemiology, EBM also emerged from an era increasingly attuned—in medical journals and government hearings alike—to the aspirations of "rational medicine" and the critique of "irrational medicine." It is perhaps no coincidence that David Sackett—the founding chair of the Department of Clinical Epidemiology and Biostatistics at McMaster University and best-known advocate of EBM—trained in internal medicine under Harry Dowling at the University of Illinois in the early 1960s. *Evidence-based medicine* was explicitly chosen by Sackett's McMaster colleague Gordon Guyatt as an inoffensive term, in a way that Guyatt's first choice, "scientific medicine"—and, one suspects, "rational medicine"—was not.[9] Nonetheless, as countless commentators have noted, EBM is far from value-neutral. At one level, it has generated a hierarchy of acceptable forms of evidence, with the randomized control trial at the top, well above observational studies; "testimonials" no longer carry the overt semantic or moral weight of half a century ago, but echoes of the debates of the 1950s and 1960s continue to reverberate in contemporary methodological discussions, as well as in discussion over the role of bias and conflict of interest in the conduct of, and formation of guidelines derived from, such studies. At another level, the guidelines derived from EBM in the United States most often find their de facto regulatory implementation not through unified policy formation but rather through their undergirding of differential local insurance reimbursement rates for particular medications or procedures.[10]

Antibiotics, in this context, are no different from most other medications governed by the decentralized findings of EBM. Decades after the first calls for antibiotic guidelines, a Pub Med search of "antibiotic guidelines" now calls forth over ten thousand articles. Yet the regulation of antibiotics in the United States remains decentralized. Even the specter of a post-antibiotic era has not overcome resistance to more centralized strategizing regarding this most central component of the therapeutic armamentarium, as the repeated legislative failure of the proposed Strategies to Address Antimicrobial Resistance (STAAR) Act has further demonstrated. Recent years, amid the "epidemic of antibiotic resistance," have seen scattered calls in the United States for more centralized oversight of antibiotic prescribing, whether at the national or state level. As one commentator put it, invoking public health rationales dating back to Philip Lee's arguments at Gaylord Nelson's Senate hearings (described in chapter 4): "A physician can be likened to a nuclear power plant to allow regulatory action to prevent a disaster. . . . In each case, the physician and the power plant are granted a license which allocates permission to the holder to conduct activities within the confines of the license. However, if the privilege is abused to the point of jeopardizing the health and safety of society, the license can be revoked or suspended."[11] Such proposals, though, have found little traction in this country. In this respect, it is useful to briefly look to several international comparisons.

From Finland to Finland

As noted in chapter 4, even the formation of local, hospital-based antibiotic approval committees has been met with resistance in the United States in the decades since Max Finland first proposed them.[12] Outside the United States, however, some countries have enacted large-scale policies of regulation or centralized education.[13] At one extreme, Czechoslovakia in the 1960s implemented strict top-down national control of antibiotics. Antibiotic Centres were established across the country, providing microbiological surveillance data, introducing new antibiotics when deemed appropriate (erythromycin, for example, was not released for "free use" until 1975, over two decades after its release in the United States), and interfacing with broader Regional Rational Chemotherapy Committees. Certain antibiotics could be prescribed freely; certain antibiotics could be prescribed "on special order forms only by doctors at hospital wards"; while others could be prescribed "by ward specialists only on recommendations of Antibiotic Centres."[14]

While Communist-era Czechoslovakia represents one end of a spectrum (and the worst fears of Arthur Sackler and Louis Lasagna realized), much of Scandinavia also implemented varying forms of centralized surveillance, encouragement, and in some respects de facto regulation. Denmark, since the mid-1960s, has engaged in centralized surveillance, nationally coordinated lectures, and the intentional shaping of antibiotic choices via nationally determined reimbursement rates for varying classes of antibiotics.[15] Sweden, through the later formation of its Strama program (Swedish Strategic Programme for the Rational Use of Antimicrobial Agents and Surveillance Resistance), utilized national public and practitioner educational measures informed by microbiological surveillance data to decrease outpatient antibiotic use by 20 percent between 1995 and 2004.[16] And Finland, in response to erythromycin resistance in group A streptococci (the agents of strep throat and cellulitis, among other infections) in the 1990s, famously conducted a national educational campaign to decrease erythromycin use, yielding a nearly 50 percent reduction in usage and a nearly 50 percent reduction in streptococcal erythromycin resistance over five years, further serving to exemplify the Darwinian selection pressures on antibiotic resistance.[17]

These are but examples, and the long-term outcomes of such interventions remain unknown, albeit suggestive. Sweden's Strama program at the very least demonstrated that it did not yield an increase in hospitalizations for mastoiditis or peritonsillar abscesses during the course of its program.[18] The applicability of such efforts to the United States, moreover, is admittedly complicated. Readers will draw their own conclusions regarding the relationship between the uptake of such measures and national or state size, population and physician distribution, political, social, and cultural history, and a host of other variables. And yet, we will need to become increasingly cognizant of such efforts. On the one hand, they serve as data points, large-scale experiments that may or may not be deemed to have the internal or external validity to be applicable to other areas. On the other hand, if we are to confront antibiotic resistance as a *global* concern, with an ever-shrinking distance between the local and the global, then such efforts affect all of us, as future clinicians or patients, regardless of where we live.

From Strictures to Structures

Finally, such Scandinavian data represent findings from highly developed, industrialized nations. Debates over the regulation versus education of clinicians in the pursuit of "rational care" in developed nations, while important in their own right, seem almost luxuries when looked at from a truly global standpoint. Since the 1970s, reformers have drawn attention to the conditions promoting antibiotic resistance in the developing world, focusing on such sites of intervention as preventing the marketing and sale of antibiotics that have not been approved in the developed world, as well as prohibiting the over-the-counter sale of antibiotics (see chapter 5). The World Health Organization (WHO) has likewise drawn attention both to the global nature of antibiotic resistance and to the need for such interventions, in many ways carrying forward the ethos behind its efforts in the 1970s and 1980s to inculcate "the rational use of drugs" more generally worldwide.[19]

But as the WHO and others have recognized, there are deeper structural impediments to the rational global delivery of antibiotics, resulting in both the undertreatment *and* overtreatment (and simply mistreatment) of common infections. "Global," in this sense, includes the United States, and we have already drawn attention to the linked structural factors (rapid visits, the nature of the reimbursement system, etc.) favoring antibiotic misuse.[20] In the developing world, the situation can entail still more fundamental barriers to rational antibiotic delivery. In Zurich in the fall of 2012, I participated in a European Science Foundation "Drugs Network" conference entitled "Is This the End? The Eclipse of the Therapeutic Revolution." I spoke on "'Multi-Resistance': Antibiotic Resistance and Delimitations of Physician Prescribing Activity" and showed my slides concerning debate in the United States over regulatory versus educational attempts to inculcate "rational" antibiotic prescribing in the United States over the past half-century. The next day, I was followed by anthropologist Kristin Peterson, who spoke on "Speculative Drug Markets: Monopolies and Derivative Life in Nigeria," detailing the flotsam and jetsam of the Nigerian pharmaceutical marketplace and Nigerian pharmacy shelves in the wake of pharmaceutical market speculation from the 1970s through the 1990s.[21] Her slides—with their depictions of unregulated pharmaceutical stores—looked quite different from mine, and numerous papers have described disturbing patterns of antibiotic self-medication (over 90% of respondents in one study) and what we would term "irrational" antibiotic prescribing (with wide and varying forms of antibiotics found in patent medicine stores) in Nigeria in recent years.[22] Nigeria is chosen here simply as an example, and several studies in the past decade have likewise drawn attention to the pervasiveness and persistence of inappropriate antibiotic prescribing in the developing world more broadly, as well as to the limits to the interventions—educational, managerial—designed to help reverse such a process.[23]

Here we may draw important lessons from those who have focused on drug-resistant tuberculosis in particular. Tuberculosis has made only occasional appearances in this book to this point. In part, this reflects the divergent interests and paths of those focused on the broad use of antibiotics in American medical offices, clinics, and hospitals for much of the latter half of the twentieth century and those interested in the treatment or eradication of tuberculosis during the same time. In part, this divergence reflects the related and relative invisibility of tuberculosis (despite its ongoing global prevalence) in the eyes of most Americans and their physicians throughout much of this time.[24] As such, tuberculosis was largely absent from the more general discourse surrounding antibiotic resistance throughout much of the era covered in this book. But such global "emerging" infections as multidrug-resistant tuberculosis (MDR-TB) and resistant pneumococci would explode in tandem in the American medical, public health, and popular consciousness from the 1990s onward.

Scholars, from René Dubos in the 1950s to Paul Farmer today, have emphasized our need to understand tuberculosis—and later, MDR-TB—not solely as biological phenomena but rather as *biosocial* phenomena.[25] Who contracts MDR-TB, as well as who is able to access and adhere to "rational" drug regimens, is shaped by economic and structural conditions that constrain individual patient agency and outcomes. And it has been all too easy to conflate patient beliefs and agency with the structural conditions that permit or inhibit the delivery of rational care. Similarly, structural conditions—economic, hygienic, nutritional, educational, to name a few—shape who gets sick, and often who develops antibiotic-resistant infections, more broadly.[26] And it is impossible to deliver "rational" antibiotic therapy, in a global sense, in the absence of conditions—access to some form of appropriate primary care and diagnostics, access to and delivery of an appropriate corresponding set of medications—that permit its application.[27]

Henry Welch, chief of the FDA's Division of Antibiotics throughout the 1950s, responded to staphylococcal antibiotic resistance at his 1956 annual antibiotics symposium by advocating the empiric administration of wide-spectrum fixed-dose combination antibiotics as a "rational" solution to the inadequate laboratory infrastructure and other structural conditions facing most outpatient clinicians in the United States (see chapter 2).[28] Welch's accommodation to the pharmaceutical industry, and the degree to which such a proposal would seemingly lead to "irrational" therapy, antagonized Max Finland, Harry Dowling, and their colleagues, setting in motion drug reforms that extended well beyond antibiotics. But Welch's sin of accommodation in this case went deeper, to an acceptance of existing conditions and a focus on an antibiotic arms race alone. Today, those concerned with the global phenomenon of antibiotic resistance would do well to continue to investigate and confront the conditions—starting with pharmaceutical marketing and patient expectations but extending to the structures of care delivery and the forces shaping such expectations—that promote antibiotic misuse and antibiotic resistance alike.[29]

Neither the developed nor the developing world can afford irrational antibiotic therapy; in this, as in most respects, we are ever more inextricably connected. The history of antibiotic reform in the United States detailed in this volume has permitted us to delve into the tensions inherent in the construction of "rational" therapy and the constraints attendant to its implementation. Such analysis can be extended to still other nations worldwide, as we continue to unpack the workings of worldwide antibiotic usage and the global antibiotic resistance crusade. Calvin Kunin wrote of the emergence of resistant organisms in 1985: "Simplistic generalizations are misleading. A separate series of equations must be established for each microbe, each drug, and each social, cultural, and environmental effect."[30] The same will hold true of our understanding of each nation's efforts to address antibiotic resistance, both individually, and, increasingly, as part of a global community of patients, microbes, healthcare personnel, academicians, industry, and regulatory bodies. For each of these nations, history will have shaped the direction and outcome of proposed measures, even as those measures intersect with those proposed elsewhere. Historical analysis is a powerful tool for helping us to place such a process in perspective, to demonstrate the choices taken and the resistances posed as we have arrived at our present position. It may, if we are thoughtful, point to alternative possibilities and trajectories as well.

ABBREVIATIONS

Archival Sources Cited

AIHP	Kremers Research Files, American Institute of the History of Pharmacy
AMAA	American Medical Association Archives
ATS	Anti-Trust Subcommittee (Drugs) Records (subset of RG 46), National Archives
CEP	Charles Edwards Papers, University of California–San Diego, Mandeville Special Collections Library
EKP	Edward Kass Papers, Francis A. Countway Medical Library
FDAD	Food and Drug Administration Division of Dockets Management, Rockville, Maryland
FDAOHC	FDA Oral History Collection, National Library of Medicine
FMIP	Félix Martí-Ibañez Papers, Yale University Library, Manuscripts & Archives
HDP	Harry Dowling Papers, National Library of Medicine
HLP	Herbert Ley Papers [Acc 2006-051], National Library of Medicine
JGP	James Goddard Papers, National Library of Medicine
IDSA	Infectious Diseases Society of America, privately held
JLP	Joshua Lederberg Papers, National Library of Medicine
JOAP	John Adriani Papers, National Library of Medicine
LLP	Louis Lasagna Papers, University of Rochester River Campus Libraries, Department of Rare Books, Special Collections & Preservation
LTWP	Louis Tompkins Wright Papers, Francis A. Countway Medical Library

MFP — Maxwell Finland Papers, Francis A. Countway Medical Library

NASD — National Academy of Sciences Archives—DESI Program

NASM — National Academy of Sciences Archives—Medical Sciences

PDC — Parke-Davis Research Laboratory Records, Archive Center, National Museum of American History, Smithsonian Institution

RAC — Commonwealth Fund Papers, Rockefeller Archive Center

RG 46 — Records of the U.S. Senate, National Archives

RG 88 — Records of the Food and Drug Administration, National Archives

RG 122 — Records of the Federal Trade Commission, National Archives

RWP — Robert Burns Woodward Papers, Harvard University Archives

SLP — Stuart Levy Papers, privately held

SWP — Selman Waksman Papers, Rutgers University Libraries, Special Collections & University Archives

WBP — William Bean Papers, National Library of Medicine

WHOA — World Health Organization Archives

WHOL — World Health Organization Library

WSP — Wesley Spink Papers, University Archives, Archives & Special Collections, University of Minnesota

Government Hearings Cited

AP — *Administered Prices in the Drug Industry* [Kefauver hearings], 1959–1960, U.S. Congress, Senate Committee on the Judiciary, Subcommittee on Antitrust and Monopoly

APAI — *Administered Prices in the Automobile Industry*, 1958, U.S. Congress, Senate Committee on the Judiciary, Subcommittee on Antitrust and Monopoly

APM — *Advertising of Proprietary Medicines*, 1971–1977, U.S. Congress, Senate Select Committee on Small Business, Subcommittee on Monopoly

AR_1 — *Antibiotic Resistance*, 1984, U.S. Congress, House Committee on Science and Technology, Subcommittee on Investigations and Oversight

AR$_2$	*Antimicrobial Resistance: Solutions for this Growing Public Health Threat*, 1999, U.S. Congress, Senate Committee on Health, Education, Labor, and Pensions, Subcommittee on Public Health
AR$_3$	*Antimicrobial Resistance*, 2000, U.S. Congress, Senate Subcommittee of the Committee on Appropriations
AR$_4$	*Antibiotic Resistance and the Threat to Public Health*, 2010, U.S. Congress, House Committee on Energy and Commerce, Subcommittee on Health
CPDI	*Competitive Problems in the Drug Industry* [Nelson hearings], 1967–1976, U.S. Congress, Senate Select Committee on Small Business, Subcommittee on Monopoly
DE	*Drug Efficacy*, 1969, U.S. Congress, House Subcommittee of the Committee on Government Operations
DIAA	*Drug Industry Antitrust Act*, 1961–1962 [Kefauver hearings], U.S. Congress, Senate Committee on the Judiciary, Subcommittee on Antitrust and Monopoly
DS	*Drug Safety*, 1964–1966, U.S. Congress, House Subcommittee of the Committee on Government Operations
EI	*Emerging Infections: A Significant Threat to the Nation's Health*, 1995, U.S. Congress, Senate Committee on Labor and Human Resources
EPI	*Examination of the Pharmaceutical Industry, 1973–1974*, U.S. Congress, Senate Committee on Labor and Public Welfare, Subcommittee on Health
ES	*Emergence of the Superbug: Antimicrobial Resistance in the United States*, 2008, U.S. Congress, Senate Committee on Health, Education, Labor, and Pensions
FMA	*False and Misleading Advertising*, 1957–1958, U.S. Congress, House Committee on Government Operations, Subcommittee on Legal and Monetary Affairs
ICDRR	*Interagency Coordination in Drug Research and Regulation*, 1962–1963, U.S. Congress, Senate Committee on Government Operations, Subcommittee on Reorganization and International Organizations
PB	*Project Bioshield: Contracting for the Health and Security of the American Public*, 2003, U.S. Congress, House Committee on Government Reform

PDAEJU — *Promoting the Development of Antibiotics and Ensuring Judicious Use in Humans*, 2010, U.S. Congress, House Committee on Energy and Commerce, Subcommittee on Health

RTP — *Review of Transitional Provisions in Drug Amendments of 1962*, 1971, U.S. Congress, House Committee on Interstate and Foreign Commerce, Subcommittee on Public Health and Environment

SEND — *The Safety and Effectiveness of New Drugs (Marketing of Fixed Combination Drugs and Unapproved New Drugs; Implementation of Drug Efficacy Findings)*, 1971, U.S. Congress, House Subcommittee of the Committee on Government Operations

SERC — *FDA Regulation of the New Drug Serc*, 1972, U.S. Congress, House Subcommittee of the Committee on Government Operations

NOTES

Introduction

1. Sharon Begley and Martha Brant, "The End of Antibiotics?" *Newsweek* (Mar. 7, 1994): 63; Sharon Begley, Martha Brant, Pat Wingert, and Mary Hager, "The End of Antibiotics," *Newsweek* (Mar. 28, 1994): 47–51; Geoffrey Cowley, John F. Lauerman, Karen Springen, Mary Hager, and Pat Wingert, "Too Much of a Good Thing" [internal story capsule], *Newsweek* (Mar. 28, 1994): 50–51.

2. Robert Bud, *Penicillin: Triumph and Tragedy* (New York: Oxford University Press, 2007).

3. Throughout the book, I frequently refer to antibiotics by their trade names, in view of the visibility and tangibility of such trade names throughout the antibiotic era. On the evolving relationship between brand and generic names of drugs, see Jeremy A. Greene, "What's in a Name? Generics and the Persistence of the Pharmaceutical Brand in American Medicine," *Journal of the History of Medicine and Allied Sciences* 66 (2011): 425–67, and *Generic: The Unbranding of Modern Medicine* (Baltimore: Johns Hopkins University Press, 2014).

4. John C. Burnham, "Medicine's Golden Age: What Happened to It?" *Science* 215 (1982): 1474–79; Allan M. Brandt and Martha Gardner, "The Golden Age of Medicine?" in Roger Cooter and John Pickstone, eds., *Medicine in the Twentieth Century* (Amsterdam: Harwood Academic Publishers, 2000), 21–37.

5. For kinematic overviews of the development of antimicrobials in the twentieth century, see Harry F. Dowling, *Fighting Infection: Conquests of the Twentieth Century* (Cambridge, MA: Harvard University Press, 1977); Wesley W. Spink, *Infectious Diseases: Prevention and Treatment in the Nineteenth and Twentieth Centuries* (Minneapolis: University of Minnesota Press, 1978); and David Greenwood, *Antimicrobial Drugs: Chronicle of a Twentieth Century Medical Triumph* (New York: Oxford University Press, 2008). For more detailed analyses of antimicrobial drug development, see Gladys L. Hobby, *Penicillin: Meeting the Challenge* (New Haven: Yale University Press, 1985), and John E. Lesch, *The First Miracle Drugs: How the Sulfa Drugs Transformed Medicine* (New York: Oxford University Press, 2007). For further attempts to place antibiotics in their social, political, and legal contexts, see Thomas Maeder, *Adverse Reactions* (New York: William Morrow, 1994), and Bud, *Penicillin*.

6. Gladys L. Hobby to Edward Kass, 4/8/86, Box 28, ff 31, EKP.

7. Harry M. Marks, *The Progress of Experiment: Science and Therapeutic Re-*

form in the United States, 1900–1990 (New York: Cambridge University Press), 1997. Marks covered streptomycin in his text but generously left the broad-spectrum antibiotics and their successors to others.

8. James Whorton, "'Antibiotic Abandon': The Resurgence of Therapeutic Rationalism," in John Parascandola, ed., *The History of Antibiotics: A Symposium* (Madison, WI: American Institute of the History of Pharmacy, 1980), 125–36.

9. Lesch, *First Miracle Drugs*; Bud, *Penicillin.*

10. See, respectively, Harry M. Marks, "Cortisone: A Year in the Political Life of a Drug," *Bulletin of the History of Medicine* 66 (1992): 419–39; David Cantor, "Cortisone and the Politics of Drama, 1949–1955," in John V. Pickstone, ed., *Medical Innovations in Historical Perspective* (Basingstoke, UK: Macmillan, 1992), 165–84; Leo B. Slater, "Industry and Academy: The Synthesis of Steroids," *Historical Studies in the Physical and Biological Sciences* 30 (2000): 443–80; Micky Smith, *A Social History of the Minor Tranquilizers: The Quest for Small Comfort in the Age of Anxiety* (New York: Pharmaceutical Products, 1991); Susan L. Speaker, "From 'Happiness Pills' to 'National Nightmare': Changing Cultural Assessment of Minor Tranquilizers in America, 1955–1980," *Journal of the History of Medicine and Allied Sciences* 52 (1997): 338–76; David Healy, *The Antidepressant Era* (Cambridge, MA: Harvard University Press, 1997); Nicolas Rasmussen, *On Speed: The Many Lives of Amphetamine* (New York: New York University Press, 2008); David Herzberg, *Happy Pills: From Miltown to Prozac* (Baltimore: Johns Hopkins University Press, 2009); David Healy, *The Creation of Psychopharmacology* (Cambridge, MA: Harvard University Press, 2002); and Jeremy A. Greene, *Prescribing by Numbers: Drugs and the Definition of Disease* (Baltimore: Johns Hopkins University Press, 2007).

11. Jeremy A. Greene and Scott H. Podolsky, "Reform, Regulation, and Pharmaceuticals—The Kefauver-Harris Amendments at 50," *New England Journal of Medicine* 367 (2012): 1481–83.

12. Daniel Carpenter: *Reputation and Power: Organizational Image and Pharmaceutical Regulation at the FDA* (Princeton: Princeton University Press, 2010), 118–227.

13. For those wondering about Bactrim (trimethoprim-sulfamethoxazole), this would be the first fixed-dose combination antibiotic approved and introduced in the United States (in 1974) *after* the DESI process had eliminated the existing fixed-dose combination antibiotics. See Scott H. Podolsky and Jeremy A. Greene, "Combination Drugs: Hype, Harm, and Hope," *New England Journal of Medicine* 365 (2011): 488–91.

14. Carpenter, *Reputation and Power*, 345–62.

15. Robert Bud, "From Epidemic to Scandal: The Politicization of Antibiotic Resistance, 1957–1969," in Carsten Timmermann and Julie Anderson, eds., *Devices and Designs: Medical Technologies in Historical Perspective* (New York: Palgrave Macmillan, 2006), 195–211; Bud, *Penicillin*, 163–212.

16. Relatedly, it appears that the American story, from the 1940s onward, had

significant worldwide consequences, whether with respect to the evolution of the FDA and its own worldwide impact or to the framing and discourse related to the "rational" usage of antibiotics worldwide over the past six-plus decades. On the developing global hegemony of the FDA throughout this era, see Carpenter, *Reputation and Power*, 686–726. On developing American hegemony in science throughout the era, see John Krige, *American Hegemony and the Postwar Reconstruction of Science in Europe* (Cambridge, MA: MIT Press, 2006).

17. Paul B. Beeson, "Infectious Diseases (Microbiology)," in John Z. Bowers and Elizabeth F. Purcell, eds., *Advances in American Medicine: Essays at the Bicentennial*, vol. 1 (New York: Josiah Macy Jr. Foundation, 1976), 100–156.

18. Chester S. Keefer, foreword [to the Maxwell Finland sixtieth birthday festschrift], *Archives of Internal Medicine* 110 (1962): 559; "Fellows in Infectious Diseases (Harvard Medical Unit) at the Boston City Hospital, 1932–1972" [part of the Maxwell Finland seventieth birthday festschrift], *Journal of Infectious Diseases* 125, suppl (1972): S4–S7; Charles S. Davidson, "Presentation of the George M. Kober Medal for 1978 to Maxwell Finland," *Transactions of the Association of American Physicians* 91 (1978): 51–62; Frederick C. Robbins, "Maxwell Finland," *Biographical Memoirs of the National Academy of Sciences* 76 (1999): 103–13; Jerome O. Klein, "Maxwell Finland: A Remembrance" [corresponding to yet another Finland festschrift, on the 100th anniversary of his birth], *Clinical Infectious Diseases* 34 (2002): 725–29.

19. Ronald A. Arky, Charles S. Davidson, Edward H. Kass, and Jerome O. Klein, "Faculty of Medicine—Memorial Minute," *Harvard Gazette* (Sept. 8, 1989): 9–10. As they continued: "At one time, when the first training grants were [awarded, we were] informed that two academic leaders, one from Columbia University and one from Western Reserve, were to come on a site visit. Finland phoned in distress to the executive officer of the study section, because both had been his former trainees. The executive replied that he was aware of that fact, but that they had run down the list of appropriate senior figures in the field and could not find any who were not."

20. Using Web of Science citation metrics, of the first ten presidents of the IDSA, one would have to combine the citations of Edward Kass with those of either John Enders or Albert Sabin to have a *pair* yield more citations than those yielded by Finland alone. And such an enumeration does not account for the several hundred unsigned editorials Finland wrote, many of them for the *New England Journal of Medicine*, or for the fact that Kass and Finland co-authored 35 papers. Web of Science, accessed Jan. 26, 2013.

21. Bud, *Penicillin*, 163–91.

22. See, e.g., Mark R. Finlay, "Reframing the History of Agricultural Antibiotics in the Postwar World: An International and Comparative Perspective" (talk given at the European Science Foundation DRUGS Research Network Programme on "Beyond the Magic Bullet: Reframing the History of Antibiotics," University of Oslo, 2011); Ulrike Thoms, "Antibiotics, Agriculture and Political Agendas: Past,

Present, and Future of German Agricultural Policy and Its Impact on Strategies against Microbiological Resistance"; and Claas Kirchelle, "Utilizing Resistance: Agricultural Antibiotics, Public Anxiety, and Expert Empowerment in Britain, 1953–2013," (talks given at the conference on "Futures Past and Present: Hopes and Fears and the History of Antibiotic Resistance," University of Oslo, 2013).

23. Brad Spellberg, *Rising Plague: The Global Threat from Deadly Bacteria and Our Dwindling Arsenal to Fight Them* (New York: Prometheus Books, 2009), 78, 195–98, 206.

24. Tanya Stivers, *Prescribing under Pressure: Parent-Physician Conversations and Antibiotics* (New York: Oxford University Press, 2007).

25. Dominique A. Tobbell, *Pills, Power, and Policy: The Struggle for Drug Reform in Cold War America and Its Consequences* (Berkeley: University of California Press, 2012).

26. G. E. R. Lloyd, *Magic, Reason, and Experience: Studies in the Origin and Development of Greek Science* (Cambridge: Cambridge University Press, 1979), 10–58; James Longrigg, *Greek Rational Medicine: Philosophy and Medicine from Alcmaeon to the Alexandrians* (New York: Routledge, 1993); John Harley Warner, *The Therapeutic Perspective: Medical Practice, Knowledge, and Identity in America, 1820–1885* (Cambridge, MA: Harvard University Press, 1986), 37–57, 235–57; Marks, *Progress of Experiment*, 17–41.

27. Ernest Jawetz, "Patient, Doctor, Drug, and Bug," *Antibiotics Annual* (1957–1958): 295; W. Clarke Wescoe, "Lewis Carroll Might Have Written It," *Food Drug Cosmetic Law Journal* 26 (1971): 463.

Chapter One: The Origins of Antibiotic Reform

Epigraph. Ernest Jawetz, "Patient, Doctor, Drug, and Bug," *Antibiotics Annual* (1957–1958): 295.

1. For the definitive account of Chloromycetin, see Thomas Maeder, *Adverse Reactions* (New York: William Morrow, 1994).

2. For the nearly immediate introduction of such military metaphors into clinical microbiological discourse, see S. L. Montgomery, "Codes and Combat in Medical Discouse," *Science as Culture* 2 (1991): 341–90, and Christoph Gradmann, "Invisible Enemies: Bacteriology and the Language of Politics in Imperial Germany," *Science in Context* 13 (2000): 9–30.

3. See Harry F. Dowling, *Fighting Infection: Conquests of the Twentieth Century* (Cambridge: Harvard University Press, 1977), 1–54; Peter Keating, "Vaccine Therapy and the Problem of Opsonins," *Journal of the History of Medicine and Allied Sciences* 43 (1988): 275–96; William C. Summers, *Félix d'Herelle and the Origins of Molecular Biology* (New Haven: Yale University Press, 1999); Scott H. Podolsky, "Cultural Divergence: Elie Metchnikoff's *Bacillus bulgaricus* Therapy and His Underlying Conception of Health," *Bulletin of the History of Medicine* 72 (1998): 1–27; Allan M. Brandt, *No Magic Bullet: A Social History of Venereal Disease in the United States since 1880* (New York: Oxford University Press,

1985); John Parascandola, *Sex, Sin, and Science: A History of Syphilis in America* (Westport, CT: Praeger, 2008); and Leo M. Slater, *War and Disease: Biomedical Research on Malaria in the Twentieth Century* (New Brunswick, NJ: Rutgers University Press, 2009), 59–83.

4. Sinclair Lewis, *Arrowsmith* (New York: Harcourt, Brace, 1925).

5. Scott H. Podolsky, *Pneumonia before Antibiotics: Therapeutic Evolution and Evaluation in Twentieth-Century America* (Baltimore: Johns Hopkins University Press, 2006).

6. John E. Lesch, *The First Miracle Drugs: How the Sulfa Drugs Transformed Medicine* (New York: Oxford University Press, 2007), esp. 197–203.

7. Gladys L. Hobby, *Penicillin: Meeting the Challenge* (New Haven: Yale University Press, 1985); Robert Bud, *Penicillin: Triumph and Tragedy* (New York: Oxford University Press, 2007), esp. 54–74.

8. On Welch's laboratory, see Simon Flexner and James Thomas Flexner, *William Henry Welch and the Heroic Age of American Medicine* (New York: Viking Press, 1941), 150–80. On the NIH, see Victoria A. Harden, *Inventing the NIH: Federal Biomedical Research Policy, 1887–1937* (Baltimore: Johns Hopkins University Press, 1986). The National Institute of Health would evolve into the National Institutes of Health in 1948, with the Division of Infectious Diseases and Division of Tropical Diseases combining with the Rocky Mountain Laboratory and Biologics Control Laboratory to form the National Microbiological Institute. In 1955, the National Microbiological Institute evolved into the National Institute of Allergy and Infectious Diseases. In *NIH Almanac, 2001* (Bethesda, MD: National Institutes of Health, 2001), 133–34.

9. The institute, founded in 1902, was renamed the McCormick Institute in 1918, after its benefactors; it was initially to have focused on the treatment of scarlet fever alone, from which the McCormicks' son had died. See Frank Billings, "The Memorial Institute for Infectious Diseases," *JAMA* 42 (1904): 1676–77. The institute closed in 1940.

10. For popular literary attention to infectious diseases, see, emblematically, Paul de Kruif, *Microbe Hunters* (New York: Harcourt, Brace, 1926). A National Library of Medicine title search of English-language books between 1890 and 1930 reveals 114 hits for "infectious disease(s)"(along with 21 for "communicable disease(s)" and 16 for "contagious disease(s)"), 698 for tuberculosis, 220 for syphilis, 109 for diphtheria, and 44 for tropical medicine.

11. New York City's was the most influential public health department; see C. E. A. Winslow, *The Life of Hermann M. Biggs: Physician and Statesman of the Public Health* (Philadelphia: Lea and Fibiger, 1929); Wade W. Oliver, *The Man Who Lived for Tomorrow: A Biography of William Hallock Park* (New York: E. P. Dutton, 1941); John Duffy, *A History of Public Health in New York City, 1866–1966* (New York: Russell Sage Foundation, 1974); David Blancher, "Workshops of the Bacteriological Revolution: A History of the Laboratories of the New York City Department of Health, 1892–1912" (PhD diss., City University of New York,

1979); and Evelynn Maxine Hammonds, *Childhood's Deadly Scourge: The Campaign to Control Diphtheria in New York City, 1880–1930* (Baltimore: Johns Hopkins University Press, 1999). Regarding the military, see Martha L. Sternberg, *George Miller Sternberg: A Biography* (Chicago: American Medical Association, 1920), and M. W. Ireland and Joseph F. Siler, *Medical Department of the United States Army in the World War, vol. 9: Communicable and Other Diseases* (Washington, DC: U.S. Government Printing Office, 1928). Regarding pharmaceutical firms, chief among these were the H. K. Mulford Company, Lederle, and Parke-Davis. On Mulford, see Jonathan Liebenau, *Medical Science and Medical Industry: The Formation of the American Pharmaceutical Industry* (Baltimore: Johns Hopkins University Press, 1987), 57–78; on Lederle, see Tom Mahoney, *The Merchants of Life: An Account of the American Pharmaceutical Industry* (New York: Harper & Brothers, 1959), 163–67; on Parke-Davis, see *A Manual of Biological Therapeutics: Sera, Bacterins, Phylacogens, Tuberculins, Glandular Extracts, Toxins, Cultures, Antigens, Etc.* (Detroit: Press of Parke, Davis, & Co., 1914), and William Haynes, *American Chemical Industry, vol. 6: The Chemical Companies* (New York: Van Nostrand, 1949), 320–24. On the Rockefeller Institute, see George Washington Corner, *A History of the Rockefeller Institute, 1901–1953: Origins and Growth* (New York: Rockefeller Institute Press, 1964).

12. On the trend to specialty formation during the first few decades of twentieth-century American medicine, see Rosemary Stevens, *American Medicine and the Public Interest*, rev. ed. (Berkeley: University of California Press, 1998), and George Weisz, *Divide and Conquer: A Comparative History of Medical Specialization* (New York: Oxford University Press, 2006). On cardiology and gastroenterology, see W. Bruce Fye, *American Cardiology: The History of a Specialty and Its College* (Baltimore: Johns Hopkins University Press, 1996), 13–51, and Joseph B. Kirsner, "One Hundred Years of American Gastroenterology," in Russell C. Maulitz and Diana E. Long, eds., *Grand Rounds: One Hundred Years of Internal Medicine* (Philadelphia: University of Pennsylvania Press, 1988), 117–57. As Joel Howell has related, the American Heart Association drew inspiration from the National Association for the Study and Prevention of Tuberculosis (renamed the National Tuberculosis Association in 1918), but no more general infectious disease society would follow suit for several decades; see Howell, "Hearts and Minds: The Invention and Transformation of American Cardiology," in Maulitz and Long, *Grand Rounds*, 243–75.

13. Barnett Cohen, *Chronicles of the Society of American Bacteriologists, 1899–1950* (Baltimore: Society of American Bacteriologists, 1950), 16. The society was renamed the American Society for Microbiology in 1960. The American Association of Pathologists and Bacteriologists, founded in 1901, was geared toward scientific research, though with obvious implications for clinical practice. See Esmond R. Long, "History of the American Association of Pathologists and Bacteriologists," *American Journal of Pathology* 77, suppl (1974): S1–S218.

14. Robert G. Petersdorf, "Whither Infectious Diseases? Memories, Man-

power, and Money," *Journal of Infectious Diseases* 153 (1986): 189–95; "Infectous Diseases" would not apper as a specialty in the AMA's *American Medical Directory* until its twenty-sixth edition; see *American Medical Directory*, 26th ed. (Chicago: American Medical Association, 1974), x.

15. Harry F. Dowling, "The Emergence of the Cooperative Clinical Trial," *Transactions and Studies of the College of Physicians of Philadelphia* 43 (1975): 20–29; Harry M. Marks, *The Progress of Experiment: Science and Therapeutic Reform in the United States, 1900–1990* (New York: Cambridge University Press, 1997), 42–70, and "'Until the Sun of Science . . . the true Apollo of Medicine has risen': Collective Investigation in Britain and America, 1880–1910," *Medical History* 50 (2006): 147–66.

16. Such data are derived from full-text searching using the terms "alternate patient(s)," "alternate case(s)," and "every other patient." Adding other terms ("alternating cases," "every other case," "every second patient," "in order of admission," etc.) adds only marginally to the totals. Of the more than four dozen alternate allocation studies reported for the first time in *JAMA* (rather than just mentioned or discussed in its pages) throughout this era, more than 80% likewise entailed the study of antimicrobial agents. See also Iain Chalmers, Estela Dukan, Scott H. Podolsky, and George Davey Smith, "The Advent of Fair Treatment Allocation Schedules in Clinical Trials During the 19th and Early 20th Centuries," *Journal of the Royal Society of Medicine* 105 (2012): 221–27 (also available at www.jameslindlibrary.org). On the statistical equivalence of alternate allocation studies and randomized controlled studies (provided that alternation remains entirely unbiased), see Iain Chalmers, "Statistical Theory Was Not the Reason That Randomization Was Used in the British Medical Research Council's Clinical Trial of Streptomycin for Pulmonary Tuberculosis," in Gérard Jorland, Annick Opinel, and George Weisz, eds., *Body Counts: Medical Quantification in Historical and Sociological Perspectives* (Montreal: McGill-Queen's University Press, 2005), 309–34.

17. Marks, *Progress of Experiment*, 1–41.

18. Maxwell Finland, "The Serum Treatment of Lobar Pneumonia," *NEJM* 202 (1930): 1244–47; Harry F. Dowling, Theodore J. Abernathy, and Clarence R. Hartman, "Should Serum Be Used in Addition to Sulfapyridine in the Treatment of Pneumococcic Pneumonia?" *JAMA* 115 (1940): 2125–28. On Dowling, see Marvin Turck, "Harry F. Dowling, M.D., 1904–2000," *Journal of Infectious Diseases* 185 (2002): 1003–4.

19. Edward H. Kass, "History of the Subspecialty of Infectious Diseases in the United States," in Maulitz and Long, *Grand Rounds*, 87–115.

20. Lesch, *First Miracle Drugs*, 207–50; Slater, *War and Disease*, 84–176. The AFEB derived from the Board for the Investigation and Control of Influenza and Other Epidemic Diseases in the Army and would not be dissolved until 1972 (and then re-formed in 1973); see Floyd W. Denny, "Atypical Pneumonia and the Armed Forces Epidemiological Board," *Journal of Infectious Diseases* 143

(1981): 305–16, and Theodore E. Woodward, ed., *The Armed Forces Epidemiological Board: The Histories of Its Commissions* (Washington, DC: Office of the Surgeon General, 1994).

21. Marks, *Progress of Experiment*, 98–128.

22. Ibid., 78–97; James Harvey Young, "The 'Elixir Sulfanilamide' Disaster," *Emory University Quarterly* 14 (1958): 230–37; Daniel Carpenter, *Reputation and Power: Organizational Image and Pharmaceutical Regulation at the FDA* (Princeton: Princeton University Press, 2010), 85–112. The diethylene glycol was found to be the toxic component of the "elixir."

23. On the implications of the Food, Drug, and Cosmetic Act of 1938 for pharmaceutical testing, see Carpenter, *Reputation and Power*, 112–56. Regarding the contemporary standards for the FDA approval of novel antibiotics (focusing on Pfizer's Terramycin), see Arthur A. Daemmrich, *Pharmacopolitics: Drug Regulation in the United States and Germany* (Chapel Hill: University of North Carolina Press, 2004), 54–58.

24. On the origins of clinical pharmacology, see Harry Gold, "Clinical Pharmacology—A Historical Note," *Journal of Clinical Pharmacology and the Journal of New Drugs* 7 (1967): 309–11, and Dominique A. Tobbell, *Pills, Power, and Policy: The Struggle for Drug Reform in Cold War America and Its Consequences* (Berkeley: University of California Press, 2012), 54–57.

25. Peter Temin, "Technology, Regulation, and Market Structure in the Modern Pharmaceutical Industry," *Bell Journal of Economics* 10 (1979): 429–46, and *Taking Your Medicine: Drug Regulation in the United States* (Cambridge, MA: Harvard University Press, 1980), 64–75; Walsh McDermott, "Pharmaceuticals: Their Role in Developing Societies," *Science* 209 (1980): 243. Government involvement in antibiotic evaluation was maintained in several respects. First, as mentioned earlier, the Armed Forces Epidemiological Board remained an important site for antimicrobial testing. Second, from the penicillin era onward, certain antibiotics were to be certified for the potency of individual batches by the Food and Drug Administration's Division of Antibiotics. Third, this division, headed by Henry Welch, would also be involved in some of the earliest laboratory studies of such emerging antibiotics as Aureomycin; see, e.g., Clifford W. Price, William A. Randall, and Henry Welch, "Bacteriological Studies of Aureomycin," *Annals of the New York Academy of Sciences* 51 (1948): 211–17, and Daemmrich, *Pharmacopolitics*, 57–58. In Welch's talks and writings on antibiotic certification from the 1950s, there is an implied notion of the efficacy and importance of ("these life-saving") antibiotic drugs, but the program was initially established to ensure the potency of individual batches of such biological products. See Henry Welch, "Government Certification of Antibiotics" (talk given to the "Division of Analytical Chemistry" in Buffalo, Mar. 1952), Box 18, ff 31, Records of the Food and Drug Administration Mixed Files, 1905–1980, RG 88; Henry Welch, "Certification of Antibiotics," *Public Health Reports* 71 (1956): 594–99.

26. John P. Swann, *Academic Scientists and the Pharmaceutical Industry:*

Cooperative Research in Twentieth-Century America (Baltimore: Johns Hopkins University Press, 1988); Nicolas Rasmussen, "The Drug Industry and Clinician Research in Interwar America: Three Types of Physician Collaborator," *Bulletin of the History of Medicine* 79 (2005): 50–80. See also Scott H. Podolsky and Jeremy A. Greene, "Academic-Industrial Relations before the Blockbuster Drugs: Lessons from the Harvard Committee on Pharmacotherapy, 1939–1943," *Academic Medicine* 86 (2011): 496–501.

27. Jeremy A. Greene, David S. Jones, and Scott H. Podolsky, "Therapeutic Evolution and the Challenge of Rational Medicine," *NEJM* 367 (2012): 1077–82.

28. On Waksman, see Selman A. Waksman, *My Life with the Microbes* (New York: Simon & Schuster, 1954), and "History of the Word 'Antibiotic,'" *Journal of the History of Medicine* 28 (1973): 284–86.

29. Mahoney, *Merchants of Life*, 176–77; Sam C. Wong and Herald R. Cox, "Action of Aureomycin against Experimental Rickettsial and Viral Infections," *Annals of the New York Academy of Sciences* 51 (1948): 290–305; Harry F. Dowling, "The History of the Broad-Spectrum Antibiotics," *Antibiotics Annual* (1958–1959): 40. "Gram-positive" versus "gram-negative" refers to how particular species of bacteria take up particular stains (helping in their identification), though such attributes also correlate with more general behaviors, including how they react to particular antibiotics. Gram-positive organisms include such well-known pathogens as staphylococci, streptococci, and pneumococci, while gram-negative organisms include such well-known pathogens as *E.coli, Salmonella, Shigella, Pseudomonas,* and *Klebsiella.* Rickettsia refer to particular obligatory intra-cellular pathogens that include the agents of typhus and Rocky Mountain spotted fever (as well as of Lyme disease, discovered after the events of this chapter).

30. "Louis Tompkins Wright, 1891–1952," *Journal of the National Medical Association* 45 (1953): 130–48.

31. Louis T. Wright, Murray Sanders, Myra A. Logan, Aaron Prigot, and Lyndon M. Hill, "Aureomycin: A New Antibiotic with Virucidal Properties," *JAMA* 138 (1948): 408–12; L. T. Wright, M. Sanders, M. A. Logan, A. Prigot, and L. M. Hill, "The Treatment of Lymphogranuloma Venereum and Granuloma Inguinale in Humans with Aureomycin," *Annals of the New York Academy of Sciences* 51 (1948): 318–30. Such access to the medication also reportedly derived from Wright's relationship with Lederle's Director of Research, Yellapragada Subbarow, who would die prematurely of a myocardial infarction in Aug. 1948. See the brief, undated, and unsigned note regarding Wright's relationship with Dr. Subbarow in Box 1, ff 5, LTWP.

32. Wright et al., "Treatment of Lymphogranuloma Venereum," 318. Note the blurry distinction between large viruses and the rickettsia at the time.

33. "Harlem Hospital Rushes Drug; Saves Peeress," *New York Age* (Jan. 15, 1949), p. 2, in Box 21, ff 4, LTWP.

34. "Miracle Drug," *Ebony* (June 1949): 13–17, in Box 21, ff 4, LTWP. The glowing *Ebony* tribute concluded, with respect to potential side effects from the

miracle drug: "Original fears that aureomycin might cause unpleasant reactions in human patients have been allayed by the discovery that it is almost toxic [*sic*] free. Tests have proved that doses much larger than necessary cause no harmful, toxic effects." The article also noted the priority dispute between Wright's group at Harlem Hospital and Perrin Long's group at Johns Hopkins regarding who first used the drug "successfully" in humans. For Long's claim of priority, see *AP*, Part 24, p. 13761.

35. Wright would go on to test Pfizer's Terramycin (oxytetracycline) as well, before his premature death in 1952. See, e.g., Louis T. Wright, James C. Whitaker, Robert S. Wilkinson, and Malcolm S. Beinfield, "The Treatment of Lymphogranuloma Venereum with Terramycin," *Antibiotics and Chemotherapy* 1 (1951): 193–97.

36. Thomas Fite Paine Jr., Harvey Shields Collins, and Maxwell Finland, "Laboratory Studies with Aureomycin," *Annals of the New York Academy of Sciences* 51 (1948): 228–30; Harvey Shields Collins, Thomas Fite Paine, and Maxwell Finland, "Clinical Studies with Aureomycin," *Annals of the New York Academy of Sciences* 51 (1948): 231–40. Lederle's Wilbur Malcolm flattered Finland, in anticipation of the meeting, as "of course . . . the star performer on the clinical side"; Malcolm to Finland, 5/12/48, Box 12, ff 44, MFP. For Dowling's presentation, see Harry F. Dowling, Mark H. Lepper, Lewis K. Sweet, and Robert L. Brickhouse, "Studies on Serum Concentrations in Humans and Preliminary Observations on the Treatment of Human Infections with Aureomycin," *Annals of the New York Academy of Sciences* 51 (1948): 241–45.

37. In 1930, the Lederle Antitoxin Laboratories had become Lederle Laboratories, a component of the American Cyanamid company. Malcolm came to Lederle in 1934 from the Massachusetts State Antitoxin Laboratories, soon beginning a swift ascent through the ranks. By 1957, Malcolm would become president of American Cyanamid, writing to Finland that "no one has contributed more to my success, and that of Lederle, than yourself." See Mahoney, *Merchants of Life*, 168–69; Maxwell Finland to Wilbur Malcolm, 7/31/46; Malcolm to Finland, 9/3/57, both in Box 12, ff 44, MFP.

38. On Lederle's support of Finland's laboratory, see Benjamin W. Carey to Maxwell Finland, 5/18/50; William B. Castle to Benjamin W. Carey, 5/9/51, both in Box 12, ff 42, MFP; on Finland's advising of Lederle regarding its fellowship program, see Maxwell Finland to Wilbur Malcolm, 10/22/52, Box 12, ff 44, MFP; Finland to Carey, 3/30/53, Box 12, ff 42, MFP. On such fellowship programs generally, see Tobbell, *Pills, Power, and Policy*, 37–58.

39. James Whorton, "'Antibiotic Abandon': The Resurgence of Therapeutic Rationalism," in John Parascandola, ed., *The History of Antibiotics: A Symposium* (Madison: American Institute of the History of Pharmacy, 1980), 125–26; Charles S. Bryan and Scott H. Podolsky, "Doctor Holmes at 200: The Spirit of Skepticism," *NEJM* 361 (2009): 846–47.

40. Jesse G. M. Bullowa, Norman Plummer, and Maxwell Finland, "Sulfapyri-

dine in the Treatment of Pneumonia," *JAMA* 112 (1939): 570; Podolsky, *Pneumonia before Antibiotics*, 97–98.

41. Of the nearly 100 papers Finland published between 1928 and 1940, all but a handful had been devoted to pneumonia or the pneumococcus. By the 1940s, with the advent of a widening array of antimicrobials, Finland likewise expanded his own research purview.

42. Maxwell Finland, Harvey Shields Collins, and Tom Fite Paine Jr., "Aueromycin: A New Antibiotic," *JAMA* 138 (1948): 946–49.

43. Harvey Shields Collins, Tom Fite Paine Jr., Edward Buist Wells, and Maxwell Finland, "Aureomycin—A New Antibiotic: Evaluation of Its Effects in Typhoid Fever, Severe Salmonella Infections, and a Case of Colon Bacillus Bacteremia," *Annals of Internal Medicine* 29 (1948): 1090. As they continued: "It can only be concluded that the effects of aureomycin in the present group of cases leave much to be desired. . . . The recent preliminary report of favorable effects from chloromycetin on typhoid fever, when carefully scrutinized, also does not give unequivocal evidence of benefits ascribable to that antibiotic" (1090–91). As early as May 1948, Finland had written to Hopkins's Emanuel Schoenbach of his fear that Perrin Long would "try to jump the gun on the A-377 [as Aureomycin was then termed]" and his "hope that nobody will come out with a half-baked blow-off which they will subsequently regret." In Finland to Schoenbach, 5/12/48, Box 4, ff 66, MFP.

44. Maxwell Finland to Benjamin W. Carey, 10/17/50, Box 12, ff 42, MFP. See also Jerome A. Marks, Louis T. Wright, and Selig Strax, "Treatment of Chronic Non-Specific Ulcerative Colitis with Aureomycin," *American Journal of Medicine* 7 (1949): 180–90. For conservative Aureomycin assessments Finland gave to Wilbur Malcolm, see Finland to Malcolm, 4/10/48; Finland to Malcolm, 5/15/48, both in Box 12, ff 44, MFP.

45. Daniel L. Shaw, "Physician-Industry-FDA Relationships: III," in *Pharmaceutical Manufacturers Association Year Book, 1965–1966* (Washington, DC: Pharmaceutical Manufacturers Association, 1966): 581; see also Tobbell, *Pills, Power, and Policy*, 127. The Chief of the FDA's own New Drug Branch, Ralph Smith, in a 1959 address, contrasted the "enthusiast" with the "ultra-conservative." In Ralph G. Smith, "F.D.A. New Drug Requirements," *Drug and Cosmetic Industry* 84 (1959): 738.

46. The term *broad-spectrum* itself appears to have entered the literature with Pfizer's Terramycin. Pfizer president John McKeen described the drug's "broad spectrum including the possibility that it may have unique value in the treatment of influenza and certain other infectious diseases not yet covered by existing antibiotics" to the New York Society of Security Analysts in March of 1950; and the University of Washington's William Kirby, at the General Scientific Meetings of the AMA in June 1950, spoke of the "broad spectrum of activity of the newer antibiotics." In the initial advertisement for Terramycin by name in July 1950,

"broad-spectrum" assumed its since-commonplace adjectival form. By Feb. 1951, Parke-Davis was describing Chloromycetin as the "broad-spectrum antibiotic of choice," while Lederle does not seem to have publicly used the term until Jan. 1952. See John E. McKeen, "Antibiotics and Pfizer & Company," *Armed Forces Chemical Journal* 3 (Apr. 1950): 38; William M. M. Kirby, "Recent Trends in Antibiotic Therapy," *JAMA* 144 (1950): 235; Terramycin advertisement, *JAMA* advertising section 143 (July 1, 1950): 10–11; Chloromycetin advertisement, *Therapeutic Notes* (Feb. 1951): back cover; "Why Is Aureomycin the Low-Cost Antibiotic in the Broad-Spectrum Field?" *Aureomycin Digest* 3 (Jan. 1952): front cover.

47. Proprietary remedies could be marketed directly to consumers, while ethical remedies were marketed only to physicians. In 1939, proprietary medication sales had outweighed ethical sales $152.4 million to $148.5 million; by 1954, three years after the passage of the Durham-Humphrey amendment formally distinguishing between over-the-counter and prescription-only medications, ethical sales outweighed proprietary sales $1.089 billion to $368.3 million. See Walter S. Meadsley, "The Pharmaceutical Industry," in Walter Adams, ed., *The Structure of American Industry* (New York: Macmillan, 1971), 162. On the increasing trend to branded drugs, evident to contemporary observers, see Joseph D. McEvilla, *Competition in the American Pharmaceutical Industry* (PhD diss., University of Pittsburgh, 1955), 103–6; *Economic Report on Antibiotics Manufacture* (Washington, DC: U.S. Government Printing Office, 1958), 125, 258–60; "Ethicals, Coming from Behind, Now Outsell Proprietaries 3-to-1," *Advertising Age* (Oct. 16, 1961): 57; Temin, *Taking Your Medicine*, 64–87. On the increasing role of drug detailing, see Jeremy A. Greene, "Attention to 'Details': Etiquette and the Pharmaceutical Salesman in Postwar America," *Social Studies of Science* 34 (2004): 271–92.

48. *Economic Report on Antibiotics Manufacture*, 67. Multiple formulations of such antibiotics were available, however.

49. "Health for Patients, Gamble for Makers," *Business Week* 25 (Mar. 25, 1950): 26; *Economic Report on Antibiotics Manufacture*, 228–30; Peter M. Costello, "The Tetracycline Conspiracy: Structure, Conduct, and Performance in the Drug Industry," *Antitrust Law and Economics Review* 1 (Summer 1968): 18–19.

50. Of the more than twenty companies producing penicillin during the wartime effort, Pfizer was by far the single largest producer; Lederle, despite its early involvement in the penicillin efforts (along with Pfizer, Merck, and Squibb), played a far smaller role, while Parke-Davis played a very minor role. See *Economic Report on Antibiotics Manufacture*, 331; Hobby, *Penicillin: Meeting the Challenge*, 104–5; and Bud, *Penicillin*, 44–45.

51. *A Review of the Clinical Uses of Aureomycin* (New York: Lederle Laboratories, 1951), 9. By the fourth issue of its *Aureomycin Digest*, Lederle could report on the use of Aureomycin for 27 types of infection (some as narrow as Boutonneuse fever, others as broad as Gram-positive infections), with the nota-

tion that "this list of infections is rapidly expanding and its ultimate extent cannot be predicted." In "The New Crystalline Aureomycin," *Aureomycin Digest* (subtitled *The Treatment of Virus and Virus-Like Diseases* for this issue) 1 (July 1950): 3. For the transforming influence of Aureomycin on Lederle, see Mahoney, *Merchants of Life*, 178.

52. Charles E. Silberman, "Drugs: The Pace is Getting Furious," *Fortune* (May 1960): 140. On the novelty of Lederle's approach, see Meadsley, "The Pharmaceutical Industry," 175–76. For contemporary advice and musings regarding such "overwhelming" advertising, see Albert J. Weisbrodt, "Advertising Pharmaceuticals," *Drug and Cosmetic Industry* 66 (1950): 528–29, 578–79, 582, 592; McEvilla, *Competition in the American Pharmaceutical Industry*, 128. Articles in popular magazines describing Aureomycin's empowering of the family doctor didn't hurt, either; see Paul de Kruif, "God's Gift to the Doctors," *Reader's Digest* 55 (July 1949): 49–52.

53. *DIAA*, Part 2, p. 783. For contemporary accounts of the general volume of pharmaceutical mailings—reported to have increased 327% between 1936 and 1954, with antibiotics driving the trend—see "Doctors and Direct Mail," *Drug and Cosmetic Industry* 75 (Sept. 1954): 307; McEvilla, *Competition in the American Pharmaceutical Industry*, 135–36.

54. Costello, "Tetracycline Conspiracy," 40; Temin, *Taking Your Medicine*, 73.

55. Mahoney, *Merchants of Life*, 68–69; Maeder, *Adverse Reactions*, 116–17.

56. Silberman, "Drugs: The Pace is Getting Furious," 269; Maeder, *Adverse Reactions*, 178.

57. See the comparatively modest "Talk on Antibiotics" given by Parke-Davis's John Ehrlich before the Annual Conventions of the North and South Dakota Pharmaceutical Associations, 6/11/51 and 6/13/51, Box 28, ff 34, PDC.

58. "Editorial: Significance of Chloromycetin," *Therapeutic Notes* 56 (July–Aug. 1949): 157. Brief mention of Chloromycetin had been made in the Mar. and May 1949 volumes in the "Therapeutic Review" sections of the journal. See also J. D. Ratcliff, "The Greatest Drug Since Penicillin," *Colliers* (Feb. 5, 1949): 26–45.

59. "Shock Treatment for Parke-Davis," *Fortune* (Sept. 1953): 212.

60. "Editorial: Present Status of Chloromycetin," *Therapeutic Notes* 57 (July–Aug. 1950): 158.

61. Indeed, the July–Aug. 1952 volume of the journal, published while the FDA was investigating Chloromycetin and its effects on the bone marrow, instead contained an "Atlas of Hematology" and a section on the "Therapy of Anemia," with no mention of Chloromycetin made. On Parke-Davis's efforts to dilute the impact of the FDA investigation, see Maeder, *Adverse Reactions*, 103–77. That said, they were certainly operating in a competitive environment in this respect. Wrote Lederle to physicians in a July 1952 "Dear Doctor" note: "Recently, a number of comments have appeared in both the lay and medical press on the subject of bone marrow depression following the administration of an antibiotic. Some

of these comments tend to link together all broad-spectrum antibiotics, or even all antibiotics, in this respect. There is no foundation for such generalizations" (Box 25, ff 13, MFP).

62. "The Facts on Chloromycetin: An Evaluation and Confirmation," *Therapeutic Notes* 60 (Jan. 1953): 22.

63. Mahoney, *Merchants of Life*, 237; Berton Roueché, "Something Extraordinary," *New Yorker* 27 (July 28, 1951): 27–47.

64. *Economic Report on Antibiotics Manufacture*, 95; Hobby, *Penicillin*, 178–90.

65. "Health for Patients, Gamble for Makers," 26. This was likely a paraphrase by the reporter; cf. McKeen, "Antibiotics and Pfizer & Company," 36–38.

66. "Pfizer and Antibiotics," *Drug and Cosmetic Industry* 66 (1950): 393.

67. Jasper H. Kane, A. C. Finley, and B. A. Sobin, "Antimicrobial Agents from Natural Sources," *Annals of the New York Academy of Sciences* 53 (1950): 226–28; see also Roueché, "Something Extraordinary," 27–47; *Terramycin: Review of the Literature* (Brooklyn, NY: Pfizer, 1951), 7–8; "Wonder-Drug Marketing," *Modern Packaging* 25 (May 1952): 138–42; John Gunther, "Inside Pfizer" [advertising supplement], *New York Times* (Mar. 23, 1958); Arthur Daemmrich, "Synthesis by Microbes or Chemists? Pharmaceutical Research and Manufacturing in the Antibiotic Era," *History and Technology* 25 (2009): 243–47.

68. Mahoney, *Merchants of Life*, 242. See also *Economic Report on Antibiotics Manufacture*, 140–41, and Terry Armstrong, "Distribution in Jig Time: The Story of Terramycin," *Sales Management* 66 (1951): 74–78.

69. Mahoney, *Merchants of Life*, 237; see also Louis Lasagna, *The Doctors' Dilemmas* (New York: Harper & Brothers, 1962), 134–36.

70. *Economic Report on Antibiotics Manufacture*, 140–42; "Pfizer Put an Old Name on a New Drug Label," *Business Week* (Oct. 13, 1951): 131–36. Pfizer's use of "teaser" ads (mentioning, e.g., "Terra firma" but not the name Terramycin itself) prior to Terramycin's release would be noted neutrally in the FTC report but fondly by fellow medical advertisers as the first teaser ad ever used. In *Medicine Avenue: The Story of Medical Advertising in America* (Huntington, NY: Medical Advertising Hall of Fame, 1999), 23.

71. On Sackler, see *DIAA*, Part 6, pp. 3064–3137; *Medicine Avenue*, 17–18; and Barry Meier, *Pain Killer: A Wonder Drug's Trail of Addiction and Death* (New York: Rodale, 2003), 193–222. On Sackler's views regarding the role of pharmaceutical marketing, see Arthur M. Sackler, "Freedom of Inquiry, Freedom of Thought, Freedom of Expression: 'A Standard to Which the Wise and the Just Can Repair': Observations on Medicines, Medicine, and the Pharmaceutical Industry," 10/18/57, in Box 5, "William Douglas McAdams," FMIP; Jeremy A. Greene and Scott H. Podolsky, "Keeping Modern in Medicine: Pharmaceutical Promotion and Physician Education in Postwar America," *Bulletin of the History of Medicine* 83 (2009): 352–55.

72. "Terramycin," *Scientific American* 183 (July 1950): 29. As Gladys Hobby

wrote to Edward Kass decades later, "Who else ever received a new drug for first animal testing during the week before Thanksgiving and got it approved through the FDA by March 22nd? Those were exciting days!" In Gladys L. Hobby to Edward Kass, 4/8/86, Box 28, ff 31, EKP.

73. F. A. Hochstein et al., "Terramycin. VII. The Structure of Terramycin," *Journal of the American Chemical Society* 74 (1952): 3708–9; F. A. Hochstein et al., "The Structure of Terramycin," *Journal of the American Chemical Society* 75 (1953): 5455–75. Woodward was originally hired in Apr. 1951 by Pfizer president John McKeen as a consultant concerning "the chemical structure of terramycin and aureomycin," with the timing of any publications derived from such a partnership to be "determined by mutual agreement." See John McKeen to Robert Woodward, 4/10/51; "Conference[s] on the Structure of Terramycin," 7/20/51–4/14/52, all in 68.10, Box 47, "Terramycin Correspondence," RWP. On the ensuing program in synthetic chemistry set up by Pfizer to explore tetracycline analogs and synthesis, see Lloyd H. Conover, "Terramycin Proposal," 8/1/52; Frederick J. Pilgrim, "Terramycin Research Program," 9/29/52, both in 68.10, Box 27, "Pfizer, 1951–1979 (5 of 6)," RWP; and Daemmrich, "Synthesis by Microbes or Chemists," 237–56. Regarding Pfizer's boast to physicians, see "Dear Doctor" letter, 8/21/52, in Box 20, ff 1, MFP. Similarly, upon the determination of the molecular structure of Chloromycetin, Parke-Davis had reported, "Here indeed was not only a new antibiotic, but a compound unique in the history of chemistry!" In "Editorial: Significance of Chloromycetin," 158.

74. "Pfizer Put an Old Name on a New Drug Label," 131; Mahoney, *Merchants of Life*, 237. As Pfizer reported of the students: "A pioneer group of 70 medical students . . . have been selected for a special course in antibiotics covering recent research and clinical developments. These young men and women are qualified to serve physicians in 36 major cities during their summer vacation and will make available reprints, abstracts, bibliographic research and other data as requested by members of the profession." In Terramycin advertising in *JAMA* advertising section 146 (July 14, 1951): 13. See also "Remarks of Mr. John E. McKeen, President, Chas. Pfizer & Company, Inc. before Meeting of the New York Security Analysts' Society, February 10, 1953," C38 (9) I, AIHP. There, McKeen also admitted that the costs of such "saturation advertising" were high and cut into profits.

75. "Remarks of Mr. John E. McKeen."

76. *Medicine Avenue*, 18.

77. Ibid., 42.

78. *Economic Report on Antibiotics Manufacture*, 132–33.

79. Ibid., 136. Advertising for pharmaceuticals overall in *JAMA* doubled during the same period. While only antibiotic-producing companies were included in the Federal Trade Commission survey, this included many of the leading pharmaceutical makers of the day, including Abbott, Commercial Solvents, Lederle, Lilly, Merck, Parke-Davis, Pfizer, Sharp and Dohme, Squibb, Upjohn, and Wyeth. Such

firms, as of 1955, would themselves account for 33% ($1.35 million worth) of *JAMA* advertising. Calculated from *DIAA*, Part 1, pp. 129, 133.

80. *Economic Report on Antibiotics Manufacture*, 133–35. Lederle, between 1950 and 1956, would increase its yearly *JAMA* broad-spectrum advertising from 0 to 138 pages; Parke-Davis, during the same period, would increase its from 14 to 76 pages.

81. Ibid., 134–35. Advertising in the eight-page *Spectrum* inserts was essentially Terramycin advertising for much of its existence. Terramycin ads appeared in every issue of *Spectrum* from the house organ's inception until Apr. 1955. Until Feb. 1954, no non-antibiotic remedy would be advertised in *Spectrum*.

82. Tetracycline would be produced by multiple companies, with the FTC launching a massive inquiry into patent and licensing arrangements among the companies by 1958. See Costello, "Tetracycline Conspiracy," 13–44; Robert Bud, "Antibiotics, Big Business, and Consumers: The Context of Government Investigations into the Postwar American Drug Industry," *Technology and Culture* 46 (2005): 329–49.

83. Sam G. Brock [Pfizer], "Regional Managers Weekly Activity Report," 7/8/55, Box 246; Edward Caso [Lederle], "Monthly Report for February, 1955," 3/9/55, Box 250, both in FTC Docket 7211, RG 122. See also Brock, "Easy Prey for Terramycin," 10/8/54, Box 246, in ibid., and Bud, *Penicillin*, 110–11. I am grateful to Robert Bud for drawing my attention to this archival treasure trove.

84. J. F. Hanifan [Lederle], "May, 1954, Report," 5/21/54, Box 244; Seymour Taylor [Pfizer], "Field Report on 'What Are Our Men Saying about Terramycin-Tetracyn?' and 'What Are Doctors' Reactions to Our Story?'" 10/13/54, Box 246; Robert H. Medlen [Pfizer], "District Managers Monthly Activity Report," Feb. 1956, Box 246; K. R. Von Fritzinger [Lederle], "Personal and Confidential," 4/15/55, Box 247; J. P. Hanifan [Lederle], "Monthly Report," [dated "Probably 1956"], Box 250; W. F. Taylor [Lederle], "Tetracycline Competitive Situation," 4/19/55, Box 250, all in FTC Docket 7211, RG 122.

85. Edward Caso [Lederle], "Report on Tetracycline Situation," 4/18/55, Box 248, FTC Docket 7211, RG 122.

86. W. Alderisio [Squibb], "Problems!!!" 12/1/55, Box 254, FTC Docket 7211, RG 122.

87. *Economic Report on Antibiotics Manufacture*, 67. Contributing to this increase, with the advent of the Korean War, was further government interest in antibiotics and the federal stimulation of a dramatic private expansion of antibiotic production facilities (57–60). However, by 1956, 27% of the antibiotic output was for animal feed supplementation (68).

88. Ibid., 269, 272. The largest yearly percentage increase in clinical antibiotic consumption occurred between 1950 and 1951 (upon the advent of all three original broad-spectrum agents), representing a 109.7% increase; the second-largest occurred the following year, a 43.1% increase (followed respectively by 24.9%, 2.9%, 16.3%, and 2.9% increases). The FTC data on production versus

consumption don't quite add up, though dramatic increases in both production and consumption are apparent in the two sets of figures.

89. *Economic Report on Antibiotics Manufacture*, 73. A brief dip in penicillin sales did occur between 1954 and 1955.

90. Ibid., 67. On erythromycin, see Henry Welch, "Erythromycin—A New Broad Spectrum Antibiotic," *Antibiotics and Chemotherapy* 2 (1952): 279–80, and David Greenwood, *Antimicrobial Drugs: Chronicle of a Twentieth Century Medical Triumph* (New York: Oxford University Press, 2008), 236–37.

91. Regarding profit margins, see *Economic Report on Antibiotics Manufacture*, 211–12. Such margins, it should be noted, did tend to decline over the period as prices were cut and marketing efforts were expanded. Regarding overall sales, see ibid., 78–79. Of such sales, the medicinal (as opposed to feed supplement) figures were $159 million and $64 million respectively.

92. See, e.g., the letterhead of the "Dear Doctor" letter, 8/21/52, in Box 20, ff 1, MFP; "Pfizer Put an Old Name on a New Drug Label," 130; and "Wonder Drugs' Wonder," *Time* 58 (Oct. 1, 1951): 91.

93. "Remarks of Mr. John E. McKeen."

94. *Economic Report on Antibiotics Manufacture*, 83. In terms of sales dollars, however, owing to Pfizer's heavy sales of antibiotics as feed supplements, Lederle continued to hold 28.1% of the market, compared to Pfizer's 23.4% (ibid., 96).

95. Ibid., 95, 199. At the same time, Lederle's and Parke-Davis's domestic antibiotic sales accounted for 18% and 18.4% of their consolidated net domestic sales, respectively.

96. Thomas J. Winn, "The Antibiotics Market," *Drug and Cosmetic Industry* 67 (1950): 472; see also Winn, "Increased Prescription Business Due to Antibiotics," *American Professional Pharmacist* 16 (1950): 984, and George Urdang, "The Antibiotics and Pharmacy," *Journal of the History of Medicine* 6 (1951): 404.

97. Winn, "Antibiotics Market," 563.

98. Silberman, "Drugs: The Pace Is Getting Furious," 140.

99. See, e.g., the rotating advertisements in *Spectrum* during Terramycin's first year of release. Lederle and Parke-Davis would use similar strategies, though in slightly more subtle fashion. Such efforts preceded the more general usage of market segmentation strategies by the late 1950s; see Lizabeth Cohen, *A Consumers' Republic: The Politics of Mass Consumption in Postwar America* (New York: Knopf, 2003), 292–344.

100. John E. McKeen, "The Antibiotic Vista," *Drug and Cosmetic Industry* 66 (1950): 652; see also Winn, "Antibiotics Market," 472.

101. McKeen, "Antibiotic Vista," 759, 764; see also Terramycin advertisement in *JAMA* advertising section 145 (Feb. 3, 1951): 14–15. Perhaps in response, after producing a somewhat cautious 1951 *Review of the Clinical Uses of Aureomycin* (arguing against "the indiscriminate use of antibiotics"), Lederle published a more

aggressive 1952 monograph. In it, Aureomycin was endorsed as "the best cold remedy which we have at present," on the basis of a study in which it was found "from two to two and a half times as effective as antihistamines"; the original study authors had themselves concluded from their study that "neither antihistamines nor aureomycin can be regarded as satisfactory treatment for the common cold." Cf. *Review of the Clinical Uses of Aureomycin*, 10; *Fifth Year of Aureomycin* (New York: Lederle Laboratories, 1952), 39; Calvin H. Chen and Robert B. Dienst, "Aureomycin and Antihistamines in the Treatment of the Common Cold," *Journal of the Medical Association of Georgia* 40 (1951): 113.

102. Roueché, "Something Extraordinary," 29. Roueché based such a proclamation on impressions "gathered on my tour of the Pfizer laboratories." The previous year, *Business Week* had similarly declared that "antibiotics alone have made it almost certain that infectious diseases will soon become a medical curiosity" ("Health for Patients, Gamble for Makers," 34).

103. Charles O. Jackson, *Food and Drug Legislation in the New Deal* (Princeton: Princeton University Press, 1970). The original agency responsible for enforcing the 1906 Act had been the United States Department of Agriculture's Bureau of Chemistry. By 1927, the Bureau of Chemistry was reorganized, resulting in a newly named Food, Drug, and Insecticide Administration; the ensuing agency would be re-named the Food and Drug Administration in 1930.

104. On such implicit attention to efficacy, see Marks, *Progress of Experiment*, 71–97; John P. Swann, "Sure Cure: Public Policy on Drug Efficacy before 1962," in George J. Higby and Elaine C. Stroud, eds., *The Inside Story of Medicines: A Symposium* (Madison: American Institute of the History of Pharmacy, 1997), 223–61; Carpenter, *Reputation and Power*, 118–227.

105. Marks, *Progress of Experiment*, 17–41; Eric W. Boyle, *Quack Medicine: A History of Combating Health Fraud in Twentieth-Century America* (Santa Barbara, CA: Praeger, 2013), 17–38.

106. "The Council on Pharmacy and Chemistry: A Twenty-fifth Anniversary," *JAMA* 94 (1930): 414; Austin Smith, "The Council on Pharmacy and Chemistry," in Morris Fishbein, *A History of the American Medical Association, 1847–1947* (Philadelphia: Saunders, 1947), 883. Advertising in *JAMA* was far more important in the pre–World War II era than it would be today. *JAMA* was both the dominant medical journal in America in the pre–World War II era and the dominant location for the pharmaceutical advertising of ethical remedies. See Jennie Gregory, "An Evaluation of Medical Periodicals," *Bulletin of the Medical Library Association* 25 (1937): 172–88; Elizabeth Knoll, "The American Medical Association and Its Journal," in William F. Bynam, Stephen Lock, and Roy Porter, eds., *Medical Journals and Medical Knowledge: Historical Essays* (London: Routledge, 1992), 146–64; and *Medicine Avenue*, 16. As late as 1960, it was still estimated that 55% of all ethical pharmaceutical journal advertising dollars were devoted to four journals (out of the 600 that carried advertisements): *JAMA, Medical*

Economics, *MD*, and *Modern Medicine*. In "Critics Fail to Inhibit Ethical Ad Growth," *Advertising Age* (Feb. 1, 1960): 34.

107. *DIAA*, Part 1, p. 129.

108. Greene and Podolsky, "Keeping Modern in Medicine," 336–46.

109. "Report on a Study of Advertising and the American Physician," *DIAA*, Part 2, p. 490.

110. Ibid., p. 491.

111. Ibid.

112. Ibid., pp. 490–91. The survey likewise included questions about the inclusion of Pfizer's *Spectrum*, of which approximately "half of the advertisers expressed a definite dislike . . . because they felt that it gave Pfizer an unfair advantage over the other advertisers and because they felt it made the AMA the publisher of the Pfizer house organ" (p. 517).

113. "JAMA Marketing Data, Reports 1 to 20," BD 991, 33–14, AMAA; see also Jeremy A. Greene, "Pharmaceutical Marketing Research and the Prescribing Physician," *Annals of Internal Medicine* 146 (2007): 742–48.

114. "Reports #2 and #3," in "JAMA Marketing Data, Reports 1 to 20," BD 991, 33-14, AMAA.

115. *DIAA*, Part 1, p. 129.

116. Robert J. Lyon to Warren Albert, 12/22/65, BD 1467, AMAA; see also *CPDI*, Part 14, p. 5808. On the study, see *DIAA*, Part 2, pp. 698–806, and Greene and Podolsky, "Keeping Modern in Medicine," 345–46.

117. "Questions for American Medical Association" [49-page preparation for the AMA appearance at the Kefauver hearings in July 1961], Box 19, ATS; *DIAA*, Part 1, pp. 114, 122–26.

118. "A.M.A. Councils Expand Their Programs," *JAMA* 157 (1955): 664–65; *DIAA*, Part 1, p. 114.

119. *DIAA*, Part 1, pp. 104–5; Harry F. Dowling, *Medicines for Man: The Development, Regulation, and Use of Prescription Drugs* (New York: Knopf, 1970), 172.

120. Minutes, AMA Board of Trustees, 10/22/53, p. 32, AMAA.

121. Karl F. Nygren to C. Joseph Stetler, 7/7/54, "Seal of Acceptance Program," AMAA.

122. Minutes, AMA Board of Trustees, 6/20/54, p. 223; Minutes, AMA Board of Trustees 11/26/54, pp. 55–55a; R. T. Stormant to Abbott Laboratory, 2/15/55, COD 15-9, all in AMAA; "A.M.A. Councils Expand their Programs," 664–65.

123. As one *JAMA* copy supervisor, quoting the comments of two "consultants," communicated to the head of the William Douglas McAdams's Terramycin team with respect to a proposed advertisement in *Spectrum*: "(1) It borders on the ridiculous in its broad implication that Terramycin is always indicated in the treatment of each of the 160 disease conditions listed, (2) it represents an over-simplification of the role of broad-spectrum antibiotics in the treatment of infections and thus

tends to foster careless and haphazard therapy, and (3) it includes numerous conditions in which the value of broad-spectrum antibiotics has not been recognized by the Council [on Pharmacy and Chemistry]." In Hazel [Eggert] to Deforest Ely, 9/27/54, 71A-5170, Box 10, "Extreme Claims e. General," RG 46. See also Robert J. Lyon to Gil Totten, 5/28/58, in ibid.

124. "Attitudes of U.S. Physicians toward the American Pharmaceutical Industry," 1959, p. iv, CD 1056, AMAA.

125. The former study, conducted through Taylor, Hawkins, & Lea, Inc., was presented to the industry as "a kind of 'pure research' which can serve as a firm foundation of strategic long term value, possibly on the policy level, in using drug samples in pharmaceutical marketing." In "A Study of Drug Sampling—Spring 1958," BD 1465, 33-14, AMAA; for the latter study, see "Attitudes of U.S. Physicians toward the American Pharmaceutical Industry," CD 1056, AMAA. Between 1950 and 1958, the AMA had commissioned Gaffin to conduct seven separate studies (ibid., appendix D-1).

126. See William B. Bean, "Vitamania, Polypharmacy, and Witchcraft," *Archives of Internal Medicine* 96 (1955): 137–41; William B. Bean to Joseph Garland [editor-in-chief, *NEJM*], 2/14/56; Bean to Garland, 2/23/56, both in Box 17, "Joseph Garland," WBP. The *Medical Letter* was likewise first developed in the context of this apparent vacuum; see Harold Aaron, "Proposal for a new publication to provide timely and accurate information on new drugs and biologicals to physicians and other health professionals," sent to Harry Dowling, 11/4/57, in Box 4, "Twixt the Cup and the Lip," HDP.

127. Maxwell Finland to Harry Dowling, 11/14/56, Box 2, ff 8, MFP.

128. Maxwell Finland to Thomas Bradley, 4/16/57, Box 4, ff 3, MFP.

129. Harry F. Dowling, "The Rise and Fall of Pneumonia-Control Programs," *Journal of Infectious Diseases* 127 (1973): 201–6; Podolsky, *Pneumonia before Antibiotics*, 53–87. Regarding the aggressive end of the spectrum concerning such oversight, see Massachusetts's Henry D. Chadwick to Lucy F. Forrer, 5/24/37, Box 184, ff 1736, Series 18.1 (Grants), RAC.

130. Podolsky, *Pneumonia before Antibiotics*, 132–36. On decreasing public health oversight of venereal disease control by the 1950s, see Brandt, *No Magic Bullet*, 174–78, and Parascandola, *Sex, Sin, and Science*, 135–38. On the closing of tuberculosis sanatoria by the mid-1950s, see Sheila M. Rothman, *Living in the Shadow of Death: Tuberculosis and the Social Experience of Illness in American History* (Baltimore: Johns Hopkins University Press, 1994), 248–50.

131. *AP*, Part 22, p. 11927; *AP*, Part 23, pp. 12634–35, 12801–2, 12983–85. Section 507 of the Food, Drug, and Cosmetic Act mandated that the FDA certify each batch of penicillin, streptomycin, aureomycin, chloramphenicol, bacitracin, "or any derivative thereof" for potency. See McEvilla, *Competition in the American Pharmaceutical Industry*, 70–71; *Economic Report on Antibiotics Manufacture*, 123–24. For antibiotics covered under such regulations, the Division of Antibiotics appears to have played a large role in determining approval for drug

marketing as well; see Welch, "Certification of Antibiotics," 596, and *AP*, Part 22, p. 11929.

132. *AP*, Part 23, pp. 12218–25, 12634.

133. Ibid., pp. 12949–50, 12804–5.

134. "Dr. Félix Martí-Ibañez Presented with the Order of Carlos J. Finlay," *Antibiotics and Chemotherapy* 5 (1955): 105–6.

135. Box 3, "CV," FMIP. Wrote Martí-Ibañez in 1943, "Among my duties as Medical Director, are the preparation and handling of Medical Propaganda, direct supervision of our house organ[,] . . . the preparation of colored mailing pieces, and all medical literature sent to physicians in Spanish America." In Félix Martí-Ibañez to O.F. Ball, 10/21/43, Box 2, "Personal, January 1943–October 1943," FMIP. For a three-page exposition of Martí-Ibañez's recommended techniques for "export marketing," see Martí-Ibañez to Arthur M. Sackler, 8/19/52, Box 5, "William Douglas McAdams," FMIP. In the midst of such efforts, Martí-Ibañez would also be offered the editorship of a proposed Spanish version of *JAMA*, an offer he would eventually decline. See Félix Martí-Ibañez to Morris Fishbein, 8/24/45; Martí-Ibañez to Fishbein 10/3/45; Martí-Ibañez to Fishbein, 11/2/45; Martí-Ibañez to Charles Grotheer, 12/21/45, all in Box 2, "Personal, August 10, 1945–December 1945," FMIP.

136. Box 3, "McAdams International," FMIP; Box 5, "William Douglas McAdams," FMIP; Félix Martí-Ibañez to Miriam Perry, 5/9/56, Box 7, "P," FMIP; "Personnel Changes," *The Advertiser* 26 (Mar. 1955): 6; "Preliminary Notes for the Introduction of Terramycin and Terramycin SF Abroad," 2/8/55, Box 99, "Ideas," FMIP. Martí-Ibañez wrote to Henry Welch in late 1954, "This has been the most hectic month in what will be the most hectic year of my life. Between the work at the [William Douglas McAdams] Agency which has again suddenly expanded so that it is now demanding a great deal of time from me, the journals with a million problems each, the promotional campaigns for Antibiotic Medicine, the forthcoming Annual which keeps us in the office nights and weekends, my columns for the Latin-American papers and the television and movie projects, I have begun to have headaches and insomnia again." In Martí-Ibañez to Welch, 12/10/54, Box 1089, "Welch, Dr. Henry, FDA Materials Correspondence (10 of 10)," RG 46.

137. *AP*, Part 23, pp. 12805–6, 12950–51.

138. "Symposia Registration," Box 1089, "Miscellaneous (2 of 6)," RG 46.

139. *AP*, Part 23, pp. 12563, 13155. At the initial 1953 symposium, 43 out of 102 papers were presented by industry; by 1958, this proportion had fallen to 23 out of 181. In "Symposia Manuscripts," Box 1089, "Miscellaneous (2 of 6)," RG 46.

140. Félix Martí-Ibañez, "Historical Perspectives of Antibiotics: Past and Present," *Antibiotics Annual* (1953–1954): 3. To his credit, Martí-Ibañez thus foreshadowed later attempts to place antibiotic use within its larger social context, as, e.g., through the New York Academy of Medicine's Institute of Social and Historical Medicine. See Iago Galdston, ed., *The Impact of the Antibiotics on Medicine and Society* (New York: International Universities Press, 1958).

141. Cf. Stephen Harbarth and Matthew H. Samore, "Antimicrobial Resistance Determinants and Future Control," *Emerging Infectious Diseases* 11 (2005): 794, with Félix Martí-Ibañez, "The Next Half Century in Antibiotic Medicine and Its Effect on the History of the Clinical Case History," *Antibiotics Annual* (1955–1956): 11–12. Indeed, even Martí-Ibañez's lone public "deplor[ing of the] abuses of this as well as of any other therapy" occurred in the midst of an intense defense of antibiotics:

> Antibiotics are constantly being accused of encouraging the development of bacteria-resistant strains, of causing superimposed infections activated by alterations in bacterial ecology, and of arousing secondary reactions. They also stand accused of changing the typical course of disease, thereby increasing diagnostic difficulties and creating new clinical pictures. . . . But the syndromes caused by antibiotics are only one more chapter in the picture of diseases caused by man. . . . The critics of antibiotics should not forget that antibiotics have won a place for themselves, which they will never lose, in the anti-infectious arsenal (ibid.).

142. Félix Martí-Ibañez, "The Philosophical Impact of Antibiotics on Clinical Medicine," *Antibiotics Annual* (1954–1955): 19.

143. Martí-Ibañez, "Next Half Century in Antibiotic Medicine," 13, and "Philosophical Impact," 22.

144. Martí-Ibañez's interests reflected emerging contemporary attention to the roles of television and other media in the dissemination of medical information, especially in graduate medical education. See Richard H. Orr [of the Institute for the Advancement of Medical Communication, formed in 1958] to Martí-Ibañez, 5/31/58, Box 10, "O," FMIP.

145. Martí-Ibañez, "Next Half Century in Antibiotic Medicine," 16.

146. Martí-Ibañez, "Historical Perspectives," 8, and "Antibiotics and the Problem of Medical Communication," *Antibiotics Annual* (1956–1957): 14.

147. Martí-Ibañez, "Historical Perspectives," 7.

148. Martí-Ibañez, "Antibiotics and the Problem of Medical Communication," 14.

149. Maxwell Finland to Selman Waksman, 11/14/56, Box 5, ff 43, MFP; Finland to Harry Dowling, 11/24/56, Box 2, ff 9, MFP.

150. Marks, *Progress of Experiment*, 1–41.

151. Whorton, "Antibiotic Abandon," 125–36.

152. Hobart A. Reimann, "Infectious Diseases: Fourteenth Annual Review of Significant Publications," *Archives of Internal Medicine* 82 (1948): 468.

153. For "antibioticist," see Hobart A. Reimann, "Infectious Diseases: Fifteenth Annual Review of Significant Publications," *Archives of Internal Medicine* 85 (1950): 157. See also Perrin H. Long, Caroline A. Chandler, Eleanor A. Bliss, and Morton S. Bryer, "The Use of Antibiotics," *JAMA* 141 (1949): 315; William M. M. Kirby, "Recent Trends in Antibiotic Therapy," *JAMA* 144 (1950): 233; Per-

rin H. Long, "The Clinical Use of Antibiotics," *Medical Clinics of North America* 34 (1950): 308; E. Jawetz, J. B. Gunnison, and R. S. Speck, "Antibiotic Synergism and Antagonism," *NEJM* 245 (1951): 966; Paul S. Rhoads, "The Antibiotics—Uses and Abuses," *GP* 5 (Feb. 1952): 67; and Caroline A. Chandler, "Antibiotic Therapy," *Yale Journal of Biology and Medicine* 25 (1953): 369. For "committees," see Wesley W. Spink, "Clinical Problems Relating to the Management of Infections with Antibiotics," *JAMA* 152 (1953): 590. For concerns regarding the ability of not only general practitioners but also infectious disease experts themselves to "keep our heads above water" and stay abreast of emerging remedies and indications (as well as to see through the "bias" of the commercial houses and their publications), see Manson Meads to Maxwell Finland, n.d. [likely spring 1949]; Finland to Meads, 4/20/49, both in Box 3, ff 47, MFP.

154. Harry F. Dowling, "Established Use of the Antibiotics," *Illinois Medical Journal* 100 (1951): 192.

155. For such an invocation of "rational" therapy, see E. Jawetz and M. S. Marshall, "The Role of the Laboratory in Antibiotic Therapy," *Journal of Pediatrics* 37 (1950): 545, 553; Ernest Jawetz, "Antibiotic Synergism and Antagonism: Review of Experimental Evidence," *Archives of Internal Medicine* 90 (1952): 308; Jawetz, "Combined Antibiotic Action: Experimental and Clinical," *Antibiotics Annual* (1953–1954): 41; and Edwin D. Kilbourne, "Rational Antibiotic Therapy," *GP* 8 (Sept. 1953): 35–40.

156. Ernest Jawetz, "Infectious Diseases: Problems of Antimicrobial Therapy," *Annual Review of Medicine* 5 (1954): 1–2. As Stephen Snelders, Charles Kaplan, and Toine Pieters have noted, Jawetz's graph was to some degree anticipated by German psychiatrist Max Seige (commenting on the psychotropic medications that had "flooded" the market) nearly four decades previously. On such "Seige cycles" of psychotropic medications, see Toine Peters and Stephen Snelders, "Mental Ills and the 'Hidden History' of Drug Treatment Practices," in Marijke Gijswitj-Hofstra, Harry Oosterhuis, Joost Vijselaar, and Hugh Freeman, eds., *Psychiatric Cultures Compared: Psychiatry and Mental Health Care in the Twentieth Century* (Amsterdam: Amsterdam University Press, 2005), 381–401, and Stephen Snelders, Charles Kaplan, and Toine Peters, "On Cannabis, Chloral Hydrate, and Career Cycles of Psychotropic Drugs in Medicine," *Bulletin of the History of Medicine* 80 (2006): 95–114.

157. Jawetz, "Patient, Doctor, Drug, and Bug," 295.

158. Wendell H. Hall, "The Abuse and Misuse of Antibiotics," *Minnesota Medicine* 35 (1952): 629. Hall had trained under antibiotic researcher Wesley Spink at the University of Minnesota.

159. Waldo E. Nelson, "Clinical Bases for Selection of Patients for Antimicrobial Therapy," *JAMA* 147 (1951): 1340.

160. Regarding the cost of "indiscriminate" use to the individual, see Nelson, "Clinical Bases," 1340. Regarding adverse reactions, see Maeder, *Adverse Reactions*, 103–51; Perrin Long, "Fatal Anaphylactic Reactions to Penicillin," *Antibiotics*

Annual (1953–1954): 35–40; Maxwell Finland and Louis Weinstein, "Complications Induced by Antimicrobial Agents," *NEJM* 248 (1953): 220–26. Regarding superinfections, see Louis Weinstein, "The Spontaneous Occurrence of New Bacterial Infections during the Course of Treatment with Streptomycin or Penicillin," *American Journal of Medical Science* 214 (1947): 56–63; Emanuel Applebaum and William A. Leff, "Occurrence of Superinfections During Antibiotic Therapy," *JAMA* 138 (1948): 119–21; Chester S. Keefer, "Alterations in Normal Bacterial Flora of Man and Secondary Infections during Antibiotic Therapy," *American Journal of Medicine* 11 (1951): 665–66; David T. Smith, "The Disturbance of the Normal Bacterial Ecology by the Administration of Antibiotics with the Development of New Clinical Syndromes," *Annals of Internal Medicine* 37 (1952): 1135–43. With respect to diagnostic sloppiness, this often took the form of warnings about missing ominous, antibiotic-unresponsive etiologies in the process; see, e.g., Louis Weinstein, "The Use and Misuse of Antibiotics," *Boston Medical Quarterly* 2 (1951): 97; W. H. Oatway Jr., "Delays in the Diagnosis of Tuberculosis from the Incautious Use of Antibiotics," *Arizona Medicine* 8 (1951): 25–28.

Antibiotic resistance is discussed at length in chapter 5. While the potential for encouraging antibiotic resistance loomed large in nearly all such enumerations, a good deal of variability characterized the priority given at the time to the various unwanted outcomes of apparent antibiotic overuse. A sampling of five general presentations of unintended consequences of antibiotics, published from 1951–1953, is illustrative:

Nelson 1951	*Hall 1952*	*Rhoads 1952*	*Spink 1953*	*Washington 1953*	*Hartung 1953*
1. Cost 2. Negative effects on lay perception of medical profession 3. Resistance 4. Diagnostic sloppiness	1. Diagnostic sloppiness/ false sense of security 2. Resistance 3. Side effects	1. Super-infection 2. Resistance 3. Side effects	1. False sense of security 2. Side effects 3. Super-infection 4. Resistance	1. Diagnostic sloppiness 2. Side effects 3. Cost 4. Effort of adminis-tration	1. Side effects 2. Resistance

SOURCES: Nelson, "Clinical Bases," 1340–41; Hall, "Abuse and Misuse," 629–32; Rhoads, "Antibiotics," 67–68; Spink, "Clinical Problems," 585–86; John A. Washington, "Some Dangers of Antibiotic Therapy," *Medical Annals of the District of Columbia* 22 (1953): 59–61; Carl A. Hartung, "Abuses in the Use of Antibiotics," *Journal of the Tennessee Medical Association* 46 (1953): 403–6.

161. Jawetz and Marshall, "The Role of the Laboratory in Antibiotic Therapy," 545–54; see also Louis Weinstein, "The Newer Antibiotics: A Brief Review of the Biological and Clinical Properties of Bacitracin, Polymixin, Aureomycin,

Chloromycetin, and Neomycin," *Ohio State Medical Journal* 46 (1950): 552, and Perrin H. Long, "Bacterial Resistance as a Factor in Antibiotic Therapy," *Bulletin of the New York Academy of Medicine* 28 (1952): 809.

162. Maxwell Finland, "Clinical Uses of the Presently Available Antibiotics," *Antibiotics Annual* (1953–1954): 11; Spink, "Clinical Problems," 586. As would ultimately be enunciated at Gaylord Nelson's hearings on the pharmaceutical industry, such notions intersected at the moment when pharmaceutical companies, in promoting drug usage, encouraged the use of "shotgun" preparations or diagnosis by therapeutic response. See *CPDI*, Part 11, 4484–85, and Maeder, *Adverse Reactions*, 191–92.

163. Allen E. Hussar, "A Proposed Crusade for the Rational Use of Antibiotics," *Antibiotics Annual* (1954–1955): 379–82.

164. Dowling, "History of the Broad-Spectrum Antibiotics," 41; Whorton, "Antibiotic Abandon," 131–32; Maeder, *Adverse Reactions*.

165. As described in chapter 4, this situation would change dramatically by the late 1960s, when would-be reformers turned their attention to the individual prescribing clinician.

166. James W. Haviland, "Advances in Antibiotic Therapy," *Annals of Internal Medicine* 39 (1953): 313–15.

167. Ethan Allan Brown, "Discussion," following Perrin H. Long, "Fatal Anaphylactic Reactions to Penicillin," 39.

168. Perrin H. Long to Robert T. Stormant, 3/8/55, COD 15-9, 1955, LL–McL, AMAA.

169. Jawetz, "Infectious Diseases: Problems of Antimicrobial Therapy," 5; see also Wesley W. Spink, "Sulfonamide and Antibiotic Therapy—After Twenty Years," *Minnesota Medicine* 40 (1957): 218–19.

170. Maxwell Finland to Benjamin W. Carey, 11/14/52, Box 12, ff 42, MFP. See also Finland, "Clinical Uses of the Presently Available Antibiotics," 17; Harry M. Marks, "Making Risks Visible: The Science and Politics of Adverse Drug Reactions," in Jean Paul Gaudillière and Volker Hess, eds., *Ways of Regulating: Therapeutic Agents between Plants, Shops, and Consulting Rooms* (Berlin: Max-Planck-Institut für Wissenschaftsgeschichte, 2009), 113–16.

171. Finland, undated response to R. Keith Cannan to Finland, 12/1/60, Box 4, ff 4, MFP. Finland may have conflated physician Albe Watkins with publisher Edgar Elfstrom; see Maeder, *Adverse Reactions*, 265–70. See also Finland to Mrs. Gordon C. Pate, 10/5/60; Finland to J. A. Bradshaw [attorney general for Parke-Davis], 2/2/62, both in Box 2, ff 24, MFP; and *AP*, Part 24, pp. 13945–50.

172. The NAS-NRC's Keith Cannan forwarded to the FDA's William H. Kessenich the 1960 committee's final recommendations, which concluded: "Almost, if not all, potent therapeutic agents cause some undesirable side effects. Therefore, it should be pointed out that Chloramphenicol is not the only antibiotic that may cause unfavorable reactions of a serious and sometimes fatal nature." In Cannan to Kessenich, 1/11/61, Box 18, ff 5, ATS.

173. See, e.g., Maxwell Finland to Emanuel Schoenbach, 5/19/50, Box 4, ff 66, MFP.

Chapter Two: Antibiotics and the Invocation of the Controlled Clinical Trial

Epigraph. Harry F. Dowling, Maxwell Finland, Morton Hamburger, Ernest Jawetz, Vernon Knight, Mark H. Lepper, Gordon Meikeljohn, Lowell A. Rantz, and Paul S. Rhoads, "The Clinical Use of Antibiotics in Combination," *Archives of Internal Medicine* 99 (1957): 537.

1. Harry M. Marks, *The Progress of Experiment: Science and Therapeutic Reform in the United States, 1900–1990* (New York: Cambridge University Press, 1997), 129–63.

2. Daniel Carpenter, *Reputation and Power: Organizational Image and Pharmaceutical Regulation at the FDA* (Princeton: Princeton University Press, 2010), 118–227.

3. Carpenter convincingly contrasts the "conceptual" influence of the FDA with its "directive" and "gatekeeping" powers (ibid., 61–65).

4. Félix Martí-Ibañez, "Historical Perspectives of Antibiotics: Past and Present," *Antibiotics Annual* (1953–1954): 6.

5. Henry Welch, "Opening Remarks," *Antibiotics Annual* (1956–1957): 2.

6. Paul Ehrlich, "Chemotherapy," in F. Himmelweit, ed., *The Collected Papers of Paul Ehrlich, vol. 3: Chemotherapy* (New York: Pergamon Press, 1960), 514.

7. Scott H. Podolsky, *Pneumonia before Antibiotics: Therapeutic Evolution and Evaluation in Twentieth-Century America* (Baltimore: Johns Hopkins University Press, 2006), 104–14.

8. Colin MacLeod, "Chemotherapy of Pneumococcic Pneumonia," *JAMA* 113 (1939): 1407.

9. J. Ungar, "Synergistic Effect of Para-aminobenzoic Acid and Sulphapyridine on Penicillin," *Nature* 152 (1943): 245–46; Joseph W. Bigger, "Synergic Action of Penicillin and Sulphonamides," *Lancet* 2 (1944): 142–45; G. Soo-Hoo and R. J. Schnitzer, "The Activity of Penicillin Combined with Other Anti-Streptococcal Agents towards β-Hemolytic Streptococci in Vivo," *Archives of Biochemistry* 5 (1944): 99–106. See, with respect to clinicians and pneumonia, Charles H. Rammelkamp to Maxwell Finland, 4/3/47, Box 4, ff 39, MFP.

10. C. W. Price, W. A. Randall, Henry Welch, and Velma Chandler, "Studies of the Combined Action of Antibiotics and Sulfonamides," *American Journal of Public Health* 39 (1949): 340–44; Harry F. Dowling, "Established Use of the Antibiotics," *Illinois Medical Journal* 100 (1951): 195.

11. Maxwell Finland, "The Present Status of Antibiotics in Bacterial Infections," *Bulletin of the New York Academy of Medicine* 27 (1951): 213.

12. Robert B. Lawson, in Perrin H. Long, Caroline A. Chandler, Eleanor A. Bliss, Morton S. Bryer, and Emanuel B. Schoenbach, "The Use of Antibiotics," *JAMA* 141 (1949): 317.

13. See, e.g., E. Jawetz, J. B. Gunnison, and R. S. Speck, "Antibiotic Synergism and Antagonism," *NEJM* 245 (1951): 968, and E. Jawetz and J. B. Gunnison, "An Experimental Basis of Combined Antibiotic Action," *JAMA* 150 (1952): 693–95.

14. C. E. Lankford and Helen Lacy, "*In Vitro* Response of Staphylococcus to Aureomycin, Streptomycin, and Penicillin," *Texas Reports on Biology and Medicine* 7 (1949): 111–24; Price et al., "Studies of the Combined Action of Antibiotics and Sulfonamides," 343–44; Sophie Spicer, "Bacteriologic Studies of the Newer Antibiotics: Effect of Combined Drugs on Microorganisms," *Journal of Laboratory and Clinical Medicine* 36 (1950): 183–91; E. Jawetz, J. B. Gunnison, and V. R. Coleman, "The Combined Action of Penicillin with Streptomycin or Chloromycetin on Enterococci *in Vitro*," *Science* 111 (1950): 254–56; Joseph W. Bigger, "Synergism and Antagonism as Displayed by Certain Antibacterial Substances," *Lancet* 2 (1950): 46–50; Jim Ruegsegger to Welsey M. Spink, 7/11/50; Spink to Ruegsegger, 7/13/50, both in Box 15, "Lederle Laboratories, 1950–1952," WSP. Such "interference" had been noted even in the pre-antibiotic era by researchers working with anti-trypanosomal chemotherapeutics; see C. H. Browning and R. Gulbranson, "An Interference Phenomenon in the Action of Chemotherapeutic Substances in Experimental Trypanosome Infections," *Journal of Pathology and Bacteriology* 25 (1922): 395–97.

15. Mark H. Lepper and Harry F. Dowling, "Treatment of Pneumococcic Meningitis with Penicillin Compared with Penicillin Plus Aureomycin: Studies Including Observations on an Apparent Antagonism between Penicillin and Aureomycin," *Archives of Internal Medicine* 88 (1951): 489–94.

16. E. Jawetz and J. B. Gunnison, "Studies on Antibiotic Synergism and Antagonism: A Scheme of Combined Antibiotic Action," *Antibiotics and Chemotherapy* 2 (1952): 243–48. For present-day molecular biological aspirations to this effect, see Tobias Bollanbach and Roy Kishony, "Resolution of Gene Regulatory Conflicts Caused by Combinations of Antibiotics," *Molecular Cell* 42 (2011): 413–25.

17. E. Jawetz, J. B. Gunnison, J. B. Bruff, and V. R. Coleman, "Studies on Antibiotic Synergism and Antagonism: Synergism among Seven Antibiotics against Various Bacteria *In Vitro*," *Journal of Bacteriology* 64 (1952): 30–31.

18. Ernest Jawetz, "Antibiotic Synergism and Antagonism: Review of Experimental Evidence," *Archives of Internal Medicine* 90 (1952): 308.

19. Jawetz and Gunnison, "Experimental Basis of Combined Antibiotic Action," 694; Jawetz, Gunnison, and Speck, "Antibiotic Synergism and Antagonism," 966; Dowling, "Established Use of the Antibiotics," 196.

20. Jawetz, "Antibiotic Synergism and Antagonism," 309; Jawetz and Gunnison, "Experimental Basis of Combined Antibiotic Action," 695.

21. Hobart A. Reimann, "Infectious Diseases: Sixteenth Annual Review of Significant Publications," *Archives of Internal Medicine* 87 (1951): 128. Max Finland stated at the time, "The pharmaceutical manufacturers are not entirely to blame, for they will concoct any combination of agents that the medical public demands or can be persuaded to buy—and that will pass the relatively minor

requirements of the licensing authorities"; "Antibiotic Combinations [unsigned editorial]," *NEJM* 245 (1951): 989.

22. "Combinations of Drugs," n.d. [likely 1952], Box 5, "William Douglas McAdams," FMIP; punctuation as in the original. Reflecting the competitive crucible out of which such ideas emerged, Lederle's director of sales likewise recommended combining Lederle antibiotic products with other forms of drugs, in the context of Pfizer's "getting a real jump on us in the antibiotic business." See Henry Wendt Jr., "Inter-Office Correspondence, Attention of Dr. W. G. Malcolm," 2/1/51, and J. H. Williams to Henry Wendt Jr., 3/16/51, both in Box 247, FTC Dockets 7211, RG 122.

23. Harry F. Dowling, "Twixt the Cup and the Lip," *JAMA* 165 (1957): 658. While Dowling attempted in the address to buttress his argument by including ointments and powders, he actually *undershot* the mark, as Welch's list included only those "combinations of antibiotics approved since publication of [Welch's] *Manual of Antibiotics* [in 1954]." Welch's *Manual*, for instance, had already demonstrated that sixteen separate companies were marketing combinations of penicillin with dihydrostreptomycin, while twenty-six were marketing combinations of pencillin with "triple sulfonamides." See Henry Welch to Harry Dowling, 11/26/56, Box 3, "W–Z," HDP, and Henry Welch, *The Manual of Antibiotics, 1954–1955* (New York: Medical Encyclopedia, 1954), 33, 55–57.

24. Maxwell Finland to Wesley Spink, 11/15/54, Box 5, ff 8, MFP; Finland to Henry Welch, 11/16/54, Box 5, ff 36, MFP; Finland to Benjamin Carey, 11/24/54, Box 12, ff 42, MFP.

25. Harry Dowling to Maxwell Finland, 11/14/55; Finland to Dowling, 11/18/55; both in Box 2, ff 8, MFP. See also *AP*, Part 23, p. 13155.

26. Maxwell Finland to Henry Welch, 11/10/55, Box 5, ff 37, MFP.

27. Henry Welch to Maxwell Finland, 5/7/56, Box 5, ff 42. In 1955, Welch and Martí-Ibañez had first published and presented *Antibiotic Medicine* as a pragmatic outgrowth of *Antibiotics and Chemotherapy* (which could thereafter be devoted to laboratory issues), portending a "journey towards the luminous shores, already outlined on the present horizon, of a future Medical Science radically transformed by the impact of antibiotics." By 1956, the further transformation in journal title and scope was to "open up many other new windows, enabling the practicing physician to view the full therapeutic panorama of our age." The widened scope of expected pharmaceutical advertising was apparent as well. See Henry Welch and Félix Martí-Ibañez, "Statement of Purposes," *Antibiotic Medicine: Journal of Clinical Studies and Practice of Antibiotic Therapy* 1 (1955): 2, and "On the Scope and Purpose of This Journal," *Antibiotic Medicine and Clinical Therapy* 3 (1956): 24; Félix Martí-Ibañez to H. C. McKenzie [Ortho Pharmaceuticals], 12/6/55; and Martí-Ibañez to E. L. Severinghaus [Hoffman-LaRoche], 4/12/56, both in Box 7, "M," FMIP.

28. Harry Dowling to Maxwell Finland, 5/11/56, Box 5, ff 42, MFP. The journals maintained an approximately 90% article acceptance rate (*AP*, Part 23, p. 12310).

29. Harold Aaron to Harry Dowling, 7/6/56; Harry Dowling to Maxwell Finland, 8/13/56, both in Box 2, ff 8, MFP. For the article, see H. Marvin Daskal, "Clinical Trial of an Antibiotic-Vitamin Combination—Oxytetracycline and Stress Formula Vitamins," *Antibiotic Medicine and Clinical Therapy* 3 (1956): 33–37. Having treated "45 unselected patients suffering from a variety of infections . . . it was the *author's impression* that the cases treated in this clinical study, as compared to other similar cases treated with an antibiotic alone, not only responded more rapidly and had a much shorter convalescent period, but also that the medication was better tolerated and produced fewer and less severe side effects than are usually seen" [italics added, quotation ordering inverted]. For the more public repudiation of the article by Boston City Hospital's Charles Davidson (an esteemed hematologist and close colleague of Max Finland), see Charles S. Davidson, "Antibiotics Plus?" *American Journal of Clinical Nutrition* 4 (1956): 687–88. Maxwell Finland, on his copy of the Mar. 7, 1955, announcement sent to physicians by J. B. Roerig & Company and Pfizer regarding the "striking pharmaceutical and therapeutic advance" of the "Stress Fortified antibiotics," had scribbled "Terra ballyhoo" (Box 20, ff 1, MFP). On the contemporary prevalence of the term *ballyhoo* (signifying the carnival barker's cry), see Nancy Tomes, "The Great American Medicine Show Revisited," *Bulletin of the History of Medicine* 79 (2005): 635–45.

30. Welch, "Opening Remarks," 1.

31. Ibid.

32. Ibid., 2. Ironically, given his mentioning of antibiotic resistance, Welch also drew attention at the very same symposium to a "relatively new" antibiotic, Eli Lilly's vancomycin, on which the first clinical report had actually been delivered at the previous year's symposium. See R. S. Griffith and Franklin B. Peck Jr., "Vancomycin, A New Antibiotic: III. Preliminary Clinical and Laboratory Studies," *Antibiotics Annual* (1955–1956): 619–22, and R. C. Anderson, H. M. Worth, P. N. Harris, and K. K. Chen, "Vancomycin, A New Antibiotic: IV. Pharmacologic and Toxicologic Studies," *Antibiotics Annual* (1956–1957): 75–81. On the limited early use—and wider later utility—of vancomycin, see Donald P. Levine, "Vancomycin: A History," *Clinical Infectious Diseases* 42 suppl 1 (2006): S5–S12, and Kevin Outterson, "The Legal Ecology of Resistance: The Role of Antibiotic Resistance in Pharmaceutical Innovation," *Cardoza Law Review* 31, no. 3 (2010): 613–78.

33. Pfizer mailed to physicians envelopes emblazoned on the outside with the words "The Third Era of Antibiotic Therapy . . . And Sigmamycin." Inside each was a "Dear Doctor" letter entitled "In the New Third Era of Antibiotic Therapy," describing the "Third Era envisioned by Dr. Welch" (Box 5, ff 5, ATS).

34. Henry Welch, "A Rational Approach to Combined Antibiotic Therapy," *Antibiotic Medicine and Clinical Therapy* 3 (1956): 377. From a mechanistic standpoint, Pfizer had first reported Sigmamycin's "synergistic" action in Aug. 1956; see Arthur R. English, Tom J. McBride, G. Van Halsema, and Michael Carlozzi, "Biologic Studies on PA 775, A Combination of Tetracycline and Ole-

andomycin with Synergistic Activity," *Antibiotics and Chemotherapy* 6 (1956): 511–22. For Max Finland's inability to reproduce such studies, see Finland to Lawrence Garrod, 1/22/57, Box 5, ff 45, MFP.

35. Sigmamycin advertisement, in *Antibiotic Medicine and Clinical Therapy* 3 (1956): 363–66. Note that the name Sigmamycin would be changed to Signemycin in 1958, owing to the similarity of its name to an existing product (likely Schering's anti-inflammatory drug Sigmagen). Sigmamycin became Pfizer's "primary product to detail for over a year and a half," generating "60 to 65" pieces of direct mail and twelve different journal advertisements. See "Testimony of John E. McKeen," 12/18/59, pp. 41–42, FTC Dockets 7487, RG 122. For Lederle's attempt to counter Sigmamycin's advent, asking its detailing force to impeach the very rationale of synergistic fixed-dose combination therapy and warning them that "from all the signs we see at the present this new move of Pfizer is not an attempt to push vigorously a new development of scientific therapeutic merit but an attempt to launch a product with high pressure promotion which has only laboratory backing to prove clinical claims," see Austin Joyner, "Confidential to All Lederle Representatives," 10/19/56, Box 15, "Lederle Laboratories, 1952–1959," WSP.

36. Harry Dowling to Maxwell Finland, 11/16/56, Box 2, ff 8, MFP.

37. C. H. Carter and M. C. Maley, "Application of Tetracycline-Oleandomycin and of Oxytetracycline-Oleandomycin in Clinical Practice," *Antibiotics Annual* (1956–1957): 52.

38. Harry F. Dowling, "A New Generation," *Antibiotic Medicine and Clinical Therapy* 3 (1956): 25–26. It could hardly have been lost on Dowling that his article happened to be nearly sandwiched between Welch and Martí-Ibañez's justification for their new journal business model and H. Marvin Daskal's article on vitamin-enhanced antibiotic preparations.

39. Vance Packard, *The Hidden Persuaders* (New York: Pocket, 1957); see also Mark Crispin Miller, introduction to Packard, *Hidden Persuaders*, 50th anniversary ed. (Brooklyn: Ig Publishing, 2007), 15; Daniel Horowitz, *Vance Packard and American Social Criticism* (Chapel Hill: University of North Carolina Press, 1994).

40. Dowling, "Twixt the Cup and the Lip," 659; see also Harry Dowling to Norris Brookens, 10/28/57, Box 4, "Twixt the Cup and the Lip," HDP. For the larger context of pharmaceutical anxiety amid which Dowling prepared his talk, see Tomes, "Great American Medicine Show Revisited," 645–54. For the still larger context of consumerism amid which he prepared the talk, see Lizabeth Cohen, *A Consumer's Republic: The Politics of Mass Consumption in Postwar America* (New York: Knopf, 2003).

41. Dowling, "Twixt the Cup and the Lip," 659. For further anxieties about such a trend's threatening to "drag the whole field of antibiotic therapy far into the mire," see also Maxwell Finland to W. A. Altemeier, 11/2/56, Box 5, ff 42, MFP.

42. Harry Dowling to Maxwell Finland, 11/6/56, Box 2, ff 8, MFP. Concerning their anger with Welch, see Maxwell Finland to A. O. Zink, 10/31/56, Box 5, ff 43, MFP, and Finland to Robertson Pratt, 2/14/57, Box 5, ff 47, MFP.

43. On the AMA and FDA, see Maxwell Finland to Thomas Bradley, 4/16/57, Box 4, ff 3, MFP, and Harry Dowling, statement, *FMA*, Part 3, p. 137. On the ensuing infectious disease expert activity, see Harry Dowling to Maxwell Finland, 11/16/56, Box 2, ff 8, MFP; Finland to Morton Hamburger, 11/17/56; and David E. Rogers to Finland, 11/19/56, both in Box 15, ff 26, MFP. At the two-day Brook Lodge Symposium on the Theory and Use of Antibiotic Combinations, hosted (ironically, ultimately) by Upjohn two weeks after the 1956 annual antibiotics symposium, "in some of the post-conference conversation [from the first night], comment was made that we were returning to the pre-Listerian days in regard to our aseptic techniques, and the age of 'nostrums' in the blind use of antimicrobial agents in the management of undiagnosed disease." See "Brook Lodge Invitational Symposium," p. 162, 71A-5170, Box 5, ff 3, RG 46.

44. "Antibiotic Combinations," *NEJM* 255 (1956): 1057. Surprisingly, this editorial does not appear in Max Finland's collected reprints. For confirmation that it was penned by Finland, see Finland to Harry Dowling, 11/14/56, Box 2, ff 8, MFP.

45. Harry F. Dowling, Maxwell Finland, Morton Hamburger, Ernest Jawetz, Vernon Knight, Mark H. Lepper, Gordon Meikeljohn, Lowell A. Rantz, and Paul S. Rhoads, "The Clinical Use of Antibiotics in Combination," *Archives of Internal Medicine* 99 (1957): 537. The paper was originally written by Morton Hamburger and Vernon Knight before being sent to others for their input. See, e.g., Morton Hamburger to Maxwell Finland, 11/15/56, Box 15, ff 26, MFP.

46. Maxwell Finland, "The New Antibiotic Era: For Better or for Worse?" *Antibiotic Medicine and Clinical Therapy* 4 (1957): 18. The paper, drafted by Finland and sent to colleagues (including Harry Dowling, Ernest Jawetz, Walsh McDermott, Perrin Long, Wesley Spink, Louis Weinstein, and W. Barry Wood Jr.) for their input, was originally to be published as a jointly authored manuscript. After resistance from Martí-Ibañez, however, the remaining 18 "authors" were reduced to a footnote on the first page of the text. See Maxwell Finland to Henry Welch, 11/23/56, and Félix Martí-Ibañez to Finland, 11/30/56, both in Box 15, ff 26, MFP.

47. Finland, "New Antibiotic Era," 18.

48. See, e.g., Irvine Loudon, "'The Vile Race of Quacks with Which This Country Is Infested,'" in W. F. Bynum and Roy Porter, eds., *Medical Fringe and Medical Orthodoxy, 1750–1850* (London: Croom Helm, 1987), 106–28, and Roy Porter, *Health for Sale: Quackery in England, 1660–1850* (New York: Manchester University Press, 1989), 21–59.

49. "Medical Testimonials," *Lancet* 33 (1839): 344; see also "Corn-Cutting.—The Testimonial System," *Lancet* 48 (1846): 136.

50. "Patronage of Quacks," *Medico-Chirurgical Review* 50 (1836): 576;

American Medical Association, *Code of Ethics of the American Medical Association, Adopted May, 1847* (Philadelphia: T. K. & P. G. Collins, 1848), 12.

51. See, e.g., James Harvey Young, *The Toadstool Millionaires: A Social History of Patent Medicines in America before Federal Regulation* (Princeton: Princeton University Press, 1961), 165–89; William H. Helfand, *Quack, Quack, Quack: The Sellers of Nostrums in Prints, Posters, Ephemera, and Books* (New York: Grolier Club, 2002); and Eric W. Boyle, *Quack Medicine: A History of Combating Health Fraud in Twentieth-Century America* (Santa Barbara, CA: Praeger, 2013), 1–16.

52. "Endorsed by Physicians," *JAMA* 34 (1900): 304.

53. The clergy were often portrayed in medical journals as the most culpable of testimonial providers. See, e.g., "Habitual Endorsers of Nostrums," *JAMA* 17 (1891): 979; Henry Bixby Hemenway, "Modern Homeopathy and Medical Science," *JAMA* 22 (1894): 376; "The Clergy and the Medical Profession; The Medical Code a Tonic," *JAMA* 22 (1894): 996; and John Madden, "The Mendacity and Filth of Quack Advertising," *JAMA* 28 (1897): 404.

54. See, e.g., Oliver Wendell Holmes, "Currents and Counter-Currents in Medical Science," in *Currents and Counter-Currents in Medical Science, with Other Addresses and Essays* (Boston: Ticknor and Fields, 1861), 15–16, and "Homœopathy and Its Kindred Delusions," in *The Works of Oliver Wendell Holmes, vol. 9: Medical Essays, 1842–1882* (Boston: Houghton, Mifflin, 1892), 37, 75–76.

55. Holmes, "Currents and Counter-Currents in Medical Science," 17–18.

56. William Osler, *The Principles and Practice of Medicine: Designed for the Use of Practitioners and Students of Medicine* (New York: D. Appleton, 1892). On the line from Holmes to Osler, see Charles S. Bryan, "'The Greatest Brahmin': Overview of a Life," in Scott H. Podolsky and Charles S. Bryan, eds., *Oliver Wendell Holmes: Physician and Man of Letters* (Sagamore Beach, MA: Science History Publications, 2009), 3–21. On the line from Osler to the infectious disease–based therapeutic rationalists of the 1950s, see James C. Whorton, "'Antibiotic Abandon': The Resurgence of Therapeutic Rationalism," in John Parascandola, ed., *The History of Antibiotics: A Symposium* (Madison: American Institute of the History of Pharmacy, 1980), 125–36.

57. O. B. Campbell, "Modern Therapeutics," *JAMA* 18 (1892): 458.

58. Ibid., 459.

59. Ibid., 458; see also "The Evolution of a Testimonial," *JAMA* 33 (1899): 553.

60. See, e.g., the back-and-forth in John N. Upshur, "The Treatment of Typhoid Fever," *JAMA* 29 (1897): 20–23; John Eliot Woodbridge, "The Treatment of Typhoid Fever," *JAMA* 29 (1897): 138–39; and J. N. Upshur, "Dr. Upshur's Final Reply to Dr. Woodbridge," *JAMA* 29 (1897): 547–48.

61. Henry P. Loomis, "Physicians and Proprietary Remedies," *JAMA* 45 (1905): 1782.

62. See the collection of relevant articles in Samuel Hopkins Adams, *The Great American Fraud: Articles on the Nostrum Evil and Quacks, in Two Series,*

Reprinted from Collier's Weekly, 4th ed. (Chicago: American Medical Association, 1907).

63. See, e.g., Arthur J. Cramp [director, Propaganda Department] to Carl Hunt, 7/31/19; Cramp to Professor Puckner, 7/10/19; Cramp to Mr. Gardiner, "In re: Testimonials," 8/8/28; Cramp to the editor, *Printers' Ink*, 10/8/29; Cramp to Pearl Kendall Hess, "In re: Sargon Testimonials," 5/29/31, all in Historical Health Fraud and Alternative Medicine Collection, "Advertising, 1887–1981," ff 27-8, AMAA; "Official Minutes, House of Delegates," *JAMA* 45 (1905): 264–65; Morris Fishbein, *A History of the American Medical Association, 1847 to 1947* (Philadelphia: W. B. Saunders, 1947), 168–69, 198–99; and James G. Burrow, *AMA: Voice of American Medicine* (Baltimore: Johns Hopkins Press, 1963), 74–75, 107–31.

64. James Harvey Young, *The Medical Messiahs: A Social History of Health Quackery in Twentieth-Century America* (Princeton: Princeton University Press, 1967), 129–32; Boyle, *Quack Medicine*, 61–89; Bliss O. Halling, "Bureau of Investigation," in Fishbein, *History of the American Medical Association*, 1034–38; "Professional Testimonial Writers: Their Pedigrees and Accomplishments," Historical Health Fraud and Alternative Medicine Collection, "Advertising, 1887–1981," ff 27-10, AMA.

65. Marks, *Progress of Experiment*, 17–41; Boyle, *Quack Medicine*, 17–38.

66. See, e.g., Arthur J. Cramp, "The Part Played by the Testimonial," *Hygeia* 1 (1923): 170–74; "The Propaganda for Reform: Whole Grain Wheat—Quackery in the Food Field," *JAMA* 84 (1925): 1441–43; and "Council on Pharmacy and Chemistry: Reports of the Council—Diampysal Not Acceptable," *JAMA* 101 (1933): 1482.

67. Harold S. Diehl, "Medicinal Treatment of the Common Cold," *JAMA* 101 (1933): 2047.

68. W. D. Sutliff, "Adequate Tests of Curative Therapy in Man," *Annals of Internal Medicine* 10 (1936): 90.

69. "Test by Testimonial," *Science* 123 (1956): 1059.

70. See especially his unsigned editorial, "The Play of the Market Place," *NEJM* 255 (1956): 528–29. Reflecting contemporary stigma attached to the "testimonial," the editorial pages of *NEJM* likewise used the term to refer to the claims of the tobacco industry and chiropractors, as well as to premature scientific and therapeutic claims. See "The Press and the Profession," *NEJM* 245 (1951): 155–56; "A Scientific Perspective," *NEJM* 250 (1954): 923–24; "Double Standard of Medicine," *NEJM* 252 (1955): 286–87; "Smoking and Health," *NEJM* 253 (1955): 480–81; and "Tobacco and Health—Volume 1, Number 1," *NEJM* 258 (1958): 99–100.

71. Comically, yet tellingly, after *JAMA*'s copy supervisor, Hazel Eggert, reported to William Douglas McAdams's Deforest Ely how Pfizer's Bonamine (meclizine, prescribed by the AMA's Robert Stormant) had helped her on a fishing trip (which she went on along with "some of the pharmaceutical men and Dr. Stormant"), Ely

responded, "We are planning a testimonial campaign on Bonamine; may we use yours?" See Hazel [Eggert] to Deforest Ely, 12/10/54; Ely to Eggert, 12/12/54, both in 71A-5170, Box 10, "Extreme Claims e. General," RG 46. The FDA actually had been noting "testimonials" for decades; see James Harvey Young, "Laetrile in Historical Perspective," in *American Health Quackery: Collected Essays* (Princeton: Princeton University Press, 1992), 240; Carpenter, *Reputation and Power*, 175–76; and Boyle, *Quack Medicine*, 127–34.

72. Maxwell Finland to Harry Dowling, 11/24/56, Box 2, ff 9, MFP. As Finland remarked of Martí-Ibañez to Selman Waksman at the same time, "His journals are being used as 'trade journals' and with as much discretion as is used by manufacturers but not by scientists"; Maxwell Finland to Selman Waksman, 11/14/56, Box 5, ff 43, MFP.

73. Finland, "New Antibiotic Era," 20.

74. The manuscript was originally to be a monologue, to which an upset Henry Welch felt compelled to write a "rebuttal." See Maxwell Finland to Walsh McDermott, 12/7/56, Box 15, ff 26, MFP.

75. Henry Welch, "Further Comments on Combined Therapy," *Antibiotic Medicine and Clinical Therapy* 4 (1957): 23, 21 [italics in original]. For Dowling and Finland's bristling at such an assignment of "blame," see Harry Dowling to Maxwell Finland, 12/10/56, Box 15, ff 26, MFP; Maxwell Finland to William P. Boger, 12/19/56, Box 5, ff 42, MFP.

76. Welch, "Further Comments on Combined Therapy," 23. Wrote Welch to Finland privately, "No matter what we say about combinations, I am convinced that they are coming and that doctors are going to use them, and it seems to me that the most helpful thing we could do is to try to establish the proper ones for those sections of the spectrum where we are going to need them"; Henry Welch to Maxwell Finland, 12/6/56, Box 15, ff 26, MFP.

77. Félix Martí-Ibañez, paragraph attached to Finland, "New Antibiotic Era," 17.

78. Maxwell Finland to Selman Waksman, 11/14/56, Box 5, ff 43, MFP. As he did with Welch, however, Finland maintained a complicated relationship in his correspondence with Martí-Ibañez throughout the era. See, e.g., the expressed good wishes in Félix Martí-Ibañez to Maxwell Finland, 11/21/55, Box 7, "N," FMIP; Finland to Joan Plaut, 3/9/57, Box 5, ff 47, MFP; and Finland to Martí-Ibañez, 10/27/58, Box 9, "F," FMIP. Atop this last letter, in which Finland expressed his admiration for Martí-Ibañez's facility for writing and for his productivity, Martí-Ibañez had scrawled, "The Mighty become human!" Attached to it was paper-clipped the note, "Material for our own use—coveted."

79. Summarized in Podolsky, *Pneumonia before Antibiotics*, 35–36.

80. Marcia L. Meldrum, "Departures from the Design: The Randomized Clinical Trial in Historical Context, 1946–1970" (PhD diss., State University of New York Stony Brook, 1994); J. Rosser Matthews, *Quantification and the Quest for*

Medical Certainty (Princeton: Princeton University Press, 1995); Marks, *Progress of Experiment*; Peter Keating and Alberto Cambrosio, *Cancer on Trial: Oncology as a New Style of Practice* (Chicago: University of Chicago Press, 2012); Suzanne White Junod, "FDA and Clinical Drug Trials: A Short History," in Madhu Davies and Faiz Kerimani, eds., *A Quick Guide to Clinical Trials* (Washington: Bioplan, Inc., 2008), 25–55; Carpenter, *Reputation and Power*.

81. Harry M. Marks, "Trust and Mistrust in the Marketplace: Statistics and Clinical Research, 1945–1960," *History of Science* 38 (2000): 344.

82. Iain Chalmers, Estela Dukan, Scott H. Podolsky, and George Davey Smith, "The Advent of Fair Treatment Allocation Schedules in Clinical Trials during the 19th and Early 20th Centuries," *Journal of the Royal Society of Medicine* 105 (2012): 221–27 (see also at www.jameslindlibrary.org).

83. Maxwell Finland, "The Serum Treatment of Lobar Pneumonia," *NEJM* 202 (1930): 1244–47; Podolsky, *Pneumonia before Antibiotics*, 21.

84. See Norman Plummer, James Liebman, Saul Solomon, W. H. Kammerer, Mennasch Kalkstein, and Herbert K. Ensworth, "Chemotherapy versus Combined Chemotherapy and Serum in the Treatment of Pneumonia: A Study of 607 Alternated Cases," *JAMA* 116 (1941): 2366–71; Maxwell Finland, "Controlling Clinical Therapeutic Experiments with Specific Serums: With Particular Reference to Antipneumococcus Serums," *NEJM* 225 (1941): 495–506; and Podolsky, *Pneumonia before Antibiotics*, 119–24.

85. Maxwell Finland to Osler Peterson, 10/8/43, Box 4, ff 22, MFP; Finland to Elias Strauss, 3/27/45, Box 5, ff 19, MFP.

86. Maxwell Finland, Elias Strauss, and Osler L. Peterson, "Sulfadiazine: Therapeutic Evaluation and Toxic Effects on Four Hundred and Forty-six Patients," *JAMA* 116 (1941): 2641–47. Finland's resistance was admittedly a bit idiosyncratic, despite his prominence. For the enthusiasm of his colleagues for conducting controlled studies of the sulfa drugs at the time, see Charles H. Rammelkamp to Finland, 12/26/43, Box 4, ff 39, MFP.

87. Maxwell Finland to John Dingle, 1/28/53, Box 2, ff 6, MFP.

88. John Dingle to Maxwell Finland, 2/3/53; Finland to Dingle, 2/16/53, both in Box 2, ff 6, MFP.

89. Monroe D. Eaton, "Chemotherapy of Virus and Rickettsial Infections," *Annual Review of Microbiology* 4 (1950): 223–46; Joseph E. Smadel, "Present Status of Antibiotic Therapy in Viral and Rickettsial Disease," *Bulletin of the New York Academy of Medicine* 27 (1951): 221–31.

90. Gordon Meikeljohn and Robert I. Shragg, "Aureomycin in Primary Atypical Pneumonia: A Controlled Evaluation," *JAMA* 140 (1949): 391.

91. Emanuel Schoenbach to Maxwell Finland, 10/4/48, Box 4, ff 66, MFP.

92. Emanuel Schoenbach to Maxwell Finland, 11/10/48; Schoenbach to Finland, 12/14/48, both in Box 4, ff 66, MFP; Schoenbach to Finland, 12/18/48, Box 5, ff 30, MFP; Schoenbach to Finland, 1/24/49, Box 5, ff 31, MFP. It does not

appear that Dingle conducted the proposed study; see John H. Dingle, "Primary Atypical Pneumonia," in H. Stanley Banks, ed., *Modern Practice in Infectious Fevers* (New York: Paul B. Hoeber, 1951), 2:626–45.

93. Maxwell Finland to Elias Strauss, 12/16/48, Box 4, ff 66, MFP.

94. In reporting in 1952 on the potential treatment of herpes zoster with broad-spectrum antibiotics, Finland and his colleagues explained that given the eventless recovery entailed in the large majority of cases, "the evaluation of a beneficial agent in herpes zoster requires observation of a sufficiently large number of treated and untreated patients to permit statistical comparison of both groups." See Edward H. Kass, Robert R. Aycock, and Maxwell Finland, "Clinical Evaluation of Aureomycin and Chloramphenicol in Herpes Zoster," *NEJM* 246 (1952): 171. They did not find that either antibiotic helped in the treatment of zoster. By 1959, Finland appears to have become comfortable with the prospect of conducting a placebo-controlled study of the treatment of atypical pneumonia with tetracycline, requesting from Lederle "a placebo that looks the same as the Declomycin but has only a bland filler—perhaps lactose or calcium phosphate or whatever. . . . If the capsule has something that tastes bitter or perhaps has a little aspirin in it, that might be helpful in making it somewhat more difficult to discern"; Finland to Stanton M. Hardy, 10/7/59, Box 12, ff 39, MFP. It does not appear that Finland conducted the study.

95. Marks, *Progress of Experiment*, 236.

96. Maxwell Finland, "Antimicrobial Treatment for Viral and Related Infections: I. Antibiotic Treatment of Primary Atypical Pneumonia," *NEJM* 247 (1952): 318.

97. Ibid., 324.

98. Ibid., 319. For the study itself, see John C. Harvey, George S. Mirick, and Isabelle G. Schaub, "Clinical Experience with Aureomycin," *Journal of Clinical Investigation* 28 (1949): 987–91.

99. John H. Dingle to Arthur M. Walker, 10/28/52, Box 5, ff 34, MFP; see also Maxwell Finland, "Pneumonia: Present Status of Diagnosis and Treatment," *Veterans Affairs Technical Bulletin, TB 10–84* (Nov. 30, 1952): 1–16, found in Maxwell Finland, *Collected Reprints, 1949–1952*, Francis A. Countway Medical Library.

100. Maxwell Finland to Arthur M. Walker, 11/10/52, Box 5, ff 34, MFP.

101. Maxwell Finland to D. H. Garrow, 2/27/57, Box 5, ff 45, MFP. Beyond his epistemological concerns, Finland was equally concerned with the ongoing logistical roadblocks to conducting trials (especially comparative effectiveness ones); see Maxwell Finland to Charles D. May, 6/26/58, Box 5, ff 40, MFP.

102. "Antihistamine Cold Cures and Modern Sales Methods," *NEJM* 242 (1950): 106; see also "Present Status of the Antihistaminic Drugs in the Common Cold," *NEJM* 242 (1950): 596–97; "Lack of Effect of Antihistaminic Drugs on Common Cold," *NEJM* 242 (1950): 765; and "Guilty or Gullible," *NEJM* 244 (1951): 156–57.

103. "Antibiotic Combinations" [1956], 1059.

104. Dowling et al., "Clinical Use of Antibiotics in Combination," 538.

105. Harry F. Dowling, "Mixtures of Antibiotics," *JAMA* 164 (1957): 45; Henry D. Brainerd, Ernest Jawetz, and Lowell A. Rantz, "Fixed Antibiotic Combinations," *California Medicine* 86 (1957): 57.

106. Maxwell Finland to Lawrence Garrod, 1/22/57, Box 5, ff 45, MFP.

107. "Oleandomycin-Tetracycline Mixture," *NEJM* 257 (1957): 289.

108. Ibid. With still more invective, Finland would write the following month: "The encouragement of the so-called 'new era of antibiotic combinations' and the manner in which it is being exploited represent a major backward step in the management of infections. . . . It is ardently hoped that this trend will be reversed as the medical profession becomes better acquainted with the very weak and treacherous foundation upon which it rests"; "Erythromycin, Oleandomycin and Spiramycin—and their Combinations with Tetracycline," *NEJM* 257 (1957): 526. For confirmation that such unsigned editorials were written by Finland, see William M. Kirby to Finland, 9/25/57, Box 5, ff 39, MFP, as well as the fact that both editorials are among Finland's bound collected reprints at the Francis A. Countway Medical Library.

109. "Oleandomycin-Tetracycline Mixture," 290. Finland concluded by countering: "This type of medical practice has gradually been giving way to the modern scientific medicine that was ushered in with the changes in medical education taking place during the last half century. Therapeutics should not return to the ancient, barbaric, and irrational era of the shotgun." Again, this transposition of former concerns regarding proprietary remedies onto the products (and marketing practices) of the ethical pharmaceutical industry has been well described in Tomes, "Great American Medicine Show Revisited."

110. "Oleandomycin-Tetracycline Mixture," 290.

111. See, e.g., Austin Bradford Hill, "The Clinical Trial," *NEJM* 247 (1952): 113–19, and "Assessment of Therapeutic Trials," *Transactions of the Medical Society of London* 68 (1953): 128–36; Donald Mainland, "The Modern Method of Clinical Trial" *Methods in Medical Research* 6 (1954): 152–59; and Marks, *Progress of Experiment*, 136–63.

112. Dominique A. Tobbell, *Pills, Power, and Policy: The Struggle for Drug Reform in Cold War America and its Consequences* (Berkeley: University of California Press, 2012), 54–57; Carpenter, *Reputation and Power*, 302–45.

113. See, e.g., Louis Lasagna, Frederick Mosteller, John M. von Felsinger, and Henry K. Beecher, "A Study of the Placebo Response," *American Journal of Medicine* 16 (1954): 770–79.

114. Louis Lasagna, "The Controlled Clinical Trial: Theory and Practice," *Journal of Chronic Diseases* 1 (1955): 353.

115. Louis Lasagna, "Statistics, Sophistication, Sophistry, and Sacred Cows," *Clinical Research Proceedings* 3 (1955): 185. Such ambivalence regarding industry would foreshadow Lasagna's later support of industry as being overregulated

by the FDA. And such a critique of those relying on "Clinical Judgment" would seem quite ironic to FDA staff in the wake of Lasagna's later defense of such judgment vis à vis the controlled clinical trial from the late 1960s onward. These turns are discussed further in chapters 3 and 4.

116. See, e.g., Wilfred F. Jones and Maxwell Finland, "Antibiotic Combinations: Antistreptococcal and Antistaphylococcal Activity of Plasma of Normal Subjects after Ingestion of Erythromycin or Penicillin or Both," *NEJM* 255 (1956): 1019–24; "Antistreptococcal and Antistaphylococcal Activity of Plasma of Normal Subjects after Oral Doses of Penicillin, Oleandomycin, and a Combination of These Antibiotics," *NEJM* 256 (1957): 115–19; "Antibacterial Action of Plasma of Normal Subjects after Oral Doses of Penicillin V, Tetracycline Hydrochloride, and a Combination of Both," *NEJM* 256 (1957): 869–75; and "Tetracycline, Erythromycin, Oleandomycin, and Spiramycin, and Combinations of Tetracycline with Each of the Other Three Agents—Comparisons of Activity in Vitro and Antibacterial Action of Blood after Oral Administration," *NEJM* 257 (1957): 481–91, 536–47.

117. As would be outlined in a 1962 Festschrift in honor of his sixtieth birthday, an analysis of Finland's then-published 500 articles (in addition to 370 unsigned editorials) revealed them to have been co-written with 53 separate research fellows working under him over the preceding three decades, along with multiple colleagues from both Harvard and across the country. In frontispiece, *Archives of Internal Medicine* 110 (1962): 558.

118. See, e.g., Finland to John Ehrlich [Parke-Davis], 10/15/52; Finland to H. E. Carnes [Parke-Davis], 10/24/55, both in Box 4, ff 20, MFP; Gladys Hobby [Pfizer] to Finland, 9/27/55; Finland to Hobby, 7/10/56, both in Box 4, ff 24, MFP; entire "Terramycin Correspondence" folder, Box 20, ff 1, MFP; Finland to Benjamin Carey [Lederle], 6/12/59, Box 3, ff 34, MFP. For "stamp of approval," see Robert G. Petersdorff, "Sulfadiazine," *JAMA* 251 (1984): 1476.

119. Warwick Anderson, *The Collectors of Lost Souls: Turning Kuru Scientists into Whitemen* (Baltimore: Johns Hopkins University Press, 2008), 91–115, 133–60.

120. "Personal Note of WWS," n.d., Box 15, "Personal Note of WWS," WSP. Spink's claim may initially appear off the mark, in view of the postwar surge in federal support for medical research, but it would buttress Dominique Tobbell's claim that while support for medical research in general may have increased after the war, support for clinical pharmacology lagged behind (*Pills, Power, and Policy*, 37–58).

121. For an overview of the exchange between Spink and Pfizer (especially with Gladys Hobby), see Gladys L. Hobby to Wesley W. Spink, 11/29/49; Spink to Hobby, 12/5/49; Spink to Hobby, 3/10/52; Spink to John McKeen, 11/30/53; Hobby to Spink, all in Box 15, "Pfizer, 1949–1957," WSP; Hobby to Spink, 2/3/55; Spink to Hobby, 2/8/55 (in which Spink balks at being asked to study stress-fortified antibiotics); Hobby to Spink, 8/21/57; and Spink to Hobby, 8/26/57, all

in Box 15, "Pfizer, 1955–1965," WSP. For an overview of the exchange between Spink and Lederle, see Spink to Benjamin Carey, 4/15/50; Spink to Carey, 3/19/51; Carey to Spink, 3/22/51, all in Box 15, "Lederle Laboratories, 1950–1952," WSP; Spink to Carey, 11/17/52, Box 4, "1952," WSP; and W. G. Malcolm to Spink, 9/30/57, Box 15, "Lederle Laboratories, 1956–1958," WSP.

122. Regarding the direct advising of his trainees in this respect, and given that in subsequent years former Max Finland fellow and eventual IDSA president Calvin Kunin would take the mantle as the most vocal and influential infectious disease–based moralist and critic of conflict of interest, see especially Finland to Calvin Kunin, 8/10/59; Finland to Kunin 9/14/59; Kunin to Finland, 9/16/59, all in Box 3, ff 34, MFP; and Finland to Kunin, 9/21/59, Box 3, ff 35, MFP.

123. Félix Martí-Ibañez, "The Great Historical Challenges in Medicine," *Antibiotics Annual* (1959–1960): 4; Austin Smith, "The Challenge of New Drugs to the Pharmaceutical Industry," *Antibiotics Annual* (1959–1960): 16.

124. Maxwell Finland, "The Challenge of New Drugs to the Clinical Investigator," *Antibiotics Annual* (1959–1960): 24.

125. Ibid., 26.

126. For interest in Finland's address and requests from members of both government agencies and industry for copies of the address, see Paul L. Day [HEW] to Finland, 11/4/59; Austin Smith [PMA] to Finland, 11/13/59; Marie Walbrecht [Lederle] to Finland, 11/5/59; G. F. Roll [Smith Kline French] to Finland, 11/13/59; Eugene N. Beesley [Eli Lilly] to Finland, 11/23/59; Andrew J. Moriarity [Upjohn] to Finland, 11/30/59; and William W. Ferguson [Michigan Department of Health] to Finland, 1/18/60; all in Box 3, ff 35, MFP.

127. See George P. Larrick, "Welcome from the Office of the Commissioner," attachment #1 to the first meeting of the FDA Medical Advisory Board, 3/2/65, Box 11, "FDA Medical Advisory Board," WSP; Sheila Jasanoff, *The Fifth Branch: Science Advisers as Policymakers* (Cambridge, MA: Harvard University Press, 1990), 152–55; Carpenter, *Reputation and Power*, 302–32. In several ways, Finland anticipated the FDA's own use of the National Academy of Sciences–National Research Council to implement the Drug Efficacy Study and Implementation (DESI) process nearly a decade later (see chapter 3).

128. Maxwell Finland to Thomas Bradley, 4/16/57, Box 4, ff 3, MFP; see also Maxwell Finland to W. A. Altemeier, 11/2/56, Box 5, ff 42, MFP.

129. As far back as 1951, Finland had remarked of Pfizer to Lederle's Wilbur Malcolm, "I have long since thought that both their original entrance into the direct sale to the medical profession and the methods they have employed since then have been a little out of line and not becoming a major pharmaceutical house"; Finland to Malcolm, 12/4/51, Box 12, ff 44, MFP. For Malcolm's denigration of Pfizer and fomenting of such irritation at the time (when Aureomycin, of course, was in direct competition with Terramycin), see Malcolm to Finland, 11/29/51, Box 12, ff 44, MFP. Several caveats must be kept in mind. First, as detailed in chapter 1, Lederle was nearly as aggressive as Pfizer in its marketing, with Fin-

land's radically different perception of the two companies likely colored to a large extent by his own relationships with them. Indeed, for all his moralizing, Finland felt comfortable sending Malcolm samples of Pfizer's competing Terramycin, in Feb. 1950, when it was still in the testing phase (Finland to Malcolm, 2/1/50, Box 12, ff 44, MFP). Second, despite such a dichotomization, Finland maintained a fairly healthy working relationships with Pfizer for much of the 1950s (and especially with renowned microbiologist Gladys Hobby, who worked in Pfizer's research division from 1943 until Feb. 1959), even to the point of meeting with its president, John McKeen, to air his concerns in early 1959. See, e.g., Gladys Hobby to Finland, 4/28/54; Hobby to Finland, 6/18/54; John E. McKeen to Finland, 1/14/59; McKeen to Finland, 1/22/59; and Hobby to Finland, 2/4/59; all in Box 4, ff 24, MFP.

130. "LCD" to H.W. Wendt Jr., n.d. [likely mid-1950s], Box 250, FTC Dockets 7211, RG 122. As Duncan wrote to Max Finland in the fall of 1957, regarding American Cyanamid president Wilber Malcolm's chairing of an American Drug Manufacturers Association (forerunner to the Pharmaceutical Manufacturers Association) committee on ethical practices in drug promotion, "We are heartily in favor of a sound code of ethical practices, particularly, of course, for the other fellow to observe!" (Duncan to Finland, 9/26/57, Box 12, ff 39, MFP). At the same time, in the detailing skirmishes, taking an explicitly low-key, low-pressure approach was yet another (though seemingly less successful) tactic; see William Alderisio [Squibb], "Tetracycline 'deals,' " 12/17/54, and A. J. Longpre [Squibb] to Clyde E. Place, 4/14/55, both in Box 254, FTC Dockets 7211, RG 122.

131. Regarding such new investigators, see Maxwell Finland to Emanuel Schoenbach, 5/19/50, Box 4, ff 66, MFP. On Finland himself being offered seemingly excessive, and at times unsolicited, quantities of Terramycin for whatever research he saw fit, see, e.g., his handwritten plea, "Can't we stop this?" on a letter accompanying the shipment to him of still more Terrramycin, in Gladys L. Hobby to Finland, 8/16/51, Box 20, ff 1, MFP. See also Michael Corlozzi to Finland, 5/20/53, in Box 20, ff 1, MFP. Regarding Pfizer's distribution of Terramycin to clinics and hospitals for testing (the actual number was closer to 100 than Finland's hyperbolized figure), see Arthur A. Daemmrich, *Pharmacopolitics: Drug Regulation in the United States and Germany* (Chapel Hill: University of North Carolina Press, 2005), 54–58. Regarding the Terramycin monograph, see Merle McNeil Musselman, *Terramycin* (New York: Medical Encyclopedia, 1956); Harry Dowling to Maxwell Finland, 11/16/56; Finland to Dowling, 11/24/56; both in Box 2, ff 9, MFP; and Maxwell Finland to Henry Welch, 11/21/56, box 5, ff 43, MFP.

132. See, especially, Finland's exchanges with his British counterparts: Maxwell Finland to Lawrence Garrod, 1/22/57, Box 5, ff 45, MFP; Finland to Garrod, 7/17/57; Finland to Hugh Clegg, 7/17/57; and Garrod to Finland, 7/25/57, all in Box 5, ff 39, MFP.

133. Maxwell Finland to Wilbur Malcolm, 10/2/57, Box 12, ff 44, MFP. Fin-

land also had expressed such concerns to Gladys Hobby of Pfizer, well before his disenchantment with the fixed-dose combination antibiotics and their marketing; see Finland to Hobby, 7/24/53, Box 4, ff 24, MFP.

134. Maxwell Finland to Wilbur Malcolm, 10/2/57, Box 12, ff 44, MFP. See also Finland, "Challenge of New Drugs to the Pharmaceutical Industry," 27.

135. Lyman C. Duncan to Maxwell Finland, 6/3/58, Box 12, ff 39, MFP; *AP*, Part 24, pp. 13932–33, 13939–40.

136. Maxwell Finland to Charles H. Mann, 6/3/58, Box 5, ff 40, MFP.

137. Tobbell, *Pills, Power, and Policy*, 69–76, 81–84.

138. Federal Trade Commission, *Economic Report on Antibiotics Manufacture* (Washington, DC: U.S. Government Printing Office, 1958), 1, 30–33; Robert Bud, "Antibiotics, Big Business, and Consumers: The Context of Government Investigations into the Postwar American Drug Industry," *Technology and Culture* 46 (2005): 329–49.

139. Blatnik's four hearings would respectively focus on filter-tip cigarettes, weight-reducing preparations (with discussions of proprietary remedies again preceding those of the ethical pharmaceutical industry), prescription tranquilizing drugs, and dentifrices.

140. *FMA*, Part 1, p. 1.

141. *FMA*, Part 3, pp. 132, 135.

142. Middleton Kiefer, *Pax* (New York: Random House, 1958). The football-themed sales pitch to the Raven detail men (pp. 250–52) vividly echoed the portrayal of John McKeen and his baseball-themed pitch to the Pfizer sales staff in "Pfizer Put an Old Name on a New Drug Label," *Business Week* (Oct. 13, 1951): 131. Middleton Kiefer was actually the pseudonym of Harry Middleton and Warren Kiefer, who had worked in public relations for Pfizer during the Sigmamycin campaign; see *AP*, Part 22, pp. 11993–12003.

143. John P. Swann, "Sure Cure: Public Policy on Drug Efficacy before 1962," in Gregory J. Higby and Elaine C. Stroud, eds., *The Inside Story of Medicines: A Symposium* (Madison, WI: American Institute of the History of Pharmacy, 1997), 223–61; Carpenter, *Reputation and Power*, 118–227.

144. *FMA*, Part 3, p. 158. A year prior, while Harry Dowling was preparing his "Twixt the Cup and the Lip" talk, Holland had communicated to him: "As you know, the term 'safety' is, in a medical sense, a comparative one and as a matter of administrative operation our New Drug Branch can and does comment more or less strongly on claims of efficacy, although often we are not in a position to institute regulatory action in court" (Albert Holland to Harry F. Dowling, 4/1/57, Box 2, "D–E–F," HDP).

145. *FMA*, Part 3, p. 158.

146. Louis Lasagna, "Across-the-Counter Hypnotics: Boon, Hazard, or Fraud," *Journal of Chronic Diseases* 4 (1956): 554.

147. Louis Lasagna, "Gripesmanship: A Positive Approach," *Journal of Chronic Diseases* 10 (1959): 464; see also Box 1, ff 13, LLP.

148. Ibid.

149. "New Investigation of Pharmaceutical Promotion Will Be Added to Govt. Pricing and Patent Probes: Result of Exposé Articles," *F-D-C Reports* (Feb. 9, 1959): 12.

150. John Lear, "Taking the Miracle out of the Miracle Drugs," *Saturday Review* 42 (Jan. 3, 1959): 35–41. Such emphasis on Sigmamycin's broad "spectrum" of utility of course paralleled the approach applied to Terramycin earlier in the decade.

151. Pfizer president John McKeen, in privately justifying the advertisement, simultaneously validated and critiqued the use of testimonials:

> In using physicians' professional cards as illustrations the [advertising] agency advises me they took care that no physician with the names used did in fact practice in the cities mentioned. On the other hand, the agency did know that physicians in these and many other cities were prescribing Sigmamycin with excellent results. . . . Traditionally, we, as well as the entire pharmaceutical industry and practicing physicians, turn to scientific papers . . . as the authority for new drugs rather than endorsements by individual physicians, which would have no weight. . . . [The advertisement] was a symbolic way of indicating that the antibiotic was in widespread use by doctors in various medical specialties. (John E. McKeen to Aaron G. Benesch [managing editor, *St. Louis Globe-Democrat*], 2/3/59, Box 5, ff 2, ATS)

McKeen's former assistant would testify a year later during a Federal Trade Commission investigation of the advertisement, "Medical ethics absolutely prohibits a physician from giving a testimonial or anything that even looks like one"; "Testimony of Ward John Haas," 11/25/59, p. 14, FTC Docket 7487, RG 122. The FTC, in May 1959, issued a formal complaint against Pfizer regarding the advertisement; the complaint was formally dismissed on May 23, 1960, just as concerns regarding the advertising of Sigmamycin were being discussed at length as part of the Kefauver hearings (see chapter 3).

152. Richard Harris, *The Real Voice* (New York: Macmillan, 1964), 18–19, and "The Real Voice," *New Yorker* (Mar. 14, 1964): 64.

153. See, e.g., Finland to Harry J. Loynd, 3/10/59, Box 4, ff 20, MFP; Finland to Charles D. May, 2/20/61, Box 15, ff 8, MFP; and Finland to Robert P. Parker, 5/8/65, Box 12, ff 39, MFP. For evidence that Finland knew Lear and supplied him with information for at least one later article, see Maxwell Finland to John Dingle, 10/3/60, Box 2, ff 6, MFP; John Lear to Finland, 4/27/62; and Finland to Lear, 5/7/62, both in Box 6, ff 7, MFP. That Finland was capable of suggesting such strategies to others is suggested by his later comments regarding Martí-Ibañez to Charles May, then-editor of *Pediatrics*: "You would do the profession a service by making a bit of a survey of his empire, how he runs it, what it is doing, and what influence it might have on industry and on the profession. You could get

food for more than one 'political' type editorial" (Finland to May, 7/14/58, Box 5, ff 40, MFP). Eventually, no love would be lost between Finland and Welch in particular. For suggestive evidence that Henry Welch suspected Finland of having provided Lear with information for later articles, see Welch to Félix Martí-Ibañez, 11/15/60, Box 95, "Henry Welch to 1966," FMIP. And as Finland would describe to a colleague the following decade, after Welch's affairs had become a national scandal and catalyzed a series of FDA changes (described in chapter 3): "It is only fair to say that 'they who have sown the wind must now reap the whirlwind.' By they, I mean the FDA as well as the responsible drug manufacturers. All of them followed a bad leader and must suffer the consequences"; Maxwell Finland to William L. Hewitt, 6/5/68, Series 2, DES Panels, "Anti-Infectives (II): General: 1966–1968, NASD. Having stated all of this, I have not found a smoking gun either in Finland's archival collection or elsewhere.

154. John Lear, "The Certification of Antibiotics," *Saturday Review* 42 (Feb. 7, 1959): 47.

155. See Richard Edward McFadyen, "Estes Kefauver and the Drug Industry" (PhD diss., Emory University, 1973), and "The FDA's Regulation and Control of Antibiotics in the 1950s: The Henry Welch Scandal, Félix Martí-Ibañez, and Charles Pfizer and Co.," *Bulletin of the History of Medicine* 53 (1979): 159–69; and Bud, "Antibiotics, Big Business, and Consumers" 329–49.

156. Harris, *Real Voice*, 19–20.

Chapter Three: From Sigmamycin to Panalba

Epigraph. Stanley Temko, oral arguments, "In the Matter of Novobiocin-Tetracycline Drugs; Calcium Novobiocin-Sulfamethizole Tablets (Upjohn Co.), Arlington, VA, August 13, 1969," p. 92, Box 88-78-36, FDAD.

1. Richard Harris, *The Real Voice* (New York: Macmillan, 1964); Arthur Maass, *Congress and the Common Good* (New York: Basic Books, 1983), 212–13.

2. Daniel Carpenter, *Reputation and Power: Organizational Image and Pharmaceutical Regulation at the FDA* (Princeton: Princeton University Press, 2010), 118–227.

3. Harry M. Marks, *The Progress of Experiment: Science and Therapeutic Reform in the United States, 1900–1990* (New York: Cambridge University Press, 1997), 71–97.

4. Harris, *Real Voice*, 3–15. See also Robert Bud, "Antibiotics, Big Business, and Consumers: The Context of Government Investigations into the Postwar American Drug Industry," *Technology and Culture* 46 (2005): 329–49, and Dominique A. Tobbell, *Pills, Power, and Policy: The Struggle for Drug Reform in Cold War America and Its Consequences* (Berkeley: University of California Press, 2012), 69–76.

5. Federal Trade Commission, *Economic Report on Antibiotics Manufacture* (Washington, DC: U.S. Government Printing Office, 1958).

6. Irene Till to John M. Blair, "Memorandum: Drug Industry," 9/23/58; Wayles

Browne to "Irene, Lucille, Ray and George," "Priority for Drug Chores," 5/4/59, both in Box 7, ff 12, ATS.

7. See, e.g., John M. Blair, "Drug Prices Obtained from Webb City," 1/13/59 [most certainly misdated as 1/13/58], Box 10, ff 12; Ray Cole, "Drugs—Philadelphia, Pa" [Jan Laboratories], 1/28/59, Box 9, ff 26; George R. Clifford, "Drugs: New York" [Premo], 1/29/59, Box 9, ff 6; Raymond C. Cole, "S. E. Massengill Company," 2/26/59, Box 9, ff 29, all in ATS. Of the detail men for Massengill, which had been responsible for the 1937 Elixir Sulfanilamide fiasco, Cole reported that "Mr. Massengill remarked that they were qualified if they were breathing."

8. John M. Blair to Paul Rand Dixon, "Proposed Points of Inquiry for Drug Hearings," 2/14/59, p. 8, Box 7, ff 12, ATS.

9. Ibid., pp. 10–11. See also *APAI*, pp. 85, 87, 123, 147, and Vance Packard, *The Waste Makers* (New York: David McKay, 1960), 78–91.

10. John Lear, "The Certification of Antibiotics," *Saturday Review* 42 (Feb. 7, 1959): 43–48; "New Investigation of Pharmaceutical Promotion Will Be Added to Govt. Pricing & Patent Probes: Result of Exposé Articles," *F-D-C Reports* (Feb. 9, 1959): 10–13.

11. John M. Blair to Paul Rand Dixon, "Proposed Points of Inquiry for Drug Hearings," 2/14/59, pp. 10, 11–12, Box 7, ff 12, ATS.

12. Martin Seidell to John Lear, 5/29/60, attached to Seidell to John M. Blair, 5/29/60, Box 3, ff 1, ATS; see also Abraham Flexner, *Medical Education in the United States and Canada* (New York: Carnegie Foundation for the Advancement of Teaching, 1910); Kenneth M. Ludmerer, *Learning to Heal: The Development of American Medical Education* (New York: Basic Books, 1985), 166–90; and Thomas Neville Bonner, *Iconoclast: Abraham Flexner and a Life in Learning* (Baltimore: Johns Hopkins University Press, 2002), 69–90.

13. On Lear, Seidell, and Blair, see John Lear to John M. Blair, 11/12/59; Blair to Martin Seidell, 11/20/59, both in Box 2, ff 20, ATS; "Memo from the Desk of M.A. Seidell [to Blair]," n.d.; and Seidell to Lear, 5/29/60, attached to Seidell to Blair, 5/29/60, both in Box 3, ff 1, ATS. See also Harris, *Real Voice*, 25.

14. Harold Aaron to John M. Blair, 10/27/59, Box 7, ff 6, ATS. The other three "outstanding physicians" with antibiotics expertise Aaron noted were Robert Wise, Mark Lepper [misspelled Leppert], and Paul Beeson. Blair had actually written to Aaron regarding experts who could weigh in on side effects, rather than efficacy; Blair to Aaron, 10/13/59, Box 7, ff 6, ATS.

15. Of Pfizer, in an undated report from 1959, a staff member had written, "In industry Pfizer believed to be the 'worst' in dealing with others & in exploitation"; "Drugs" [handwritten report], Box 2, ff 9, ATS. However, by the spring of 1961, Irene Till could conclude that "in our examination of the 20 largest companies, we have been unable to distinguish one that is much superior in practices or ethics over its fellows" (Till to Harold Aaron, 3/24/61, Box 7, ff 6, ATS). Sackler is discussed later in the chapter. On the AMA, see especially the later "Questions for American Medical Assocation" [49-page preparation for their appearance at

the Drug Industry Antitrust Hearings, July 1961], 71A-5170, Box 19, untitled folder, RG 46. Therein, staff had written of AMA advertising and Pfizer's house organ, *Spectrum*, designed by McAdams: "[*Spectrum*] triggered the change. An examination of the McAdams advertising materials in these issues of the JAMA shows what a revolution was occurring in the former, vivid colors, extravagant advertising claims, and the like. All the other companies adopted the new advertising methods."

16. Harris, *Real Voice*, 41. For a more cynical take on the investigation and hearings, viewing them as political fodder for Kefauver, see "Inside the Kefauver Sub[committee]: Staff on Brink of Crack-up as 'Ideological Zealots' Find It Harder to Get Along with Pros," *F-D-C Reports* (Sept. 4, 1961): 3–6; C. Joseph Stetler [president, Pharmaceutical Manufacturers Association], "The Congressional Hearing," *Pharmaceutical Manufacturers Association Year Book, 1968–1969* (Washington, DC: Pharmaceutical Manufacturers Association, 1969), 89–103.

17. Richard Edward McFadyen, "Estes Kefauver and the Drug Industry" (PhD diss., Emory University, 1973), 254–55.

18. *AP*, Part 14, p. 7837.

19. Ibid., pp. 8136–37. As staff wrote in an undated (likely mid-1960) "Outline for Drug Report": "Non-price competition the key factor in determining market shares. Importance due not only to general tendency of producers who do not compete in terms of price to compete on non-price basis but also to unique feature of this industry in that he who orders does not buy" (Box 7, ff 11, ATS).

20. On contemporary Arthritis and Rheumatism Foundation interest in discrediting such quackery, see Barbara Yuncker, "Arthritis: A Vast Racket in Human Agony," *New York Post* (Nov. 10, 1959): 7; Ruth Walrad, *The Misrepresentation of Arthritis Drugs and Devices in the United States* (New York: Arthritis and Rheumatism Foundation, 1960); and Eric Boyle: *Quack Medicine: A History of Combating Health Fraud in Twentieth-Century America* (Santa Barbara, CA: Praeger, 2013), 135.

21. *AP*, Part 14, pp. 7977, 7994.

22. Ibid., p. 7995.

23. Ibid., pp. 8140–41. "They are supposed to deal with toxicity of compounds and pay little attention to claims of efficacy. They in fact do pay attention to efficacy, they have to, but in a strange sort of extralegal way." This was of course similar to Lasagna's talk before the Pharmaceutical Manufacturers Association in Feb. 1959 (detailed in chapter 2). See Louis Lasagna, "Gripesmanship: A Positive Approach," *Journal of Chronic Diseases* 10 (1959): 464; Box 1, ff 13, LLP; cf. the depiction of Lasagna at the hearings in Carpenter, *Reputation and Power*, 189–90.

24. *AP*, Part 16, p. 9003.

25. Haskell J. Weinstein to Estes Kefauver, 12/9/59, Box 3, ff 2, ATS. For a more sanitized version of such sentiments, see *AP*, Part 18, pp. 10251–52.

26. *AP*, Part 18, p. 10266.

27. Ibid., p. 10347. Five years previously, Bean had foretold a therapeutic "reversion to the polypharmacy which condoned combinations of puppy dog fat, powdered unicorn's horns, dried mosquito wings, and spider webs, obtained from the graveyard in the dark of the moon, and brewed by witches as panacea for real and imagined ills of every kind"; William B. Bean, "Vitamania, Polypharmacy, and Witchcraft," *Archives of Internal Medicine* 96 (1955): 141.

28. *AP*, Part 18, pp. 10371, 10375. By 1961, Console would become much firmer in his support of the efficacy requirement; see A. Dale Console to the editor, *AMA News* (Aug. 17, 1961), in Box 2, ff 13, ATS. In contrast, pharmacologist (and historian of medicine) Chauncey Leake, who followed Console, demurred when asked by Senator Roman Hruska (R-Nebraska) about such an FDA efficacy requirement. Hruska served as chief opposition to Kefauver throughout the duration of the hearings. See *AP*, Part 18, p. 10441. See also Leake, "To the Members of the American Society for Pharmacology and Experimental Therapeutics," 5/25/60, Box 2, ff 13, ATS, in which Leake criticized Kefauver and expressed his admiration for Hruska, stating that "we already have too much government regulation as it is, and that free men in a free society such as ours should have and be expected to have the responsibility of regulating themselves."

29. *AP*, Part 22, pp. 11890, 11918–23. Kefauver had investigated the mob in the early 1950s, and his staff, from Mar. through May 1960, had traced what seemed to them a web of nefarious relationships among Welch and Martí-Ibañez, on the one hand, and Arthur Sackler and the William Douglas McAdams agency, on the other. At one point, Sackler—at the head of an apparently vertically integrated "empire" of pharmaceutical companies, clinical testing groups, public relations outfits, and journals—seemed the larger catch, but this route ultimately became a tantalizing dead end for Blair and his staff. See John M. Blair to Paul Rand Dixon, "Memorandum: Sackler Brothers," 3/16/60; Blair to Dixon, "Drafts of Subpoenas to Dr. Welch and M.D. Publications," 4/6/60, both in Box 1087, "Miscellaneous (1 of 6)," RG 46; Blair to Dixon, "Further Information Concerning M.D. Publications and the Sackler Brothers," 71A-5171, Box 19, RG 46; and charts of the Sackler "Empire," Box 5, ff 2, ATS. For the later, public representation of such concerns, see John Lear, "Drugmakers and the Govt.—Who Makes the Decisions?" *Saturday Review* 43 (July 2, 1960): 37–42, and "The Struggle for Control of Drug Prescriptions," *Saturday Review* 45 (Mar. 3, 1962): 35–39.

30. *AP*, Part 22, pp. 11940–46. Entailed in such a critique was one concerning the FDA's own laxity in reining in Henry Welch's external activities. Ironically, it appears that Connors had first heard of Welch's publication arrangements from the FDA's Albert ("Jerry") Holland at a 50th anniversary celebration of the FDA, the proceedings of which Welch and Martí-Ibañez would publish. See "The Drug Industry Inquires About Dr. Welch's Outside Activities" and "Inquiry by John Connor, American Drug Manufacturers Association, Respecting Outside Activities of Dr. Welch and Response of FDA Thereto," both in Box 3, ff 13, ATS. See also Henry Welch and Félix Martí-Ibañez, eds., *The Impact of the Food and Drug*

Administration on Our Society: A Fiftieth Anniversary Panorama (New York: M.D. Publications, 1956).

31. *AP*, Part 22, pp. 11947–48; "M.D. Publications, Computations of Payments to Dr. Welch, 1953 through 1959," Box 1087 "Miscellaneous (2 of 6)," RG 46; "U.S. Drug Aide Got $287,142 on the Side," *New York Times* (May 19, 1960): 1; "FDA Official Is Fired after Pay Disclosure," *Washington Post* (May 19, 1960): A2.

32. Félix Martí-Ibañez to E. W. Whitney, 12/22/25, in *AP*, Part 23, pp. 12493–94.

33. Henry Welch to Arthur M. Sackler, 2/23/56, Box 3, ff 13, ATS.

34. *AP*, Part 22, p. 11958.

35. For Welch's defiant resignation letter, in which he challenged "anyone to search the journals [he had edited] and come up with any article, paragraph, or sentence which reflects a lack of editorial or scientific integrity," see Henry Welch to George P. Larrick, 5/19/60, Box 5, ff 7, ATS. By mid-1963, a grand jury investigation would dismiss pursuing a case against Welch; see "Henry Welch Grand Jury Dismissed June 12 without Handing Down an Indictment after 17-Month Intensive Investigation of the Ex-FDA-Er and Industry Execs," *F-D-C Reports* (June 17, 1963): 3–8. Welch was left with lasting resentments against Finland, Kefauver, and even Frances Kelsey (who he felt "got her accolade by sitting on her fat fanny procrastinating"). As Welch wrote of Kefauver, on the senator's death two months after the grand jury announcement: "It would be pure hypocrisy for me to say I am sorry. . . . Although there will be other 'investigators' and 'investigations,' few will reach the peak of mad . . . vituperation and character assassination than those instituted and perpetuated by the man from Tennessee." See Henry Welch to Félix Martí-Ibañez, 11/15/60; 8/11/63; and 9/21/[likely 1965], all in Box 95, "Henry Welch to 1966," FMIP.

36. *AP*, Part 22, p. 11962.

37. Ibid., p. 11963.

38. Ibid., pp. 11968–69. See also Gideon Nachumi to John Blair, n.d. [though shortly after his testimony], Box 5, ff 2, ATS. A copy of the letter was also sent to the Federal Bureau of Investigation.

39. *AP*, Part 22, pp. 11999, 12013. For the internal Pfizer drama surrounding the insertion of the phrase into Welch's speech, see the statements of Warren Kiefer (5/3/60), Morton B. Stone (5/10/60), and Joseph R. Hixson (5/30/60), all in Box 1089, "Welch, Henry, FDA Materials, Correspondence (10 out of 10)," RG 46.

40. For later depictions of such capture, see Morton Mintz (who thanked Moulton in his acknowledgments), *The Therapeutic Nightmare* (Boston: Houghton Mifflin, 1965), and Philip J. Hilts, *Protecting America's Health: The FDA, Business, and One Hundred Years of Regulation* (New York: Alfred A. Knopf, 2003), 117–28. Daniel Carpenter critiques such depictions in *Reputation and Power*, 26–27, 40–43.

41. *AP*, Part 22, p. 12021.

42. As Moulton related to Blair's staff in private: "New Drug is now staffed by young doctors who use their salary . . . to pay their expenses while they establish their practice. . . . Indeed some work only half time. As they are not career minded, they are not inclined to 'rock the boat' [and] will go along with big drug companies 'who know better' "; B. B. Howard to John Blair, "Notes on Interview with Barbara Moulton," 3/21/60, Box 1087, "Miscellaneous (6 of 6)," RG 46.

43. *AP*, Part 22, p. 12025.

44. Ibid., pp. 12033–34. For Moulton's pre-existing concern with antibiotic overuse, see Moulton, "Antibiotics in the Treatment of Virus Diseases," *Antibiotics Annual* (1955–1956): 719–26. For her characterization of Welch as the opinionated and domineering "King Henry," see "Notes on Interview with Barbara Moulton." For Welch's actual instructions to his staff, in the wake of the 1956 Antibiotics Symposium, that "all efforts must be engaged in the study of antibiotic mixtures particularly for the discovery of those with synergistic power," see "Record of Antibiotic Research Meeting, December 7, 1956," Box 5, ff 4, ATS.

45. *AP*, Part 22, p. 12035. As Moulton stated, Sigmamycin was initially evaluated by the Division of Antibiotics, rather than by the New Drug Branch, since tetracycline was covered by Section 507 of the Federal Food, Drug, and Cosmetic Act (as described in chapter 1); ibid., p. 12034.

46. Ibid., pp. 12035–36.

47. Ibid., p. 12040.

48. Carpenter, *Reputation and Power*, 125–57. Harry Dowling had communicated to Moulton a less "destructive" view of the FDA than she had articulated. Moulton admitted her pessimism in response, feeling that Ralph Smith himself "duck[ed] the issue—not attempting to influence policy"; Dowling to Moulton, 9/7/60; Moulton to Dowling, 9/28/60, both in Box 10, "Kefauver Subcommittee Hearings" [cf. title of folder with that in finding aid], HDP.

49. *AP*, Part 22, p. 12080. Flemming referred to the New Drug Branch as the New Drug Division. The Division of Antibiotics as a semi-autonomous entity within the Bureau of Medicine was not dissolved until 1967, when a Bureau of Medicine reorganization created a Division of Anti-Infective Drugs that served as a subcomponent of the Office of New Drugs, no more autonomous than the Division of Cardiopulmonary and Renal Drugs, for example. See Robert J. Robinson to James L. Goddard, "Proposed Reorganization of the Bureau of Medicine," 6/6/66, Box 2, "FDA Miscellaneous Records," HLP, and Donald F. Simpson, "Statement of Organization, Functions, and Delegations of Authority," *Federal Register* 32 (Nov. 15, 1967): 15721–23.

50. *AP*, Part 22, pp. 12128–29.

51. McFadyen, "Estes Kefauver and the Drug Industry," 213; 106 Cong. Rec. (June 15, 1960): S12664–65.

52. The members of the committee were permitted to review the FDA evaluations of those drugs "they considered to be of especial significance." Tellingly, Sigmamycin was the first drug chosen, while chloramphenicol was not chosen

at all (the latter to the surprise and dismay of Kefauver staff members Irene Till and Lucile Wendt); Till and Wendt, "Memorandum to Mr. Paul Rand Dixon," 11/1/60, Box 4, ff 8, ATS.

53. "Report of Special Committee Advisory to the Secretary of Health, Education, and Welfare to Review the Policies, Procedures, and Decisions of the Division of Antibiotics and the New Drug Branch of the Food and Drug Administration," 9/20/60, Box 2, ff 24, MFP. On following the direction established by Larrick's congressional testimony, see "Notes for Meeting of FDA Committee, 28 June 1960"; for HEW Secretary Flemming's concurrence with the committee's efficacy proposal, see "Secretary Flemming's Comments, 7 October 1960," both in Divisions of the NRC, Medicine, "Committee to Review Policies Procedures & Decisions of Division of Antibiotics & New Drug Branch of FDA: Special: Advisory to Secretary of HEW," NASM.

54. *AP*, Part 24, p. 13941. For Dowling's defense of the proposed efficacy clause, see pp. 14172, 14179–80.

55. 107 Cong. Rec. (Apr. 12, 1961): S5638–42; Jeremy A. Greene and Scott H. Podolsky, "Reform, Regulation, and Pharmaceuticals—The Kefauver-Harris Amendments at 50," *NEJM* 367 (2012): 1481–83.

56. *DIAA*, Part 1, pp. 44–45.

57. Ibid., p. 45.

58. Ibid.

59. Maxwell Finland to Louis Goodman, 6/1/61, Box 2, "G," HDP. For the intense back-and-forth among such witnesses prior to the hearings, and especially concerning the apparent disjunction between the views of the AMA's Board of Trustees and the views of such AMA Council on Drugs members as Harry Dowling and Louis Goodman, see also Harry F. Dowling to Louis Goodman, 6/6/61; Goodman to Finland, 6/7/61; Goodman to Dowling, 6/20/61; and Dowling to Goodman, 6/30/61; all in Box 2, "G," HDP. See also C. Joseph Stetler [director, Legal and Socio-Economic Division of the AMA] to Dowling, 6/16/61, Box 2, "S," HDP.

60. *DIAA*, Part 1, p. 267. As Bean wrote to Harry Dowling, hoping to compare notes in anticipation of the hearings, "I don't really enjoy tangling with the A.M.A. people but if one looks at what their avowed policy is now and what they said about these very things ten years ago[,] any notion that they are either forward looking or omniscient must disappear in a hurry"; William B. Bean to Harry F. Dowling, 7/7/61, Box 13, "Dowling, Harry F., 1952–1975," WBP. Dowling, who had communicated with Hugh Hussey directly that May, considered Hussey more "liberal" than the remainder of the board of trustees, lamenting to Louis Goodman in June that while Hussey was "patiently trying to change certain attitudes, I don't know how far he is getting" (Dowling to Goodman, 6/30/61, Box 2, "G," HDP). See also Dowling to Hugh Hussey, 5/24/61, and Hussey to Dowling, 5/30/61, both in Box 2, "H," HDP.

61. Richard McFadyen notes in his extensive account of the hearings that

while the AMA opposed the bill, "other significant spokesmen for the medical profession" were more supportive. He cites a series of seven anonymous editorials in *NEJM* from the fall of 1961 to support this claim; as it turns out, each of the editorials was written by Max Finland. See McFadyen, "Estes Kefauver and the Drug Industry," 230–31, and Maxwell Finland, *Collected Reprints, 1960–1964*, Francis A. Countway Medical Library.

62. Most comprehensively in McFadyen, "Estes Kefauver and the Drug Industry," 207–385.

63. "Drug Amendments of 1962: Public Law 87–781; 76 Stat.780," in *U.S. Code, Congressional and Administrative News* (Oct. 10, 1962), 87th Congress, part 1, Section 102 (d), p. 911.

64. John P. Swann, "Sure Cure: Public Policy on Drug Efficacy before 1962," in Gregory J. Higby and Elaine C. Stroud, eds., *The Inside Story of Medicines: A Symposium* (Madison, WI: American Institute of the History of Pharmacy, 1997), 223–61; Carpenter, *Reputation and Power*, 118–227.

65. Milton Silverman and Philip R. Lee, *Pills, Profits, and Politics* (Berkeley: University of California Press, 1974), 121–22; Dominique A. Tobbell, "Allied Against Reform: Pharmaceutical Industry-Academic Physician Relations in the United States, 1945–1970," *Bulletin of the History of Medicine* 92 (2008): 878–912; Carpenter, *Reputation and Power*, 345–48; Tobbell, *Pills, Power, and Policy*, 121–35.

66. "AMA-ers & FDA-ers Hold Discussion to Avoid Probable Conflicts on Drug Evaluations Growing out of Govt.'s New Legal Authority on Effectiveness," *Drug Research Reports* (Aug. 14, 1963): 20.

67. Mintz, *Therapeutic Nightmare*, 214–29.

68. See Robert C. Toth, "Drug Agency Proposes a Ban on Antibiotics in Cold Remedies," *New York Times* (Aug. 18, 1963): 1; Nate Haseltine, "Ban on Antibiotics in Cold Remedies Stirs Protests of Working Doctors," *Washington Post* (Sept. 20, 1963): A1.

69. "Antibiotic in Combination with Analgesic Substances, Anti-Histaminics, and Caffeine," *Federal Register* 28 (Aug. 17, 1963): 8471.

70. For condemnation of the remedies, see, e.g., Louis Weinstein, "The Chemoprophylaxis of Infection," *Annals of Internal Medicine* 43 (1955): 287–98; A. E. Feller, "Common Respiratory Disease," in Tinsley R. Harrison et al., eds., *Principles of Internal Medicine*, 3rd ed. (New York: McGraw-Hill, 1958), 1048. For sales figures, see *ICDRR*, Part 4, p. 1521.

71. *AP*, Part 18, pp. 10487–88.

72. Garb had first written to Lederle to complain on Dec. 9, 1959, the first day of the hearings. See Solomon Garb to President, Lederle, 12/9/59, and Charles J. Masur to Solomon Garb, 2/1/60, both in 71A-5170, Box 18, "Achrocidin," RG 46.

73. Charles J. Masur to Solomon Garb, 2/1/60, 71A-5170, Box 18, "Achrocidin," RG 46.

74. Charles J. Masur to Solomon Garb, 1/8/60, ibid.

75. *ICDRR*, Part 4, p. 1510.

76. "Conference of Consultants to Food and Drug Administration Considering Prophylactic Use of Antimicrobial Agents," 6/6/62, 71A-5170, Box 10, "FDA: Antibiotics," RG 46.

77. Dowling's panel, citing other known instances in which prophylaxis was appropriate (e.g., in contacts of patients with meningococcal infections), did state that patients with "significant chronic pulmonary disease" may benefit from prophylaxis with tetracyclines, thereby opening the door for later industry claims concerning the utility of the fixed-dose combination antibiotic/symptom-relief remedies as prophylaxis against the transformation of upper respiratory tract infections into more serious illnesses. See the text of their report, later reprinted in "Anti Antibiotic Prophylaxis Viewpoint, across the Spectrum, Stated by Dowling Panel; Provided Basis for FDA Proposal to Ban Cold Combinations," *Drug Research Reports* (Aug. 21, 1963): 1–3.

78. *ICDRR*, Part 4, p. 1300. See also Hobart A. Reimann, "Infectious Diseases: Eleventh Annual Review of Significant Publications," *Annals of Internal Medicine* 76 (1945): 114.

79. *ICDRR*, Part 4, p. 1300.

80. Ibid., p. 1301.

81. Ibid., p. 1513.

82. "Antibiotic in Combination with Analgesic Substances, Anti-Histaminics, and Caffeine," 8471–72.

83. 109 Cong. Rec. (Aug. 26, 1963): H15825. On such "cold war" pharmaceutical rhetoric, see Dominique A. Tobbell, "Who's Winning the Human Race? Cold War as Pharmaceutical Political Strategy," *Journal of the History of Medicine and Allied Sciences* 64 (2009): 429–73, and *Pills, Power, and Policy*, 89–120.

84. *ICDRR*, Part 4, p. 1514; see also "Proposed Notice: Antibiotic Combinations for Prophylaxis in Colds," 7/17/63, in General Subject Files, 1938–1974, Decimal File 051.11, Box 3418, RG 88.

85. "Gap between Practicing MDs & 'Academic Medicine' Shows in More than 100 Letters Sent to FDA Protesting Proposed Ban on Antibiotic Combinations," *F-D-C Reports* (Sept. 16, 1963): 7–8; John Troan, "FDA Turns on Heat and AMA Yelps Like Stuck Kid," *Washington Daily News* (Nov. 12, 1963): 16; also in *ICDRR*, Part 4, pp. 1526–67. See also Mintz, *Therapeutic Nightmare*, 224–25.

86. Benjamin W. Carey [medical director, Lederle Laboratories] to Harry F. Dowling, 9/6/63, Box 2, "B–C," HDP; Robert P. Parker [general manager, Lederle Laboratories] to Dowling, 9/9/63, Box 2, "P–R," HDP; Dowling to George P. Larrick, 9/13/63; Maxwell Finland to Larrick, 9/13/63, both in Box 2, "L," HDP. The "Pink Sheet" reported, "The willingness of the U. of Ill. and Harvard professors to come forward with a reconsideration proposal reflects a degree of flexibility and open-mindedness not often credited to the men in 'academic medicine' by those

who use the phrase—'town and gown'—with a sneer"; "Dowling and Finland Suggest Reconsideration of Proposed Ban on Rx Antibiotic Combinations; Events Point to Relabeling as Final FDA Decision—in 1964," *F-D-C Reports* (Sept. 23, 1963): 6.

87. *ICDRR*, Part 4, pp. 1520–21. The University of Virginia's William Jordan, for one, was not convinced by the survey data, writing to Harry Dowling that he noted "with dismay that 50% of these products are sold in the South. The data on utilization proved nothing but good salesmanship. . . . It is apparent that two-thirds of these products are still used for undifferentiated acute respiratory illnesses" (Jordan to Dowling, 9/23/63, Box 2, "I–J–K," HDP.

88. For "cold war," see William A. MacColl, "The Witch's Brew Gets Official Sanction: The Cold War," M.D. Column, *Group Health Association of America* [newsletter] (Nov. 1963), in *ICDRR*, Part 4, p. 1525.

89. A. Dale Console, "Let's You and Him Fight" [letter to the editor], *Medical World News* 4 (Oct. 11, 1963): 31; also in *ICDRR*, Part 4, p. 1523.

90. Gerard Marder to Maxwell Finland, 10/5/63, Box 2, "D–E–F," HDP [*sic*]. On choice architecture, see Richard H. Thaler and Cass R. Sunstein, *Nudge: Improving Decisions about Health, Wealth, and Happiness* (New Haven: Yale University Press, 2008).

91. "Use and Misuse of Antibiotics," *JAMA* 185 (1963): 315. "Clearly violating the principles of rational therapy are the complex combinations of antipyretics, antihistaminics, or vitamins, with minimal amounts of the antibiotic."

92. *ICDRR*, Part 4, p. 1524.

93. Ibid. Not surprisingly, the Pharmaceutical Manufacturers Association came out with a statement concurring with the AMA statement. See "PMA Protest on FDA's Proposed Ban for Antibiotic Cold Combos Hits Method of Choosing Expert Panels; Representative of Industry Should be Included," *F-D-C Reports* (Nov. 18, 1963): 18.

94. Troan, "FDA Turns on Heat and AMA Yelps Like Stuck Kid," 16. Similarly, a clinician from Iowa wrote: "Since you cannot legislate away all possibilities of my making an error in judgement [*sic*], why not leave me this one too? I'll write again when I find I need you to protect me from myself" ("Gap between Practicing MDs and 'Academic Medicine,' " 8).

95. Tobbell, "Allied Against Reform," 889–97, and *Pills, Power, and Policy*, 122–25.

96. Louis C. Lasagna, "Drug Panic and Its Aftermath," *Medical Tribune* 3 (Aug. 27, 1962): 15. By 1962, Lasagna already seems to have developed some reservations regarding the "circus atmosphere of the hearings"; Louis Lasagna, *The Doctors' Dilemmas* (New York: Harper & Brothers, 1962), 167–68. However, the thalidomide episode appears to have represented the key transition point in his career and stance, from industry-wary proponent of FDA empowerment to FDA-wary critic of drug lags and industry stultification. As Lasagna, who had studied thalidomide and had publicly recommended it as "a drug worth trying

in patients whose therapeutic needs cannot be met by older drugs," concluded his initial editorial concerning the thalidomide episode: "On the basis of the best available information, Marilyn Monroe [who had died three weeks previously] would be alive today if she had been given thalidomide instead of barbiturates for sedation. This fact is one which illustrates difficulties in dealing glibly with the advantages and disadvantages of useful drugs; a consideration of it might lend perspective to journalists who wish to provide their readers with something more than lurid prose." See also Louis Lasagna, "Thalidomide—A New Nonbarbiturate Sleep-Inducing Drug," *Journal of Chronic Diseases* 11 (1960): 627–31, and "New FDA Regulations: A Forum through the Mails," *Clinical Research* 10 (1962): 376–78. For the ensuing debate between the *Saturday Review*'s John Lear and Lasagna on the topic of new drug investigation and patient protection, see Lear, "Human Guinea Pigs and the Law," *Saturday Review* 45 (Oct. 6, 1962): 55–57, and Lasagna, "Human Guinea Pigs [letter to the editor]," *Saturday Review* 45 (Nov. 3, 1962): 65–66.

97. Regarding "cripple," see Henry Beecher's telegram to Estes Kefauver, 9/25/62, 71A-5170, Box 10, "Correspondence 1962," RG 46. As Harry Gold, arguably the founder of American clinical pharmacology, cabled Kefauver: "I have been actively engaged in testing drugs in man (human pharmacology) for about forty years. Its appearance is innocent but my experience leaves me with a deep conviction that so called informed consent is a snare and delusion it is for the most part impossible to achieve and is certain to do more harm than good. . . . One rule must prevail, namely, a human participating as a subject in experimentation must not be injured and consent or no consent if he suffers injury the experimenter pays the price which may even amount to the loss of privilege to continuing human experimentation" (Gold to Kefauver, 9/24/62, 71A-5170, Box 10, "Correspondence 1962," RG 46). Over two dozen such telegrams were sent to Kefauver over the span of several days. See also "Experts Advise 'Go Slow' on Drug Regulations," *Medical Tribune* 3 (Sept. 3, 1962): 1, 27, and "Top Pharmacologists Hit Test Consent Provision," *Medical Tribune* 3 (Oct. 8, 1962): 1, 10.

98. On informed consent and the role of Senator Jacob Javits in congressional debate on the matter, see Ruth R. Faden, Tom L. Beauchamp, and Nancy M. King, *A History and Theory of Informed Consent* (New York: Oxford University Press, 1986), 202–5, and David J. Rothman, *Strangers at the Bedside: A History of How Law and Bioethics Transformed Medical Decision Making* (New York: Basic Books, 1991), 63–67.

99. Edwin D. Kilbourne to George P. Larrick, 10/30/63, Box 2, "L," HDP.

100. Harry F. Dowling to Norris L. Brookens, 9/25/63, Box 2, "B–C," HDP. For a concurring (and perhaps influencing) opinion, see William S. Jordon to Dowling, 9/23/63, Box 2, "I–J–K," HDP. See also Harry F. Dowling to Hubert H. Humphrey, 10/14/63, Box 2, "H," HDP. As Max Finland wrote fatalistically to Cecil Sheps (at the University of Pittsburgh School of Public Health) at the same time: "All the members of the Committee involved in the deliberations upon which

the F.D.A. based its recent action have been flooded with correspondence, some from respectable physicians and investigators concerning the F.D.A.'s action. . . . Our duty was done and now if the Commissioner wishes to reconvene the same or any other groups to consider any problem he wishes to raise he may do so. I suppose then any action (or inaction) on anybody's part would be subject to criticism from one quarter or another"; Finland to Sheps, 10/3/63, Box 6, ff 9, MFP.

101. *ICDRR*, Part 4, p. 1519.

102. "Washington Wire," *Wall Street Journal* (Nov. 1, 1963): 1; see also *ICDRR*, Part 4, p. 1526.

103. "7th Meeting, Bureau of Medicine—Medical Advisory Board," 6/23–24/66, Box 6, "Bureau of Medicine—Medical Advisory Board—Meetings, 1965–1968," HDP.

104. Joseph F. Sadusk Jr. to Harry F. Dowling, 2/25/66, Box 2, "S," HDP.

105. Tobbell, "Allied against Reform," 898–902, and *Pills, Power, and Policy*, 130–35. On the FDA's increasing use of external panels of experts throughout this era, see Carpenter, *Reputation and Power*, 303–32.

106. By implication, this depiction also lends nuance to Tobbell's otherwise impressive analysis, which has focused on DESI's industry-favorable interpretation of therapeutic equivalence and generic drugs. See Tobbell, "Allied against Reform," 902–9, and *Pills, Power, and Policy*, 163–76.

107. "Conference with Dr. Goddard and Dr. Don Estes (Management Consultant)," 3/26/66; "Chronology: 1968," both in Series 1, DES, "Beginning of the Program: 1966," NASD; *Drug Efficacy Study: A Report to the Commissioner of Food and Drugs* (Washington, DC: National Academy of Sciences, 1969), 2–4; *DE*, Part 2, pp. 229–30; James L. Goddard, "Shared Responsibilities" [address delivered at the Annual Meeting of the Pharmaceutical Manufacturers Association], 4/6/66, Box 3, "Speeches by Dr. Goddard, January 17–May 25, 1966," JGP. The FDA likewise conducted an internal efficacy review of over-the-counter antibiotic remedies (chiefly lozenges and troches) in early 1966, but this effort would be dwarfed by the larger DESI process (though would face similar issues concerning proof of efficacy). See J. Lamar Callway to James L. Goddard, 3/29/66; Robert J. Robinson, "Task Force for Antibiotic Evaluation," 4/19/66; R. A. Brooks to James L. Goddard, 4/25/66; R. N. Palmer, "Memorandum of Meeting," 5/6/66; B. Harvey Minchew, "Troches Containing Antibiotics," 6/10/66, all in General Subject Files, 1938–1974, Decimal File 512x, "Antibiotic Combinations for Prophylaxis in Colds," Vol. 3, Box 4141, RG 88; "Dr. Minchew's Report of Task Force, Presented to Med Advisory Board," 6/23/66, Box 5, "FDA—Antibiotics and Insulin" [not in finding aid], HDP; *Federal Register* 31 (Mar. 9, 1966): 4128; *Federal Register* 32 (Feb. 2, 1967): 1172–73.

108. "Doubt Cast on Future of All Drug Combos—Proprietary & Rx by NAS/NRC Judgment by Academic Panel That Mysteclin F & Panalba Are 'Ineffective,'" *F-D-C Reports* (Jan. 6, 1969): 28.

109. *Drug Efficacy Study*, 2–3, 21.

110. "Efficacy Review of Pre-1962 Drugs," 3/31/66, Series 1, DES, "Beginning of the Program: 1966," NASD. After some back-and-forth, the four categories were changed to the more optimistic "effective," "probably effective," "possibly effective," and "ineffective."

111. "Verbatim Transcript—Executive Session, Agenda Item XIV," 4/1/66, Series 1, DES, "Beginning of the Program: 1966," NASD.

112. R. Keith Cannan, "The Drug Efficacy Study of the National Academy of Sciences-National Research Council" [talk given at the Sixth Annual Briefing in Science of the Council for the Advancement of Science Writing], 11/11/68, Series 5, DES Name Files, "Cannan RK, 1963–1968," NASD. This subjective aspect would be characterized as "the informed judgment and experience of the members of the panels" in the final report of the study. See *Drug Efficacy Study*, 45, and *DE*, p. 230.

113. "Minutes of Second Meeting [of the Policy Advisory Committee of the Drug Efficacy Study], 12 January 1967," Series 3, DES Policy Advisory Committee (PAC), "Meetings: Second: 12 Jan 1967," NASD. James Goddard, in anticipation of the meeting, had actually proposed the term, "partially effective as a combination"; "Selected Questions for Discussion Meeting with Commissioner James L. Goddard, 16 December 1966," Series 5, DES Name Files, "Goddard JL: 1966–1968," NASD.

114. *Drug Efficacy Study*, 7.

115. "Some Thoughts on the Efficacy Review for FDA of 3000 Drugs Approved during the Period of 1938 to 1962" [Apr. 1966]; "Discussion with Dr. Louis Lasagna, 16 April [1966]"; Robert J. Robinson, "NAS Program for Review of Drug Efficacy," 4/18/66, all in Series 1, DES, "Beginning of the Program: 1966," NASD.

116. "Categorization of Drugs by the FDA, Drug Efficacy Study as of 5 August 1966," Series 1, DES, "Beginning of the Program: 1966," NASD.

117. Harry Dowling had been asked to head panel II but declined owing to his being on sabbatical and thus "required not to take on outside positions"; Hewitt was hence asked to head the panel. See Dowling to R. Keith Cannan, 7/8/66, Series 2, DES Panels, "Anti-Infectives (I): 1966–1969," NASD. The fifth panel, largely concerned with anti-tuberculosis and anti-parasitic drugs, was headed by William B. Tucker.

118. *CPDI*, Part 12, p. 5024.

119. Ibid., p. 5053. Four decades later, Kunin would recall: "The demise of the fixed-dose combination of antibiotics was a result of the recommendations of the committee chaired by me and Bill Hewitt. We did this even though it was not part of our charge. We owed this to Max Finland"; Calvin Kunin to the author, 3/15/10, e-mail.

120. R. Keith Cannan to William M. M. Kirby, 4/19/68; Kirby to Cannan, 4/25/68, both in Series 3, DES Policy Advisory Committee (PAC), "Executive Committee: Meeting to Consider Classification of Anti-Infective Combination Drugs: 24 May 1968," NASD.

121. Herbert L. Ley Jr. to R. Keith Cannan, 4/17/68; William M. M. Kirby to Cannan, 4/25/68; William B. Castle to Duke C. Trexler, 5/17/68; Kirby to Jay H. Winemiller, 5/15/68, all in Series 3, DES Policy Advisory Committee (PAC), "Executive Committee: Meeting to Consider Classification of Anti-Infective Combination Drugs: 24 May 1968," NASD.

122. Leighton E. Cluff to Duke C. Trexler, 5/17/68, ibid.

123. "Executive Committee, Policy Advisory Committee, DES: Meeting to Consider Classification of Anti-Infective Combination Drugs, 24 May 1968," ibid.

124. *Drug Efficacy Study*, 7. Of the drugs studied by the NAS-NRC panels, two-thirds were single-entity drugs, 20% were two-drug combinations, and the remainder contained from three to twelve component drugs (5).

125. James L. Goddard, "Patterns of Change" [address delivered before the Rhode Island Medical Society], 5/8/68, Box 4, "Speeches by Dr. Goddard—April 10–May 8, 1968," JGP.

126. Cannan, "Drug Efficacy Study of the National Academy of Sciences."

127. Ibid. The FDA's successful extensive use of expert panels in the DESI process was considered "one of the more enduring of the Study's contributions"; "Drug Efficacy Study Annual Report, June 1967–June 1968," Series 1, DES, "Progress Reports: 1966–1969," NASD.

128. *DE*, Part 2, p. 195. On the paper's rejection by *JAMA*, see *CPDI*, Part 12, p. 5038. On *JAMA* editor John Talbot's denial that the paper had been rejected (stating he had only received the paper in draft form), see "NRC Disputes *JAMA* on Antibiotic Combo 'White Paper,'" *F-D-C Reports* (May 12, 1969): 17. For the published paper, see "Fixed Combinations of Antimicrobial Agents: National Academy of Sciences—National Research Council Division of Medical Science Drug Efficacy Study," *NEJM* 280 (1969): 1149–54.

129. Upjohn Company v. Robert H. Finch and Herbert L. Ley, Jr., 303 F.Supp. 241 (1969); Morton Mintz, "FDA and Panalba: A Conflict of Commercial, Therapeutic Goals?" *Science* 165 (1969): 876.

130. Rody P. Cox, E. L. Foltz, Samuel Raymond, and Richard Drewer, "Novobiocin Jaundice," *NEJM* 261 (1959): 139–41; "Reversing the Tide of Antibiotic Resistance" [unsigned editorial by Maxwell Finland], *NEJM* 262 (1960): 578–79.

131. *CPDI*, p. 5006; Mintz, "FDA and Panalba," 876.

132. See, e.g., *Medical Times* (Jan. 1962): 197a.

133. Charles D. May, "Selling Drugs by 'Educating' Physicians," *Journal of Medical Education* 36 (1961): 4–6. For a sample Panalba handout Upjohn representatives gave to clinicians, describing in technical jargon concerning bacterial genetics the drug's utility in preventing emerging drug resistance (and citing Harvard's Bernard Davis in support, despite Davis's own printed misgivings regarding "shotgun therapy" and "prefabricated mixtures"), see the handout accompanying C. G. Sheppard to Wesley W. Spink, 10/11/63, Box 5, "1963," WSP; see also Bernard D. Davis, "Principles of Chemotherapy: Drug-Parasite Interactions," in René

J. Dubos, ed., *Bacterial and Mycotic Infections of Man*, 3rd ed. (Philadelphia: J. B. Lippincott, 1958), 686–87.

134. "Novobiocin-Tetracycline Combination Drugs," *Federal Register* 33 (Dec. 24, 1968): 19204.

135. For "explosive," see *DE*, Part 2, p. 174.

136. "Doubt Cast on Future of All Drug Combos," 26–29. Herbert Ley would meet that May with the AMA's Council on Drugs, whose members (not surprisingly) communicated their endorsement of the NAS-NRC and FDA positions; see "AMA Council on Drugs," 5/16/69, General Subject Files, 1938–1974, Decimal File 505.52, Box 4247, RG 88.

137. Upjohn had met with FDA officials in January to request a six-month extension and was granted 120 days on Jan. 24, 1969. See "Memorandum of Conference," 1/16/69; Henry L. Ley Jr. to Murray B. Welch Jr., 1/24/69, both in Box 2, Panalba correspondence, HLP; and "Panalba Case: Chronology from Efficacy Report to Decertification Decision," *F-D-C Reports* (May 19, 1969): 13. On suspicions that such letters were incited, if not written, by pharmaceutical representatives, see "Responses (as of 2/10/69) to Federal Register Announcements (12/24/68) Concerning Certain Combination Antibiotic Drugs," 2/19/69, Box 2, Panalba correspondence, HLP; B. Harvey Minchew [speaking on behalf of Herbert Ley], "Drug Development and the FDA" [remarks before the Basic Science Section of the New Jersey Academy of Medicine], 3/12/69, Box 1, "Dr. Ley's Speeches, Volume I, March 1967–March 27, 1969," HLP; *DE*, Part 2, pp. 240–41; and *CPDI*, Part 12, pp. 5175–76. Many of the letters were written on behalf of Squibb's Mysteclin-F, a combination of tetracycline and the antifungal amphotericin-B, whose proposed withdrawal had similarly been announced in the Dec. 24, 1968, *Federal Register*, alongside Panalba's. For examples of the letters, and an enumeration of their state of origin, see *CPDI*, Part 12, pp. 5188–5237. At the same time, by mid-February, Upjohn's representatives "developed a protocol for a clinical study to determine the efficacy of Panalba," expected to entail 30,000 patients and take two years to complete (and which FDA staffers would tear apart); see "Memorandum of Meeting," 2/11/69; "Protocol: Panalba Efficacy Study"; and "Evaluation of Proposed Protocol (Panalba Efficacy Study) Submitted by the Upjohn Company," 2/14/69, all in Box 2, Panalba correspondence, HLP. The self-contradictory Upjohn strategy of stating that existing data were already sufficient while calling for more time to collect "substantial evidence" would not be lost on the FDA and its legal team. See William Goodrich, "Memorandum in Opposition to Petitioner's Motion for a Stay [Upjohn v. Finch and Ley]," pp. 9–10, 59–62, Box 88-78-36, FDAD.

138. *Federal Register* 34 (Apr. 2, 1969): 6004–9.

139. *DE*, Part 2, pp. 173–90; Mintz, "FDA and Panalba," 879–80.

140. *DE*, Part 2, p. 337.

141. *CPDI*, Part 12, p. 5006. William Kirby, who rendered this characterization, had chaired another of the anti-infective NAS-NRC panels. Such a metaphor

had first been voiced at the Brook Lodge Invitational Symposium on the "Theory and Use of Antibiotic Combinations" in the fall of 1956. See "Brook Lodge Invitational Symposium," p. 162, 71A-5170, Box 5, ff 3, RG 46.

142. *CPDI*, Part 12, p. 5069. As early as during the formulation of Estes Kefauver's proposed drug bill in 1960, John Blair had noted that "as the patent on a given product approached the end of its span, the product would be combined with another drug and an application would be filed for a patent on the new combination, then on another combination, and so on *infinitum*"; Blair to Paul Rand Dixon, "Proposed Substantive Provisions of a Drug Bill," 11/7/60, Box 18, ff 17, ATS.

143. *CPDI*, Part 12, pp. 5011, 5057, 5068.

144. Ibid., pp. 5029–30, 5068.

145. Ibid., p. 5019.

146. Ibid., p. 5096.

147. John Adriani to Harry F. Dowling, 1/3/69, Box 32, ff 10, JOAP.

148. *CPDI*, Part 12, p. 5108.

149. Ibid., pp. 5020, 5026, 5164, 5253. For Herbert Ley's repeated and similar characterization of their use, from February and March of that year, see Herbert L. Ley Jr., "The Physician's Information" [remarks before the New York Pharmaceutical Advertising Club], 2/13/69; Minchew [speaking on behalf of Ley], "Drug Development and the FDA"; Ley, "Mutual Concerns and Opportunities" [remarks before the School of Pharmacy, St. John's University], 3/17/69, all in Box 1, "Dr. Ley's Speeches, Volume I, March 1967–March 27, 1969," HLP; and "Unnecessary Use of Antibiotic Combos 'Irrational, Illogical & Unscientific,' Ley Tells NY Admen; Says Drug Reactions Send 1.5 Mil. to Hospitals Annually," *F-D-C Reports* (Feb. 17, 1969): 6–7.

150. Jeremy A. Greene and Scott H. Podolsky, "Keeping Modern in Medicine: Pharmaceutical Promotion and Physician Education in Postwar America," *Bulletin of the History of Medicine* 83 (2009): 339–52.

151. "The Fond du Lac Study," in *DIAA*, Part 2, p. 789.

152. *CPDI*, Part 12, pp. 5009–10, 5017, 5153.

153. James Harvey Young, *The Medical Messiahs: A Social History of Health Quackery in Twentieth-Century America* (Princeton: Princeton University Press, 1967), 131.

154. *CPDI*, Part 12, pp. 5045–46. Dominique Tobbell has likewise described the contemporary (1967) envisioning of a "Therapeutic Consultants Program" by the Drug Research Board, though the program was never launched; see Tobbell, *Pills, Power, and Progress*, 144–47. Academic detailing would later be independently developed and advanced by Jerry Avorn. See, e.g., Jerry Avorn and Stephen B. Soumerai, "Improving Drug-Therapy Decisions through Educational Outreach: A Randomized Controlled Trial of Academically Based 'Detailing,' " *NEJM* 308 (1983): 1457–63. Ironically, it appears that "counter detail" was ini-

tially an industry term used to describe when the detail men of one company criticized the products of another; see testimony of John E. McKeen, 12/18/59, pp. 75–76, FTC Dockets 7487, RG 122.

155. *CPDI*, Part 12, p. 5025; see also Ley, "Physician's Information."

156. *DE*, Part 2, p. 338.

157. Ibid., pp. 348, 387. Upjohn appears to have first developed this line of reasoning in mid-February; see R. T. Parfet Jr. [president, Upjohn] to Herbert L. Ley Jr., 2/18/69, Box 2, Panalba correspondence, HLP.

158. *DE*, Part 2, p. 346.

159. Ibid., p. 374.

160. Leonard Engel, *Medicine Makers of Kalamazoo* (New York: McGraw-Hill, 1961), 212–13. This passage from Engel's book would later be noted by FDA lawyers as well. See "Respondents' Supplemental Brief, U.S. Court of Appeals [Upjohn v. Finch and Ley]," Box 88-78-36, FDAD; "Upjohn's Basic Error in Panalba Case Revolves around Who Has Proof Burden, FDA Says; Five Main Issues in Contention in Court Case; FDA Refutes Lasagna," *F-D-C Reports* (Dec. 1, 1969): 10.

161. "Memorandum on Cooperative Study to Evaluate Hydrocortisone in Severe Infections," 7/17/56, Box 6, ff 53, EKP; Ivan Bennett et al., "The Effectiveness of Hydrocortisone in the Management of Severe Infections," *JAMA* 183 (1963): 462–65. In their published 1963 paper, the authors describe the "study group" as having been formed "some 4 yr ago," though the group actually formed in 1956.

162. Burton A. Waisbren to Edward Kass, 3/8/57; Waisbren to Kass, 2/8/57, both in Box 6, ff 53, EKP.

163. Andrew J. Moriarty to Edward H. Kass, 6/27/57, Box 6, ff 53, EKP. As Moriarty continued in a subsequent letter, "Management here is afraid that some other company, in collaboration with Henry Welch, will break with an antibiotic-steroid combination at the Fall [Antibiotics] Symposium"; Moriarty to Kass, 7/18/57, Box 6, ff 53, EKP.

164. Carpenter, *Reputation and Power*, 269–92.

165. Austin Smith, "The Drug Amendments of 1962—An Industry Appeal" [address at the XLIV Conference of the Federal Bar Association, Washington, DC], 6/27/63, Box 34, ff 14, JOAP; see also "Who Likes the New Drug Regulations?" *JAMA* 185 (1963): 34–35.

166. Harold L. Upjohn was actually present at the first meeting, commenting on difficulties inherent in a double-blind trial. In "Advisory Committee on Investigational Drugs Meeting," 6/13/63, Box 34, ff 12, JOAP. On Upjohn's presence at the FDA's open hearing on the draft Investigational New Drug rules in Feb. 1963, see Carpenter, *Reputation and Power*, 283.

167. "The manual published by the I.R.S. (Your Income Tax) was suggested as a model"; Walter Modell to George P. Larrick, 6/17/63, Box 34, ff 13, JOAP.

168. Frances O. Kelsey, "Problems Raised for the FDA by the Occurrence

of Thalidomide Embryopathy in Germany, 1960–1961" [presented at the 91st Annual Meeting, American Public Health Association], 11/14/63, Box 34, ff 14, JOAP.

169. "Drug Evaluation Manual," pp. 12, 15–18, accompanying Matthew J. Ellenhorn to John Adriani, 10/2/63, Box 34, ff 14, JOAP.

170. See C. Gordon Zubrond, in "Summary of Proceedings, Third Meeting of the Advisory Committee on Investigational Drugs," 10/24/63, Box 34, ff 12, JOAP; "Priority Order on Effectiveness Review for 13 Categories of Marketed Drugs Disclosed by Medical Director Sadusk in First Regulatory-Type Speech," *F-D-C Reports* (Oct. 12, 1964): 14–17.

171. *DE*, Part 2, p. 247.

172. "Guidelines Employed in the Bureau of Medicine for Substantial Evidence of Efficacy," 3/4/66, Series 1, DES, "Guidelines: General: 1966–1967," NASD.

173. *Drug Efficacy Study*, 45.

174. *DE*, Part 2, p. 274.

175. Ibid., p. 243.

176. Herbert L. Ley Jr. to R. T. Parfet Jr., 2/7/69, Box 2, Panalba correspondence, HLP; *DE*, Part 2, p. 335.

177. *CPDI*, Part 12, p. 5057. In the final DESI report, however, in noting the paucity of clinical trial data for many widely used drugs, the report's authors admitted:

> This situation presented the panels with a very difficult problem. How much weight should they give to the opinion of the marketplace? The final arbiter of the value of a drug is the consensus of the experience of critical physicians in its use in the practice of medicine over a period of years. Approval of a new drug for release to the market is only a license to seek this experience. When the panels were faced with this situation, they have sought to grant liberty but to restrain license by assigning a rating of "Probably effective" or "Possibly effective" on the basis of their own clinical experience with the drug and their evaluation of the opinions of their peers (*Drug Efficacy Study*, 9).

It is unclear why Upjohn's lawyers did not focus more attention on this critical hedge.

178. *DE*, Part 2, pp. 181, 220–23.

179. Ibid., p. 408.

180. Ibid., p. 250.

181. Ibid., p. 279. For prior Fountain hearings that included an extended discussion of the Measurin case, see *DS*, Part 5, pp. 2049–68. For the contemporary aspirations of the Drug Research Board to augment the funding of well-trained clinical pharmacologists, see Tobbell, *Pills, Power, and Policy*, 135–42.

182. *DE*, Part 2, pp. 260–331; see also "MD's, Ethics, and Stock Options," *Medical World News* (Nov. 14, 1969): 16–18. On later Fountain hearings devoted

to Serc's eventual withdrawal, see *SERC*, and Carpenter, *Reputation and Power,* 627–29.

183. *Federal Register* 34 (May 15, 1969): 7687.

184. Herbert L. Ley Jr., "Emerging Patterns" [prepared for the annual meeting of the Pharmaceutical Manufacturers Association], 5/20/69, Box 1, "Dr. Ley's Speeches, Volume II, March 27, 1969–June 23, 1969," HLP; "FDA's Ley Fires Sharpest Barbs at Industry in PMA Speech That Wasn't Delivered; Criticizes Abbott on Parenteral Situation, Says Ads Regs Will Test Cooperation," *F-D-C Reports* (May 26, 1969): 13. Of expected concerns about the "ivory tower" nature of the recommended withdrawal, Ley noted: "This kind of thinking has been demolished long ago. If there were any ivory towers left on our campuses, they probably would have been occupied today by protesting students." Of note, Ley did not deliver his remarks in person (owing to his having to testify on legislation at the same time), but they were delivered on paper.

185. At the same time, they spoke with National Academy of Sciences officials, as well as with both Max Finland and Calvin Kunin, to see if they would be supportive of being granted an FDA "hearing" on the subject; Finland and Kunin were reportedly supportive at the time. "Visit of Dr. E. Gifford Upjohn and Dr. Harold Upjohn," 5/26/69, Series 2, DES Panels, "Anti-Infectives (IV): 1966–1969," NASD; W. B. Rankin, "Memorandum of Telephone Conversation between Duke Trexler and W.B. Rankin," 5/26/69, General Subject Files, 1938–1974, Decimal File 505.52, Box 4247, RG 88.

186. "Objection to Order of the Commissioner of Food and Drugs and Request for a Hearing, In the Matter of Novobiocin-Tetracycline Combination Drugs; Calcium Novobiocin-Sulfamethizole Tablets," 6/14/69, p. 10, Box 86-78-36, FDAD. Upjohn had first developed this stance in their winter dealings with the FDA; see "Statement [by the Upjohn Company]," 1/16/69, Box 2, Panalba correspondence, HLP. And, indeed, upon completion of the Drug Efficacy Study review in Jan. 1969, the panelists were polled regarding such issues as drug combinations, the quality of labeling, and generic therapeutic equivalence. Responded one panelist, "Although it is easy to damn fixed drug combinations on theoretical grounds, and we allow few of them in the hospital, in practice it is often a matter of considerable convenience for the physician and cheaper and less confusing for the patient to obtain one prescription than two" (*Drug Efficacy Study*, 73).

187. "Objection to Order of the Commissioner of Food and Drugs and Request for a Hearing, In the Matter of Novobiocin-Tetracycline Combination Drugs," p. 10.

188. Ibid., p. 15.

189. Ibid., pp. 3–6. See also William Goodrich, "Memorandum in Opposition to Petitioner's Motion for a Stay [in Upjohn v. Finch and Ley]," p. 9, Box 88-78-36, FDAD.

190. *Upjohn Company*, 303 F.Supp. 241.

191. "Novobiocin-Tetracycline Combination Drugs; Calcium Novobiocin-

Sulfamethizole Tablets: Order Ruling on Upjohn's Objections and Request for a Hearing," *Federal Register* 34 (Aug. 9, 1969): 12958–68. Among additional information obtained from Upjohn was unpublished data showing reduction of novobiocin and tetracycline blood levels in the combination pill versus when the drugs were given singly. See "Review of Records Obtained by FDA Inspector, Mr. Roy D. Sanberg[,] from The Upjohn Company on March 7, 1969," 4/29/69, Box 2, Panalba correspondence, HLP; "Data Submitted by Upjohn in Support of Certification," Vol. 2, p. 131; Vol. 3, pp. 394, 397, Box 88-78-36, FDAD. There also appeared a letter from Upjohn's James A. Duggar to Henry Welch, 8/22/57, requesting that Welch's division perform absorption studies on a new mixture "we call Panalba. . . . I hope you can do these studies on a minimum of ten persons and would appreciate this extra special fast, if possible." An exclamation point—likely dating from the spring of 1969 inquiry—appears next to this statement (ibid., Vol. 1, p. 69).

192. "Novobiocin-Tetracycline Combination Drugs," 12968.

193. Stanley Temko, oral arguments "In the Matter of Novobiocin-Tetracycline Drugs; Calcium Novobiocin-Sulfamethizole Tablets (Upjohn Co.), Arlington, VA, August 13, 1969," pp. 18, 22–23, Box 88-78-36, FDAD.

194. Ibid., pp. 92, 98.

195. Temko, oral arguments, "Novobiocin-Tetracycline Drugs," pp. 84–85.

196. Ibid., p. 12.

197. Ibid., p. 70. See also Thomas Maeder, *Adverse Reactions* (New York: William Morrow, 1994), 86–92.

198. Herbert L. Ley Jr., "Partners in Therapeutics" [presented before the American Academy of Dermatology], 12/4/67, Box 1, "Dr. Ley's Speeches, Volume 1, March 1967–March 27, 1969," HLP.

199. Ley, "Physician's Information."

200. Ley, "Emerging Patterns"; see also "The 'New' Ley Bares FDA Teeth," *Medical World News* 10 (June 27, 1969): 19–20.

201. As Ley would report that September, three days before his official response to Upjohn would appear in the *Federal Register*: "As a physician in full accord with the long-standing conviction of medical experts that *fixed combination* antibiotic prescribing is not good medicine, I am privileged to treat this problem as one would an abscess before incision and drainage—namely, bring it to a head. I believe we have done that. The problem is now out-in-the-open"; Herbert L. Ley Jr., "The Art and the Science" [presentation before the Academy of Medicine of Columbus Ohio and Franklin County, Ohio], 9/16/69, Box 1, "Dr. Ley's Speeches, Volume III, July 1969–," HLP.

202. Herbert L. Ley Jr., "Antibiotic Drugs: Procedural and Interpretative Regulations," *Federal Register* 34 (Sept. 19, 1969): 14596–98. See also "Early Draft Signed 18 Aug. 1969 for Final Order Repealing Regulation and Revoking Certification," Box 88-78-36, FDAD. The timing here would seem to contradict the notion that such a regulatory framework would derive from testimony by the

FDA's William Beaver in the ensuing *Pharmaceutical Manufacturers' Association v. Finch and Ley* case, which would commence in response to the regulations themselves. Cf. Curtis L. Meinert and Susan Tonascia, *Clinical Trials: Design, Conduct, and Analysis* (New York: Oxford University Press, 1986), 8; Carpenter, *Reputation and Power*, 354–55. It is of course possible that William Beaver or other FDA staff members contributed to Ley's formulations that summer.

203. Herbert L. Ley Jr., "Novobiocin," *Federal Register* 34 (Sept. 19, 1969): 14598–99; "Early Draft Signed 18 Aug. 1969." To give further context to Ley's musings, in mid-July, American Home Products had filed a motion for an injunction against the FDA's withdrawal of its own penicillin-streptomycin and penicillin-sulfonamide combinations. The same day that Ley edited the draft version of the resolutions, and echoing the first *Upjohn v. Finch and Ley* case, the courts found in the company's favor. See American Home Products Corporation v. Robert H. Finch and Herbert L. Ley, Jr., 303 F.Supp. 448 (1969).

204. James R. Phelps, "After Panalba, Whither," *Food Drug Cosmetic Law Journal* 26 (1971): 188.

205. Pharmaceutical Manufacturers Association v. Robert H. Finch and Herbert L. Ley, Jr., 307 F.Supp. 858 (1970). Such opportunity for comment still differed from an evidentiary hearing, however. See also "Drug Efficacy and the 1962 Drug Amendments," *Georgetown Law Journal* 60 (1971–1972): 185–224.

206. While Ley's resignation took place in the context of fallout from both the Panalba case and a more proximate dispute over the toxicity of cyclamates, Ley would in posterity consider himself to have fallen on his sword over Panalba in particular. See Herbert L. Ley Jr. to Cole Palmer Werble, 12/12/89, Box 2, "FDA Resignation Letters (Mine)," HLP; Ronald T. Ottes and Robert A. Tucker, "Interview with Herbert L. Ley, M.D.," 12/15/99, FDAOHC.

207. Charles C. Edwards, "Hearing Requests on Refusal or Withdrawal of New Drug Applications and Issuance, Amendment or Repeal of Antibiotic Drug Regulations and Describing Scientific Content of Adequate and Well-Controlled Investigations," *Federal Register* 35 (Feb. 17, 1970), 3073–74; Charles C. Edwards, "Remarks ['To be presented at a meeting with the Drug Research Board of the NAS/NRC, February 20, 1970']," Box 2, ff 1, CEP.

208. William Goodrich, "Respondent's Supplemental Brief, U.S. Court of Appeals," p. 75, Box 88-78-36, FDAD. See also Minutes of the 19th, 20th, and 21st Bureau of Medicine Advisory Meetings, 6/26–27/69, 10/2–3/69, and 12/18–19/69, Box 6, "Bureau of Medicine—Medical Advisory Board—Meetings, 1968–1969," HDP.

209. For the list of affadavits, see Box 88-78-36, FDAD. Sadly, the affadavits themselves are missing from the box. Lasagna's name is not on this list, though he clearly supplied an affidavit. Goodrich's supplemental brief would serve as a pointed attack on Lasagna in particular, with a profound sense of betrayal suffusing the document. Lasagna's earlier support of the controlled clinical trial over the play of the marketplace was cited no fewer than five times, alongside the lament that "the doctor's current advocacy stands in bold contradiction to every scientific

precept he has advocated publicly and to his profession." See also "Upjohn's Basic Error," p. 8. On Lasagna's later ostracizing by his fellow reformers, see Carpenter, *Reputation and Power*, 330–32.

210. "Combination's Efficacy Backed, Hearing Urged: Dr. Lasagna Disputes FDA on Panalba Ruling," *Medical Tribune* (Dec. 15, 1969): 1, 18. Lasagna had been part of the DESI process and had long been privy to the antibiotic combination discussions; while he did not support the use of fixed-dose combination drugs in general, he felt that manufacturers should be given "a reasonable and realistic amount of time to accumulate data to buttress the claims or implied claims" about them. See "Discussion with Dr. Louis Lasagana, 16 April [1966]," Series 1, DES, "Beginning of the Program," NASD; Louis Lasagna to Duke C. Trexler, 11/29/68, Series 4, DES Reports, "White Papers: Anti-Infective Combinations: 1968–1969," NASD.

211. Upjohn Company v. Robert H. Finch and Herbert L. Ley, Jr., 422 F.2d 944 (6th Cir., 1970). For earlier reporting of the proceedings, see "Sixth Circuit Judges Indicate They Would Order Panalba Hearing If Upjohn Has Evidence FDA Has Not Reviewed; Combo Therapy, MD Prescribing Probed," *F-D-C Reports* (Dec. 15, 1969): 6–7.

212. Upjohn Co. v. Finch, Secretary of Health, Education, and Welfare, et al., 397 U.S. 970 (1970); 90 S.Ct. 1085; 25 L.Ed. 2d 274.

213. Charles C. Edwards, "Hearing Regulations and Regulations Describing Scientific Content of Adequate and Well-Controlled Clinical Investigations," *Federal Register* 35 (May 8, 1970): 7253.

214. Pharmaceutical Manufacturers Association v. Elliot L. Richardson and Charles C. Edwards, 318 F.Supp. 301 (1970).

215. Carpenter, *Reputation and Power*, 355–62.

216. Pfizer, Inc. v. Elliot L. Richardson and Charles C. Edwards, 434 F.2d 536 (1970). American Home Products also again filed—this time unsuccessfully—in 1971 for relief against the FDA on behalf of its combination drug Wycillin; see American Home Products Corporation v. Elliot L. Richardson and Charles C. Edwards, 328 F. Supp. 612 (1971).

217. American Cyanamid Company v. Elliot L. Richardson, 456 F.2d 509 (1971); Charles C. Edwards, "Proposed Statement Amplifying Policy on Drugs in Fixed Combinations," *Federal Register* 36 (Feb. 18, 1971): 3126–27, and "Fixed-Combination Prescription Drugs for Humans," *Federal Register* 36 (Oct. 15, 1971): 20037–38. See also congressional hearings held in May 1971, in *SEND*.

218. Maxwell Finland, "Combinations of Antimicrobial Drugs: Trimethoprim-Sulfamethoxazole," *NEJM* 291 (1974): 624–27; Scott H. Podolsky and Jeremy A. Greene, "Combination Drugs: Hype, Harm, and Hope," *NEJM* 365 (2011): 488–91.

Chapter Four: "Rational" Therapeutics and the Limits to Delimitation

Epigraph. W. Clarke Wescoe, "Lewis Carroll Might Have Written It," *Food Drug Cosmetic Law Journal* 26 (1971): 463.

1. G. E. R. Lloyd, *Magic, Reason, and Experience: Studies in the Origin and Development of Greek Science* (Cambridge: Cambridge University Press, 1979), 10–58; James Longrigg, *Greek Rational Medicine: Philosophy and Medicine from Alcmaeon to the Alexandrians* (New York: Routledge, 1993).

2. For an example regarding later (if implicit) attempts to contrast such a grounded (and learned) system of surgery with the empiricism of the tradesman, see Michael McVaugh, *The Rational Surgery of the Middle Ages* (Florence: Sismel, Edizioni del Galluzzo, 2006).

3. John Harley Warner, *The Therapeutic Perspective: Medical Practice, Knowledge, and Identity in America, 1820–1885* (Cambridge, MA: Harvard University Press, 1986), 37–57, 235–57.

4. Harry M. Marks, *The Progress of Experiment: Science and Therapeutic Reform in the United States, 1900–1990* (New York: Cambridge University Press, 1997), 17–41.

5. On the increasing attention to the influences on physician prescribing behavior in the 1950s and 1960s, see Jeremy A. Greene and Scott H. Podolsky, "Keeping Modern in Medicine: Pharmaceutical Promotion and Physician Education in Postwar America," *Bulletin of the History of Medicine* 83 (2009): 331–77.

6. Robert Moser, *Diseases of Medical Progress* (Springfield, IL: Charles C. Thomas, 1959); see also James Whorton, "'Antibiotic Abandon': The Resurgence of Therapeutic Rationalism," in John Parascandola, ed., *The History of Antibiotics: A Symposium* (Madison, WI: American Institute of the History of Pharmacy, 1980), 125–36.

7. U.S. Task Force on Prescription Drugs, *Final Report* (Washington, DC: U.S. Government Printing Office, 1969), xxi, 21, and *The Drug Prescribers* (Washington, DC: U.S. Government Printing Office, 1968), 3. For an articulation of such a definition of "rational" therapeutics in congressional testimony, see Charles C. Edwards (who was assistant secretary of health at HEW, having been promoted from FDA chief in 1973), in *EPI*, Part 2, p. 566. For his invocation of the notion of "rational" therapeutics throughout the early 1970s, see Edwards, "Positive and Rational Drug Therapeutics" [presented at the Symposium on Drugs—Hospital Pharmacists], 9/26/70, Box 2, ff 1, CEP, and "The Ways We Regulate You" [presented before the annual meeting of the American Society of Internal Medicine], 4/7/73, Box 4, ff 2, CEP.

8. The theme of "use and abuse" dated back to the sulfa era. See, e.g., Wesley W. Spink, "The Use and Abuse of Chemotherapy," *Minnesota Medicine* 25 (1942): 988, and Richard A. Kern, "Abuse of Sulfonamides in the Treatment of Acute Catarrhal Fever," *United States Naval Medical Bulletin* 44 (1945): 686–94.

9. Whorton, "Antibiotic Abandon," 125–36.

10. "Use and Abuse of the Antibiotics," *Rocky Mountain Medical Journal* 49 (1952): 581.

11. Maurice A. Schnitker, "Some Abuses of the Antibiotic Agents," *Ohio State Medical Journal* 43 (1947): 1140; see also Karl H. Pfuetze and Marjorie M. Pyle,

"The Use and Misuse of Streptomycin in the Treatment of Tuberculosis," *Journal-Lancet* 68 (1948): 434; Paul S. Rhoads, "The Antibiotics—Uses and Abuses," *GP* 5 (Feb. 1952): 67–78; and John F. Waldo, "Antibiotics—Their Use and Abuse," *Rocky Mountain Medical Journal* 50 (1953): 879–82.

12. A. L. Tatum, "Misuse of Antibiotics," *Wisconsin Medical Journal* 51 (1952): 881.

13. Henry C. Sweany, "The Use and Misuse of Antimicrobials," *Missouri Medicine* 53 (1956): 1068.

14. Dr. Covode, in "Use and Abuse of the Antibiotics," 582.

15. Dr. Schaffer, in "Use and Abuse of Antibiotics: A Panel Discussion," *Journal of the South Carolina Medical Association* 52 (1956): 35.

16. Erwin Neter, "Use and Abuse of Antibiotics," *Virginia Medical Monthly* 81 (1954): 362.

17. For acknowledgments of patient pressure on the physician, see E. C. Drash, in William E. Apperson, "Streptomycin in Tuberculosis: Its Use and Abuse," *Virginia Medical Monthly* 77 (1950): 215; Wendell H. Hall, "The Abuse and Misuse of Antibiotics," *Minnesota Monthly* 35 (1952): 629; Neter, "Use and Abuse of Antibiotics," 362. For similar structural pressures influencing antibiotic prescribing in Britain throughout this era, see John T. MacFarlane and Michael Worboys, "The Changing Management of Acute Bronchitis in Britain, 1940–1970: The Impact of Antibiotics," *Medical History* 52 (2007): 47–72.

18. George R. Fisher, "The Use and Abuse of Antibiotics," *Journal of the Iowa State Medical Society* 50 (1960): 245.

19. William A. Nolen and Donald E. Dille, "Use and Abuse of Antibiotics in a Small Community," *NEJM* 257 (1957): 33–34. Nolen and Dille generously defined "definite" indications as "any temperature elevation of 1°F above normal and all the following, even though there was no temperature elevation: cellulitis; abscess; lymphangitis; prophylaxis in any wounds; prophylaxis in patients with history of rheumatic fever; croup (diathesis in children); history of recurrent otitis; laboratory evidence of cystitis; and any venereal disease."

20. Thomas Maeder, *Adverse Reactions* (New York: William Morrow, 1994), 343–47.

21. F. Dennette Adams, foreword to Moser, *Diseases of Medical Progress*, vii.

22. *AP*, Part 24, p. 13770. See also the testimony of Hobart Reimann, in *ICDRR*, Part 4, 1963, p. 1303.

23. William S. Jordan Jr., "Colds, Drugs, and Doctors," *Antibiotics and Chemotherapy* 11 (1961): 372.

24. Paul D. Stolley and Louis Lasagna, "Prescribing Patterns of Physicians," *Journal of Chronic Diseases* 22 (1969): 396.

25. Maeder, *Adverse Reactions*, 282–90.

26. Paul D. Stolley et al., "Drug Prescribing and Use in an American Community," *Annals of Internal Medicine* 76 (1972): 538; *APM*, Part 3, p. 1169. Stolley would report at the Senate hearings that based on the same 1968 prescription

data set, chloramphenicol was the fifteenth-most-prescribed antibiotic, and that three of the six most-prescribed antibiotics (and six of the top twenty) were fixed-dose combination antibiotics.

27. Their collaborators included the American College of Physicians, the American College of Surgeons, the American Hospital Association, and the Southwestern Michigan Hospital Council. See Robert S. Myers, Vergil N. Slee, and Richard P. Ament, "Antibiotic Study Shows Need for Therapy Audit in Hospitals," *Bulletin of the American College of Surgeons* 48 (1963): 61–63; Robert S. Myers, Vergil N. Slee, and Robert G. Hoffman, "The Medical Audit: Protects the Patient, Helps the Physician, and Serves the Hospital," *Modern Hospital* 85 (Sept. 1955): 77–83. For an explicit reference to Ernest Codman's surgical follow-up studies from the 1910s (what would come to be considered the first "outcome studies"), see Robert S. Myers and Virgil N. Slee, "Basic Ingredients of a Medical Audit," *Modern Hospital* 86 (Apr. 1956): 62–63.

28. Winifred B. Harm, "A Medical Audit Is a Medical Education," *Modern Hospital* 85 (Nov. 1955): 96.

29. Myers, Slee, and Ament, "Antibiotic Study Shows Need for Therapy Audit," 63. See also "Antibiotic Usage and Bacterial Antibiotic Sensitivity Tests (Antibiograms)," *The Record* 1 (Mar. 15, 1962): 1; "Minimum Use of Antibiotics," *The Record* 1 (July 30, 1962): 1; "Therapeutic Use of Antibiotics," *The Record* 1 (Aug. 22, 1962): 1.

30. Don W. Branham, "The Internal Medical Audit," *Journal of the Oklahoma State Medical Association* 53 (1960): 805–7; see also Stanley Ferber and Nancy Kaye, "It's Coming Fast via Medical Audits," *Medical Economics* 40 (Aug. 12, 1963): 75–79.

31. *ICDRR*, Part 4, pp. 1297–99; Hobart Reimann and Joseph D'Ambola, "The Use and Cost of Antimicrobics in Hospitals," *Archives of Environmental Health* 13 (1966): 631–36. For an earlier, rough overview of hospital antibiotic use at a community hospital in New England, stimulated by concerns regarding antibiotic resistance, see Wei-Ping Loh and Russell B. Street, "Study of the Use of Antimicrobial Agents in a Community Hospital," *NEJM* 251 (1954): 659–60.

32. William E. Scheckler and John V. Bennett, "Antibiotic Usage in Seven Community Hospitals," *JAMA* 213 (1970): 264–67. Such studies were published shortly before John Wennberg and his colleagues published their own classic studies of therapeutic variation; see Wennberg and Alan Gittelsohn, "Small Area Variations in Health Care Delivery," *Science* 182 (1973): 1102–8.

33. Andrew W. Roberts and James A. Visconti, "The Rational and Irrational Use of Systemic Antimicrobial Drugs," *American Journal of Hospital Pharmacy* 29 (1972): 828–34. They included fourteen categories of "irrational" prescribing, summarizing that "therapy was judged irrational . . . in two major cases: where antimicrobial therapy was warranted but the specific therapy used was inappropriate and when therapy was unnecessary and unwarranted." For a similar contemporary study, see Charles W. Gibbs, J. Tyrone Gibson, and David S.

Newton, "Drug Utilization Review of Actual Versus Preferred Pediatric Antibiotic Therapy," *American Journal of Hospital Pharmacy* 30 (1973): 892–97.

34. For such cost data, see Henry E. Simmons and Paul D. Stolley, "This Is Medical Progress? Trends and Consequences of Antibiotic Use in the United States," *JAMA* 227 (1974): 1024.

35. Richard A. Gleckman and Morton A. Madoff, "Environmental Pollution with Resistant Microbes," *NEJM* 281 (1969): 677; Maxwell Finland, "Changing Patterns of Susceptibility of Common Bacterial Pathogens to Antimicrobial Agents," *Annals of Internal Medicine* 76 (1972): 1009–36.

36. Maxwell Finland, "Prophylactic Antimicrobial Agents," in *Proceedings of the International Conference on Nosocomial Infections, Center for Disease Control, August 3–6, 1970* (Chicago: American Hospital Association, 1971), 312–14.

37. See Elizabeth Siegel Watkins, *On the Pill* (Baltimore: Johns Hopkins University Press, 1998), 103–28.

38. See *CPDI*, Part 6 (1967); Part 11 (1969). Similarly, clindamycin usage would be reduced from 6.1 million prescriptions in 1973 to 0.8 million prescriptions in 1977, consequent to concerns and hearings regarding its role in producing infectious colitis (what would today be considered *Clostridium dificile* colitis). See *CPDI*, Part 27 (1975), esp. pp. 12467–68; Marion J. Finkel, "Magnitude of Antibiotic Use," *Annals of Internal Medicine* 89 (1978): 791.

39. *APM*, Part 3, p. 989. Note that Nelson's hearings on the *Advertising of Proprietary Medicines*—intended to examine the "effect of promotion and advertising of over-the-counter drugs on competition, small business, and the health and welfare of the public"—ran in parallel to his hearings on *Competitive Problems in the Drug Industry*. Antibiotic misuse made it into both sets of hearings, in the *APM* case on account of its relation to the use of over-the-counter cough and cold remedies.

40. *APM*, Part 3, p. 1084.

41. Ibid., p. 1104.

42. Ibid., pp. 1164–91.

43. Simmons and Stolley, "This Is Medical Progress," 1023–28.

44. See, e.g., testimony of Perrin Long, in *AP*, Part 24, pp. 13765, 13767; "Gap between Practicing MDs & 'Academic Medicine' Shows in More Than 100 Letters Sent to FDA Protesting Proposed Ban on Antibiotic Combinations," *F-D-C Reports* (Sept. 16, 1963): 7–8; John Troan, "FDA Turns on Heat and AMA Yelps Like Stuck Kid," *Washington Daily News* (Nov. 12, 1963): 16; the testimonies of William Kirby, Heinz Eichenwald, and William Hewitt, in *CPDI*, Part 12, pp. 5003, 5009, 5020, 5048; and Morton Mintz, "FDA and Panalba: A Conflict of Commercial, Therapeutic Goals?" *Science* 165 (1969): 876. At an FDA Board of Medicine advisory board meeting in 1969, concern was broached "that there will soon be a polarization of the academic world against the practicing world in medicine." In "19th Bureau of Medicine Advisory Board Meeting," 6/26–27/69, Box 6, "FDA Bureau of Medicine Advisory Board Meetings, 1968–1969," HDP.

45. *CPDI*, Part 12, p. 5048.

46. *APM*, Part 3, p. 1214; Calvin M. Kunin, Thelma Tupasi, and William A. Craig, "Use of Antibiotics: A Brief Exposition," *Annals of Internal Medicine* 79 (1973): 557–58. For an earlier invocation of "fear" as motivation for indiscriminate prescribing, see Ryle A. Radke, "Use and Abuse of Antibiotics," *Medical Bulletin of the United States Army Far East* 2 (1954): 2.

47. Dominique A. Tobbell, *Pills, Power, and Policy: The Struggle for Drug Reform in Cold War America and Its Consequences* (Berkeley: University of California Press, 2012), 147–92.

48. Dominique Tobbell, "Plow, Town, and Gown: The Politics of Family Practice in 1960s America," *Bulletin of the History of Medicine* 87 (2013): 648–80.

49. W. Clarke Wescoe, "Lewis Carroll Might Have Written It," *Food Drug Cosmetic Law Journal* 26 (1971): 463. Wescoe was speaking with respect to the Drug Efficacy Study and Implementation process. For similar authority-wary sentiments, using a similar Carroll trope, see Brian Boru, "An Adventure with Alice in Wonderland—Indeed," *JAMA* 184 (1963): 247–50, and Arthur M. Sackler, "In Medicineland, U.S.A.," *Hospital Tribune* 9 (Nov. 24, 1975): 11.

50. Julius Michaelson, in *RTP*, p. 259; see also Charles E. Edwards, in ibid., pp. 6–7.

51. See AMA House of Delegates Proceedings, 1971, pp. 277, 366, now accessible online at the AMA historical archive website, through the Digital Collection of Historical AMA Documents. Note that W. Clarke Wescoe, in his aforementioned talk, actually (and, it seems, incorrectly in two respects) stated that the AMA had asked the FDA to stop using the term *rational*.

52. Nea D'Amelio, "Are Family Doctors Prescribing Too Many Antibiotics?" *Medical Times* 102 (1974): 53–54. At the same time, 10,000 medical residents were polled (of whom only 2,358 responded): 65.7% felt that antibiotics were being overprescribed (and 91% of them thought they were being overprescribed in private practice), while 45.5% agreed that the average person didn't require antibiotics more than once every five to ten years (ibid.).

53. Michael Halberstam, *The Pills in Your Life* (New York: Grosset & Dunlap, 1972), 67–68.

54. *APM*, Part 3, pp. 1211–12; Calvin Kunin, review of *The Pills in Your Life*, *American Society of Microbiology News* 38 (1972): 696, wherein Kunin concluded, "Shame on you Dr. Halberstam!"

55. D'Amelio, "Are Family Doctors Prescribing Too Many Antibiotics," 56. Similarly, a resident responded: "I prefer the occurrence of resistant strains rather than see people suffering from glomerulonephritis and . . . rheumatic heart disease. New antibiotics usually come up in due time and I think time itself really causes the formation of new strains"; Nea D'Amelio, "What Young Doctors Told Us about Antibiotic Overkill," *Medical Times* 102 (1974): 148. Regarding the history of such faith in "fighting resistance with technology," see Robert Bud, *Penicillin: Triumph and Tragedy* (New York: Oxford University Press, 2007), 116–39.

56. D'Amelio, "Are Family Doctors Prescribing Too Many Antibiotics," 57–59. Among the polled medical residents, this could take the form of a more generalized stab against the grasping aspirations of therapeutic rationalism: "With some residents it seems to be a matter of pride not to use antibiotics. They must feel that they get 'Brownie' points. Some would rather have both the patients and themselves dead in a ditch before they would use an antibiotic without getting back the culture and sensitivity"; D'Amelio, "What Young Doctors Told Us," 148.

57. D'Amelio, "Are Family Doctors Prescribing Too Many Antibiotics," 58.

58. The *Medical Tribune* was apparently distributed freely to 168,000 physicians weekly. See Barry Meier, *Pain Killer: A "Wonder" Drug's Trail of Addiction and Death* (Emmaus, PA: Rodale, 2003), 195.

59. Arthur M. Sackler, "Freedom of Inquiry, Freedom of Thought, Freedom of Expression: 'A Standard to Which the Wise and the Just Can Repair': Observations on Medicines, Medicine, and the Pharmaceutical Industry," 10/18/57, in Box 5, "William Douglas McAdams," FMIP.

60. See Louis Lasagna to Senator Estes Kefauver, 1/26/62, in *DIAA*, Part 7, p. 3632.

61. "Dr. Lasagna Writes on Colds and Antibiotics," *Medical Tribune* 16 (Nov. 19, 1975): 1.

62. Arthur M. Sackler, "The Common and Not-So-Common Cold," *Medical Tribune* 16 (Nov. 19, 1975): 1, 3; see also "The Good Drugs Do to Better Your Health: Why Your Life Will Be Healthier, Happier, and Longer," *Medical Tribune* 16 (Nov. 26, 1975): 9–16.

63. Arthur M. Sackler, "Medicine Is Not a Rigid, Fixed Dogma," *Medical Tribune* 16 (Nov. 26, 1975): 20.

64. Louis Lasagna, "An Open Letter on Colds and Antibiotics," *Medical Tribune* 17 (Nov. 10, 1976): 33.

65. For his earlier musings, see Louis Lasagna, "The Controlled Clinical Trial: Theory and Practice," *Journal of Chronic Diseases* 1 (1955): 353.

66. Louis Lasagna, "An Open Letter on Colds and Antibiotics, Part II," *Medical Tribune* 17 (Nov. 17, 1976): 1. Lasagna, as it turns out, was quite prescient with respect to dental prophylaxis. See the overturning of established wisdom in this respect in Walter Wilson et al., "Prevention of Infective Endocarditis: Guidelines from the American Heart Association: A Guideline from the American Heart Association Rheumatic Fever, Endocarditis, and Kawasaki Disease Committee, Council on Cardiovascular Disease in the Young, and the Council on Clinical Cardiology, Council on Cardiovascular Surgery and Anesthesia, and the Quality of Care and Outcomes Research Interdisciplinary Working Group," *Circulation* 116 (2007): 1736–54.

67. Louis Lasagna, "Statistics, Sophistication, Sophistry, and Sacred Cows," *Clinical Research Proceedings* 3 (1955): 185–87.

68. Lasagna, "Open Letter on Colds and Antibiotics, Part II," 23. On the use of cold war metaphors by those who would defend therapeutic autonomy from

FDA or reformist intervention, see Dominique A. Tobbell, "Who's Winning the Human Race? Cold War as Pharmaceutical Political Strategy," *Journal of the History of Medicine and Allied Sciences* 64 (2009): 429–73, and *Pills, Power, and Policy*, 89–120.

69. "On 'Colds' and Antibiotics, Drs. Lasagna and Sackler," *Medical Tribune* 17 (Mar. 24, 1976): 14; "Internist Advocates Antibiotic Rx Despite Experts' 'Science,'" *Medical Tribune* 17 (Nov. 17, 1976): 23.

70. "Our Readers Write about the President's Cold, Dr. Lasagna's Letter, and Dr. Sackler's View," *Medical Tribune* 16 (Dec. 17, 1975): 35.

71. Ibid.; "Our Readers Still Write about the President's Cold, Dr. Lasagna's Letter, and Dr. Sackler's View," *Medical Tribune* 17 (Feb. 25, 1976): 21; "Physicians Report Experiences with 'Colds' and Antibiotics, Add Their Comments on Views of Drs. Lasagna and Sackler," *Medical Tribune* 17 (Mar. 3, 1976): 12; "On 'Colds' and Antibiotics," 14.

72. Lasagna, "Open Letter on Colds and Antibiotics, Part II," 1, 23; see also "Internist OK's Empiric Use of Antibiotics," *Medical Tribune* 17 (Nov. 24, 1976): 1, 12; "Expert: Gear Antibiotics to Family Practice Reality," *Medical Tribune* 18 (Jan. 12, 1977): 25, 27, 29. Such algorithms actually paralleled how febrile neutropenic oncology patients were being reconsidered in the 1970s and 1980s, though with a different risk-benefit calculus entailed. The degree to which Stead analogized from such patients in his thinking is unclear. See Stephen C. Schimpff and Joseph Aisner, "Empiric Antibiotic Therapy," *Cancer Treatment Reports* 62 (1978): 673–80, and Philip A. Pizzo, K. J. Robichaud, Fred A. Gill, and Frank G. Witebsky, "Empiric Antibiotic and Antifungal Therapy for Cancer Patients with Prolonged Fever and Neutropenia," *American Journal of Medicine* 72 (1982): 101–10.

73. Collin Baker, in "Our Readers Still Write about the President's Cold," 21. On essential medicines, see Jeremy A. Greene, "Making Medicines Essential: The Emergent Centrality of Pharmaceuticals in Global Health," *BioSocieties* 6 (2011): 10–33. Later that year, researchers from Duke University reported the impressive degree of "inappropriate" antibiotic prescribing already prevalent at Duke University Medical Center (in a study conducted using 1973 data); see Mary Castle, Catherine Wilfert, Thomas R. Cate, and Suydam Osterhout, "Antibiotic Use at Duke University Medical Center," *JAMA* 237 (1977): 2819–22.

74. William J. Holloway, in "Our Readers Still Write about the President's Cold," 21; Albert B. Sabin, "'Uncommon Colds' and Antibiotics: Not Good Enough for Either the President or the Citizens of the United States," *Hospital Tribune* 10 (April 19, 1976): 18. See also Sabin, "Control of Infectious Diseases," *Journal of Infectious Diseases* 121 (1970): 91–94.

75. "Physicians Report Experiences with 'Colds' and Antibiotics," 12.

76. "On 'Colds' and Antibiotics," 14.

77. Howard R. Seidenstein, "Trends of Antibiotic Use in the United States" [letter to the editor], *JAMA* 228 (1974): 1098–99.

78. Ibid., 1099.

79. On the PSRO program and its challenges, see *EPI*, Part 2, p. 575; William F. Jessee, William B. Munier, Jonathan E. Fielding, and Michael J. Goran, "PSRO: An Educational Force for Improving Quality of Care," *NEJM* 292 (1975): 668–71; Anthony M. Komaroff, "The PSRO, Quality-Assurance Blues," *NEJM* 298 (1978): 1194–96; T. S. Jost, "Medicare Peer Review Organizations," *Quality Assurance in Health Care* 1 (1989): 235–48; Anita J. Bhatia et al., "Evolution of Quality Review Programs for Medicare: Quality Assurance to Quality Improvement," *Health Care Financing Review* 22 (Fall 2000): 69–74; and George Weisz et al., "The Emergence of Clinical Practice Guidelines," *Milbank Quarterly* 85 (2007): 691–727. The PSRO program was ended in 1982.

80. *CPDI*, Part 6, pp. 2408, 2467. Restrictive measures along such lines had been suggested to the 1960 NAS-NRC Committee on the hematological effects of chloramphenicol; see William H. Kessenich to R. Keith Cannan, 10/28/60, and Cannan to Wesley W. Spink, 12/19/60, both in Box 11, "Special Advisory Committee on FDA to HEW Secretary Flemming, 1960," WSP. However, the committee felt the drug should not be restricted to the hospital and that "sometimes there are entirely proper indications to use Chloramphenicol on patients in the home." See Cannan to Kessenich, 1/11/61, Box 18, ff 5, ATS. Such proposed restrictions had likewise been put before the California State Senate in late 1961, leading to extensive discussion (among representatives from Parke-Davis, the FDA, the state medical society, the state department of public health, and others) regarding the pros and cons of such measures. While a Senate Fact Finding Committee concluded that "the medical profession should act to improve its self-discipline in the use of various therapeutic agents in order to better safeguard the patients," the majority opinion was also that "control of the use of chloromycetin should be left as the responsibility of the medical profession . . . [and] that restriction of the use of chloromycetin to hospitals was not in the best public interest." See *"Particularly Chloromycetin": A Study of Antibiotic Drugs* [report of Senate Fact Finding Committee on Public Health and Safety] (Sacramento: California State Department of Public Health, 1963), 74–76; Maeder, *Adverse Reactions*, 280–90.

81. "Charge to the Ad Hoc Committee on Chloramphenicol," Box 34, ff 10, JOAP.

82. Ibid.

83. Ibid. On the patient package insert for the pill, see Watkins, *On the Pill*, 103–28

84. "Food and Drug Commission Ad Hoc Committee on Chloramphenicol," 2/26/68, Box 34, ff 10, JOAP.

85. "Telephone Discussion with Dr. Joseph A. Sadusk, 8 April 1968," Series 1, DES, "General: 1966–1970," NASD.

86. Ibid.

87. "Charge to the Sub-Committee on Chloramphenicol of the Medical Advisory Board," Box 34, ff 10, JOAP.

88. "Food and Drug Administration Bureau of Medicine Sub-Committee on Chloramphenicol of the Medical Advisory Board," 2/20/69; "Response of the Ad Hoc Committee on Chloramphenicol to the Charges Placed by the Commissioner at the Meeting on February 20, 1969," both in Box 34, ff 10, JOAP.

89. *APM*, Part 3, pp. 1041–42, 1124.

90. Ibid., p. 1132.

91. Ibid., pp. 1131, 1155.

92. Ibid., p. 1132.

93. Ibid., pp. 1125–26, 1171–72. For discussion about the "cost-benefit ratio" of such measures, see pp. 1172–74.

94. Calvin M. Kunin, Thelma Tupasi, and William A. Craig, "Use of Antibiotics: A Brief Exposition of the Problem and Some Tentative Solutions," *Annals of Internal Medicine* 79 (1973): 555–60; George Gee Jackson, "Perspective from a Quarter Century of Antibiotic Usage," *JAMA* 227 (1974): 634–37. Kunin made his first—and first unsuccessful—proposal for the IDSA to establish "councils or a committee to prepare position papers which could be released as news releases and/or articles" in May 1975; "Minutes, Board of Directors," 5/5/75, IDSA 1970–1977, IDSA. On later IDSA reluctance to proceed with Society-sponsored guidelines on antibiotic usage, see "Minutes, Board of Directors," 5/3/76, IDSA 1970–1977, IDSA.

95. Daniel Carpenter and Dominique A. Tobbell, "Bioequivalence: The Regulatory Career of a Pharmaceutical Concept," *Bulletin of the History of Medicine* 85 (2011): 121–26; A. James Lee, Dennis Hefner, Allen Dobson, and Ralph Hardy Jr., "Evaluation of the Maximum Allowable Cost Program," *Health Care Financing Review* 4 (1983): 71–82; William A. Craig et al., "Hospital Use of Antimicrobial Drugs: Survey at 19 Hospitals and Results of Antimicrobial Control Program," *Annals of Internal Medicine* 89 part 2 (1978): 793–95; "A New Wave of Antibiotics Builds," *Science* 214 (1981): 1225–28.

96. *APM*, Part 3, p. 1042.

97. Ibid., pp. 1082–83.

98. "Food and Drug Administration Board of Drugs, 9th Meeting," 12/14–15/72, p. 5, Box 6, "FDA Board of Drugs—Meetings, 1972–1974," HDP.

99. John C. Ballin to Edward H. Kass, 1/12/73; Leighton E. Cluff to Kass, 2/8/73, both in Box 25, ff 36, EKP. Thirty-seven participants attended the meeting. After Kass was unable to attend, George Gee Jackson chaired the meeting; "Minutes of the Conference on Antibiotic Utilization" [3/12/73], accompanying John C. Ballin to participants, 4/3/73, Box 25, ff 36, EKP.

100. "Minutes of the Conference on Antibiotic Utilization."

101. Ibid.

102. Ibid.

103. Simmons and Stolley, "This Is Medical Progress," 1027. Note that the Council on Drugs had been dissolved as part of the AMA's 1973 budget process, though an overarching Department of Drugs remained. See House of Delegates

Proceedings, Clinical Convention, 1972, p. 71; House of Delegates Proceedings, Clinical Convention, 1973, p. 45, both now accessible online at the AMA historical archive website, through The Digital Collection of Historical AMA Documents. For publicly aired concerns that such dissolution stemmed from further AMA-industry accommodation, see *CPDI*, Part 23, pp. 9608–9, and Morton Mintz, "AMA 'Captive' of Drug Firms, Top Aides Say," *Washington Post* (Feb. 7, 1973): A2.

104. John C. Ballin et al., "In Comment," *JAMA* 227 (1974): 1029–30. In the actual minutes of the meeting, sent to participants by Ballin himself, appears the line that when the participants were asked, in speaking for themselves and not their organizations, about antibiotic overuse, "There was a unanimous vote that there appears to be an inappropriate use of antibiotics and a massive overuse." Five of the six authors of the counter-response appear to have been present at the meeting. "Minutes of the Conference on Antibiotic Utilization," Box 25, ff 36, EKP.

105. Ballin et al., "In Comment," 1029–30. They continued: "Scientific knowledge increases rapidly and arises from a consensus of the scientific community. It cannot be laid down by the long, drawn out, and difficult-to-reverse process of regulatory fiat, however enlightened the latter may be."

106. Edward H. Kass and Katherine Murphey Hayes, "A History of the Infectious Diseases Society of America," *Reviews of Infectious Diseases* 10, suppl. (1988): 28–29; "Kay—Discussion AMA Dallas June 28 [1976]," Box 32, ff 48, EKP.

107. Edward C. Rosenow Jr. to Edward H. Kass, 2/14/73, Box 25, ff 23, EKP. The project's grant title, "Development of Standards and Criteria in the Use of Antimicrobial Agents, Particularly as Related to PSRO," reflected such PSRO-inspired origins. See "The ISCAMU Project," Box 26, ff 22, EKP.

108. Edward H. Kass to Edward C. Rosenow, 4/5/74, Box 25, ff 23, EKP.

109. "November 20 [1974] ISCAMU," pp. 1, 15, Box 26, ff 22, EKP.

110. Ibid., pp. 1, 7.

111. Ibid., p. 21.

112. Ibid., p. 23.

113. "The ISCAMU Project"; see also Edward H. Kass, "Antimicrobial Drug Usage in General Hospitals in Pennsylvania," *Annals of Internal Medicine* 89 part 2 (1978): 800–801, and Mervyn Shapiro, Timothy R. Townsend, Bernard Rosner, and Edward H. Kass, "Use of Antimicrobial Drugs in General Hospitals: Patterns of Prophylaxis," *NEJM* 301 (1979): 351–55.

114. "Antibiotic Use 'Remarkably Good' in Pa. Hospitals," *Medical Tribune* 17 (May 5, 1976): 7.

115. "Kay—Discussion AMA Dallas."

116. Calvin M. Kunin and Herman Y. Efron, "Guidelines for Peer Review," *JAMA* 237 (1977): 1001–2.

117. Calvin M. Kunin, "Guidelines and Audits for Use of Antimicrobial Agents

in Hospitals," *Journal of Infectious Diseases* 135 (1977): 335. On the tensions inherent in the production and impact of professional guidelines, see Stefan Timmermans and Marc Berg, *The Gold Standard: The Challenge of Evidence-Based Medicine and Standardization in Health Care* (Philadelphia: Temple University Press, 2003).

118. For an attempt to implement such an audit in a community hospital in Wisconsin as "an educational process and not a policing function," see Rocco Latorraca and Ronald Martins, "Surveillance of Antibiotic Use in a Community Hospital," *JAMA* 242 (1979): 2585–87. See also William R. Fifer, "Antibiotic Utilization Audit: Basic Considerations," *Quality Review Bulletin* 5 (1979): 9–12, and Gerald J. Jogerst and Stephen E. Dippe, "Antibiotic Use among Medical Specialties in a Community Hospital," *JAMA* 245 (1981): 842–46.

119. David McL. Greeley to Maxwell Finland, 4/10/57; June M. Cardullo to Finland, 5/25/57, both in Box 5, ff 45, MFP; M. J. Leitner to Finland, 10/25/57; Finland to Leitner, 10/28/57, both in Box 5, ff 47, MFP; Ernest Jawetz, "Patient, Doctor, Drug, and Bug," *Antibiotics Annual (*1957–1958): 292; David Greenwood, *Antimicrobial Drugs: Chronicle of a Twentieth-Century Medical Triumph* (New York: Oxford University Press, 2008), 237. On the concerns of Eli Lilly about having its Ilotycin (erythromycin) "held in reserve," see Kenneth G. Kohlstaedt to Wesley W. Spink, 3/29/55, Box 15, "Eli Lilly, 1955–1972," WSP. At the Brook Lodge Invitational Symposium on the "Theory and Use of Antibiotic Combinations" held in the fall of 1956, brief discussion developed regarding other nations' attempts to delimit particular antibiotics at the national level. Max Finland remarked that "they are considering this in other places, like Sweden and Denmark where they can control everything, including what's given to patients." See "Brook Lodge Invitational Symposium," p. 172, 71A-5170, Box 5, ff 3, RG 46. On such efforts in Great Britain during this period, see E. M. Tansey and L. A. Reynolds, eds., *Post-penicillin Antibiotics: From Acceptance to Resistance?* Wellcome Witnesses to Twentieth-Century Medicine, vol. 6 (London: Wellcome Trust, 2006), 36–37.

120. John E. McGowan Jr. and Maxwell Finland, "Usage of Antibiotics in a General Hospital: Effect of Requiring Justification," *Journal of Infectious Diseases* 130 (1974): 165–68; see also John E. McGowan and Maxwell Finland, "Effects of Monitoring the Usage of Antibiotics: An Interhospital Comparison," *Southern Medical Journal* 69 (1976): 193–95.

121. See, e.g., Craig et al., "Hospital Use of Antimicrobial Drugs," 793–95, and Rose A. Recco, Jules L. Gladstone, Sandor A. Freedman, and Edward H. Gerken, "Antibiotic Control in a Municipal Hospital," *JAMA* 241 (1979): 2283–86.

122. L. E. Burney, "Staphylococcal Disease: A National Problem," *Proceedings of the National Conference on Hospital-Acquired Staphylococcal Disease* (Atlanta: U.S. Department of Health, Education, and Welfare, 1958), 7.

123. As John McGowan noted in 1979, while "the JCAH [today the Joint

Commission] requires that hospitals receiving accreditation review usage of antibiotics, the Commission does not mandate any specific means for implementing this procedure"; McGowan, "Continuing Education for Improving Antibiotic Usage," *Quality Review Bulletin* 5 (1979): 32.

124. Craig et al., "Hospital Use of Antimicrobial Drugs," 795; "Expert: Gear Antibiotics to Family Practice Reality," 27. Such *Medical Tribune* grumblings, of course, emerged in a medical newspaper published by Arthur Sackler, who had been battling the likes of Max Finland for over two decades. For calls at the time to subject the utility of such restrictive programs to controlled study, see Merle A. Sande, "The Need for Controlled Clinical Studies in Antimicrobial Therapy," *Annals of Internal Medicine* 89 part 2 (1978): 858, and George Gee Jackson, "Antibiotic Policies, Practices and Pressures," *Journal of Antimicrobial Chemotherapy* 5 (1979): 1–5. For ongoing debate regarding the utility of such programs two decades later, see John E. McGowan Jr., "Do Intensive Hospital Antibiotic Control Programs Prevent the Spread of Antibiotic Resistance?" *Infection Control and Hospital Epidemiology* 15 (1994): 478–83, and Scott H. Podolsky, "Pharmacological Restraints: Antibiotic Prescribing and the Limits of Physician Autonomy," in Jeremy A. Greene and Elizabeth Siegel Watkins, eds., *Prescribed: Writing, Filling, Using, and Abusing the Prescription in Modern America* (Baltimore: Johns Hopkins University Press, 2012), 63–65.

125. Of course, as articulated by McGowan and Finland, the restrictive program not only could serve "as a deterrent against abuse of certain antibiotics" but also provided "continuing education of the hospital staff (and the consultants) in the proper use of those agents in the management of their patients"; McGowan and Finland, "Usage of Antibiotics in a General Hospital," 165.

126. C. Henry Kempe, "A Rational Approach to Antibiotic Therapy of Childhood Infections," *Postgraduate Medicine* 24 (Oct. 1958): 339.

127. *APM*, Part 3, 1052–53. For the further hesitation of FDA commissioner Alexander Schmidt to interfere in antibiotic prescribing by 1975, see *CPDI*, Part 27, p. 12460.

128. *EPI*, Part 2, p. 619.

129. On the clinical pharmacy movement, see Dominique A. Tobbell, "'Eroding the Physician's Control of Therapy': The Postwar Politics of the Prescription," and Elizabeth Siegel Watkins, "Deciphering the Prescription: Pharmacists and the Patient Package Insert," both in Greene and Watkins, *Prescribed*, 84–89, 101–5.

130. *EPI*, Part 2, p. 622. As even the audit-supporting Calvin Kunin responded to a question regarding physician resistance to pharmacist input: "I don't believe it's wise to establish a drug utilization review that requires the pharmacist to directly control the prescribing patterns of physicians. The pharmacist who places himself in such a role is biting off more than he can chew. . . . The patient didn't come to the hospital to see the pharmacist"; Calvin Kunin [with Frank G. Sabatino], "Antibiotic Usage Surveillance: An Overview of the Issues," *Quality Review Bulletin* 5 (1979): 8.

131. Leighton E. Cluff, "The Prescribing Habits of Physicians," *Hospital Practice* 2 (Sept. 1967): 104.

132. Reynold Spector and Allen H. Heller, "Education vs. Restriction in Rational Use of Antibiotics," *Hospital Formulary* 13 (Apr. 1978): 273.

133. Wayne A. Ray, Charles F. Federspiel, and William Schaffner, "Prescribing of Chloramphenicol in Ambulatory Practice: An Epidemiologic Study among Tennessee Medicaid Recipients," *Annals of Internal Medicine* 84 (1976): 266–70. Regarding the emergence of such databases as sources of analysis regarding "rational" prescribing at the time, see Jeremy A. Greene, "The Afterlife of the Prescription: The Sciences of Therapeutic Surveillance," in Greene and Watkins, *Prescribed*, 247–50.

134. Wayne A. Ray, Charles F. Federspiel, and William Schaffner, "Prescribing of Tetracyline to Children Less than 8 Years Old: A Two-Year Epidemiologic Study among Ambulatory Tennessee Medicaid Recipients," *JAMA* 237 (1977): 2069–74; William Schaffner, Wayne A. Ray, and Charles F. Federspiel, "Surveillance of Antibiotic Prescribing in Office Practice," *Annals of Internal Medicine* 89 part 2 (1978): 796–99.

135. William Schaffner, Wayne A. Ray, Charles F. Federspiel, and William O. Miller, "Improving Antibiotic Prescribing in Office Practice: A Controlled Trial of Three Educational Methods," *JAMA* 250 (1983): 1728–32. In contrast to the aspirations of Visconti and the clinical pharmacy movement, the Vanderbilt team found that physician educators had a more dramatic impact on prescribing habits than did pharmacist educators, concluding, à la Marshall McLuhan, that "the messenger appears to be more important than the message."

136. Stephen R. Jones et al., "The Effect of an Educational Program upon Hospital Antibiotic Use," *American Journal of the Medical Sciences* 273 (1977): 79–85; James W. Smith and Stephen R. Jones, "An Educational Program for the Rational Use of Antimicrobial Agents," *Southern Medical Journal* 70 (1977): 215–18.

137. *APM*, Part 3, p. 1220. Kunin actually used the term in this instance to contrast such an effort with the more "glamorous" war on cancer. He would in fact be critical of the potential for waste in such early 1970s categorical programs. See p. 1216, and Kunin, "Impact of Infections and Antibiotic Use on Medical Care," *Annals of Internal Medicine* 89 part 1 (1978): 717. A 2005 Cochrane Collaboration assessment of such educational measures continued to point to the benefits of such "multi-faceted interventions where educational interventions occur on many levels" over individual components. The report's authors did not use the terms *synergy* or *fixed-dose combination educational therapy* and in fact pointed to the need for interventions to be tailored to local conditions. See Sandra R. Arnold and Sharon E. Straus, "Interventions to Improve Antibiotic Prescribing Practices in Ambulatory Care," *Cochrane Database of Systematic Reviews* (2005), Issue 4: CD003539, p. 2.

138. See, e.g., *APM*, Part 3, p. 1199.

139. "'SK-Line' Antibiotics," in *EPI*, Part 1, p. 1182.

140. Calvin Kunin, "In Comment [on 'This Is Medical Progress?']," *JAMA* 227 (1974): 1031; see also Kunin, "Clinical Investigators and the Pharmaceutical Industry," *Annals of Internal Medicine* 89 part 2 (1978): 842–45.

141. Kunin, "In Comment," 1030–32.

142. Charles V. Sanders, "Memorial: Jay P. Sanford, 1928–1996," *Transactions of the American Clinical and Climatological Society* 112 (2001): 44–46.

143. On academic counter-detailing (and for the source of the quote), see Leighton E. Cluff, "The Prescribing Habits of Physicians," *Hospital Practice* 2 (Sept. 1967): 103. See also Calvin Kunin, in *CPDI*, Part 12, pp. 5045–46, and "The 'Counter-Detail' Man" [talk given at the University of Kentucky Medical Center], 1971, in Dr. Kunin's possession; Cluff, in *APM*, Part 3, p. 1199; Jerry Avorn and Stephen B. Soumerai, "Improving Drug-Therapy Decisions through Educational Outreach: A Randomized Controlled Trial of Academically Based 'Detailing,'" *NEJM* 308 (1983): 1457–63; and Tobbell, *Pills, Power, and Policy*, 144–47. On more generalized concern regarding CME during this era, see Greene and Podolsky, "Keeping Modern in Medicine," 363–75.

144. Kunin, "Impact of Infections and Antibiotic Use on Medical Care," 717.

145. William R. Barclay, "Prescribing Antibiotics: Not Entirely Bad, but Could Be Better," *JAMA* 245 (1981): 849. See also Jogerst and Dippe, "Antibiotic Use among Medical Specialties in a Community Hospital," 842–46.

146. See Jeremy A. Greene, *Generic: The Unbranding of Modern Medicine* (Baltimore: Johns Hopkins University Press, 2014); Tobbell, "Eroding the Physician's Control of Therapy," 66–90; David Herzberg, "Busted for Blockbusters: 'Scrip Mills,' Quaalude, and Prescribing Power in the 1970s," in Greene and Watkins, *Prescribed*, 207–31; and Nancy Tomes, "Patients or Health-Care Consumers? Why the History of Contested Terms Matters," in Rosemary A. Stevens, Charles E. Rosenberg, and Lawton R. Burns, eds., *History and Health Policy in the United States: Putting the Past Back In* (New Brunswick, NJ: Rutgers University Press, 2006), 83–110.

147. Kunin, "Impact of Infections and Antibiotic Use on Medical Care," 717.

148. Ibid.

Chapter Five: Responding to Antibiotic Resistance

Epigraph. Allen E. Hussar, "A Proposed Crusade for the Rational Use of Antibiotics," *Antibiotics Annual* (1954–1955): 381.

1. Allen E. Hussar, "A Proposed Crusade for the Rational Use of Antibiotics," *Antibiotics Annual* (1954–1955): 381.

2. Ibid.

3. Martin Wood, "The Politicization of Antimicrobial Resistance," *Current Opinion in Infectious Diseases* 11 (1998): 649–51.

4. Regarding the surfacing of scientific and medical issues, see, e.g., the essays in Robert N. Proctor and Londa Schiebinger, eds., *Agnotology: The Making*

and Unmaking of Ignorance (Stanford: Stanford University Press, 2008). Randall Packard and his colleagues offer a useful "processual model" by which "emerging illnesses" surface. However, while "communities of suffering" (and activists grounded in such) play a large role in their model, patients played no role in surfacing antibiotic resistance as a clinical concern, and they have played a minimal role in maintaining its visibility, to the point where the IDSA and its leaders have implored patients to come forth with their stories (e.g., about MRSA) in hopes of influencing the political process regarding budgetary allotments (though there are such emerging groups as the MRSA Survivors Network, founded in 2003). See Randall M. Packard, Peter J. Brown, Ruth L. Berkelman, and Howard Frumkin, "Introduction: Emerging Illness as Social Process," in *Emerging Illnesses and Society: Negotiating the Public Health Agenda* (Baltimore: Johns Hopkins University Press, 2004), 1–35; Brad Spellberg, *Rising Plague: The Global Threat from Deadly Bacteria and Our Dwindling Arsenal to Fight Them* (Amherst, NY: Prometheus Books, 2009), 195–208. Regarding the prioritization of scientific issues as policy concerns, see, e.g., Julius B. Richmond and Milton Kotelchuck, "Political Influences: Rethinking National Health Policy," in Christine H. McGuire, Richard P. Foley, Alan Gorr, and Ronald W. Richards, eds., *Handbook of Health Professions Education* (San Francisco: Jossey-Bass, 1983), 386–404, and "Co-ordination and Development of Strategies for Public Health Promotion in the United States," in Walter W. Holland, Roger Detels, and George Knox, eds., *Oxford Textbook of Public Health*, 2nd ed. (New York: Oxford University Press, 1991), 1:441–54; and John W. Kingdon, *Agendas, Alternatives, and Public Policies*, 2nd ed. (New York: Longman, 2003). Regarding "politicization" itself, see Martin Wood, "The Politicization of Antimicrobial Resistance," 649–51; Robert Bud, "From Epidemic to Scandal: The Politicization of Antibiotic Resistance, 1957–1969," in Carsten Timmermann and Julie Anderson, *Devices and Designs: Medical Technologies in Historical Perspective* (New York: Palgrave MacMillan, 2006), 195–211; and L. A. Reynolds and E. M. Tansey, eds., *Superbugs and Superdrugs: A History of MRSA*, Wellcome Witnesses to Twentieth-Century Medicine, vol. 32 (London: Wellcome Trust, 2008), 75–79.

5. Lawrence Besserman, "The Challenge of Periodization: Old Paradigms and New Perspectives," in *The Challenge of Periodization: Old Paradigms and New Perspectives* (New York: Garland, 1997), 3–27.

6. L. E. Burney, "Staphylococcal Disease: A National Problem," in *Proceedings of the National Conference on Hospital-Acquired Staphylococcal Disease* (Atlanta: U.S. Department of Health, Education, and Welfare, 1958), 5.

7. William C. Summers, "Microbial Drug Resistance: A Historical Perspective," in Richard G. Wax, Kim Lewis, Abigail A. Salyers, and Harry Taber, eds., *Bacterial Resistance to Antimicrobials*, 2nd ed. (Boca Raton, FL: CRC Press, 2008), 1–9; Christoph Gradmann, "Magic Bullets and Moving Targets: Antibiotic Resistance and Experimental Chemotherapy, 1900–1940," *Dynamis* 31 (2011): 305–21.

8. Lowell Rantz to Wesley W. Spink, 10/17/44, Box 32, "Hemolytic Streptococcus Commission, 1943–1946," WSP; Medical Department, U.S. Army, *Preventive Medicine in World War II, vol. 4: Communicable Diseases Transmitted Chiefly through Respiratory and Alimentary Tracts* (Washington, DC: Office of the Surgeon General, Dept. of the Army, 1958), 233–34, and *Internal Medicine in World War II, vol. 2: Infectious Diseases* (Washington, DC: Office of the Surgeon General, Dept. of the Army, 1963), 412–13; John E. Lesch, *The First Miracle Drugs: How the Sulfa Drugs Transformed Medicine* (New York: Oxford University Press, 2007), 228, 236–38, 277.

9. Alexander Fleming, "Penicillin" [Nobel lecture, Dec. 11, 1945], in *Nobel Lectures in Physiology or Medicine, 1942–1962* (Singapore: World Scientific Publishing Company, 1999), 93.

10. On the origins of this trope, see Allan M. Brandt, *No Magic Bullet: A Social History of Venereal Disease in the United States since 1980* (New York: Oxford University Press, 1985).

11. Roy Fraser to Wesley Spink, 7/31/45, Box 32, "Penicillin Resistance, 1945–1963," WSP. For similar musings, see Leslie A. Falk, "Will Penicillin Be Used Indiscriminately?" [correspondence], *JAMA* 127 (1945): 672.

12. As Peter Temin has identified, it appears that most sulfa drugs and antibiotics sold in the United States were obtained via prescription, even prior to the 1951 passage of the Durham-Humphrey amendment and its formal distinction between prescription-only and over-the-counter medications. See Temin, *Taking Your Medicine: Drug Regulation in the United States* (Cambridge, MA: Harvard University Press, 1980), 46–57.

13. Hobart A. Reimann, "Infectious Disease," *JAMA* 132 (1946): 969.

14. Wesley W. Spink, "Staphylococcal Infections and the Problem of Antibiotic-Resistant Staphylococci," *Archives of Internal Medicine* 94 (1954): 174.

15. Mary Barber, "Staphylococcal Infection Due to Penicillin-Resistant Strains," *British Medical Journal* 2 (1947): 863; Mary Barber and Mary Rozwadowska-Dowzenko, "Infection by Penicillin-Resistant Staphylococci," *Lancet* 255 (1948): 641; Paul M. Beigelman and Lowell A. Rantz, "The Clinical Importance of Coagulase-Positive, Penicillin-Resistant *Staphylococcus Aureus*," *NEJM* 242 (1950): 353–58.

16. "Penicillin-Resistant Staphylococci" [unsigned editorial by Maxwell Finland], *NEJM* 242 (1950): 382.

17. See, e.g., Manson Meads to Maxwell Finland, 11/2/48, Box 3, ff 47, MFP, and [Finland], "Penicillin-Resistant Staphylococci," 383.

18. On the rise in staphylococcal infections, see Maxwell Finland, "The Present Status of Antibiotics in Bacterial Infections," *Bulletin of the New York Academy of Medicine* 27 (1951): 215–16. On the rise in penicillin-resistant staphylococci, see James W. Haviland, "Advances in Antibiotic Therapy," *Annals of Internal Medicine* 39 (1953): 311. On the increasing resistance of staphylococci to broad-spectrum agents, see Maxwell Finland, Paul F. Frank, and Claire Wil-

cox, "*In Vitro* Susceptibility of Pathogenic Staphylococci to Seven Antibiotics," *American Journal of Clinical Pathology* 20 (1950): 325–34; Maxwell Finland and Thomas H. Haight, "Antibiotic Resistance of Pathogenic Staphylococci," *Archives of Internal Medicine* 91 (1953): 157; Mark H. Lepper, Harry F. Dowling, George Gee Jackson, and Marvin M. Hirsch, "Epidemiology of Penicillin- and Aureomycin-Resistant Staphylococci in a Hospital Population," *Archives of Internal Medicine* 92 (1953): 48–50; Spink, "Staphylococcal Infections and the Problem of Antibiotic-Resistant Staphylococci," 174; and Wesley W. Spink to Benajmin Carey, 2/3/53, Box 15, "Lederle Laboratories, 1952–1959," WSP.

19. Maxwell Finland to Herbert R. Morgan, 12/11/50, Box 3, ff 57, MFP; Finland to Elmer M. Purcell, 11/6/51, Box 4, ff 36, MFP; Finland to Wesley Spink, 1/12/53, Box 5, ff 9, MFP.

20. [Finland], "Penicillin-Resistant Staphylococci," 383.

21. Finland, "Present Status of Antibiotics in Bacterial Infections," 208–10. For Finland's own unsigned editorial concerning his talk, see "Is It Being Overdone?" *NEJM* 249 (1951): 920. For his enumeration of "omnibiotics," see Finland, "Clinical Uses of the Presently Available Antibiotics," *Antibiotics Annual* (1953–1954): 25.

22. Finland, "Present Status of Antibiotics in Bacterial Infections," 211.

23. Alfred W. Bauer and William M. M. Kirby, "Drug Usage and Antibiotic Susceptibility of Staphylococci," *JAMA* 173 (1960): 475–80. For dissenting data and implications, see Burton Armin Waisbrin, "The Sensitivity of Staphylococci to Antibiotics: A Five-Year Study of Sensitivity, Usage, and Cross Resistance in a General Hospital," *Archives of Internal Medicine* 101 (1958): 397–406.

24. Maxwell Finland, Wilfred F. Jones, and Mildred W. Barnes, "Occurrence of Serious Bacterial Infections since Introduction of Antibacterial Agents," *JAMA* 170 (1959): 2188–97; David E. Rogers, "The Changing Pattern of Life-Threatening Microbial Disease," *NEJM* 261 (1959): 677–83.

25. Harry F. Dowling, Mark H. Lepper, and George G. Jackson, "Clinical Significance of Antibiotic-Resistant Bacteria," *JAMA* 157 (1955): 327–30; Ernest Jawetz, "Antibiotics Revisited: Problems and Prospects after Two Decades," *British Medical Journal* 2 (1963): 952–53.

26. René Dubos, "The Unknowns of Staphylococcal Infection," *Annals of the New York Academy of Sciences* 65 (1956): 243–46, and *The Mirage of Health: Utopias, Progress, and Biological Health* (New York: Harper & Brothers, 1959), esp. chap. 3; Carol L. Moberg, "René Dubos, a Harbinger of Microbial Resistance to Antibiotics," *Perspectives in Biology and Medicine* 42 (1999): 559–80.

27. Lindsey W. Batten, in "Discussion on the Use and Abuse of Antibiotics," *Proceedings of the Royal Society of Medicine* 48 (1955): 360.

28. Hobart A. Reimann, "The Misuse of Antimicrobics," *Medical Clinics of North America* 45 (1961): 855.

29. See, e.g., Edward Ellis, "The Answer to Drug-Resistant Germs: Science Fights Back by Blending Antibiotics and by Looking for New Wonder Drugs,"

New York World-Telegram and The Sun (Feb. 28, 1953), pp. 12–13, in Box 31, ff 2, SWP. See also Robert Bud, *Penicillin: Triumph and Tragedy* (New York: Oxford University Press, 2007), 116–39, and "From Germophobia to the Carefree Life and Back Again: The Lifecycle of the Antibiotic Brand," in Andrea Tone and Elizabeth Siegel Watkins, eds., *Medicating Modern America* (New York: New York University Press, 2007), 31–32.

30. "Remarks of Mr. John E. McKeen, President, Chas. Pfizer & Company, Inc. before Meeting of the New York Security Analysts' Society, February 10, 1953," C38 (9) I, AIHP. With no sense of irony, McKeen's comment that "it has been possible to expand the sales of the Antibiotic Division rapidly only by using vigorous promotional techniques" appeared in the very next paragraph of his talk. On such opportunism at Bayer, see Christoph Gradmann, "Reinventing Infectious Disease: Antibiotic Resistance and Drug Development at the Bayer Company, 1940–1980" (talk given at the European Science Foundation DRUGS Research Network Programme on "Beyond the Magic Bullet: Reframing the History of Antibiotics," University of Oslo, 2011).

31. Félix Martí-Ibañez, "The Philosophical Impact of Antibiotics on Clinical Medicine," *Antibiotics Annual* (1954–1955): 19.

32. John C. Krantz, "The Use and Abuse of the Antibiotics," *Pennsylvania Medical Journal* 58 (1955): 383.

33. Selman Waksman, "Man's War against Microbes" [undated talk, likely early 1960s], Box 39, ff 7, SWP.

34. Lawrence P. Garrod, "The Reactions of Bacteria to Chemotherapeutic Agents," *British Medical Journal* 1 (1951): 210.

35. Hans Molitor to Wesley W. Spink, 2/21/56, Box 29, "The Antibiotic-Resistant Staphylococci," WSP.

36. Selman Waksman, "Antibiotics—20 Years Later," *Bulletin of the New York Academy of Medicine* 37 (1961): 212; Jawetz, "Antibiotics Revisited," 955.

37. Henry Welch, "The Antibiotic-Resistant Staphylococci," *Antibiotics and Chemotherapy* 3 (1953): 561.

38. Ibid., 567.

39. "Staphylococcic Infections," *JAMA* 166 (1958): 1205; Stuart Mudd, "Staphylococcic Infections in the Hospital and Community," *JAMA* 166 (1958): 1178. For the argument that physicians—as opposed to other hospital staff, let alone representatives from the public health system—should take the lead on such activities, see John W. Brown, "Hygiene and Education within Hospitals to Prevent Staphylococcic Infections," *JAMA* 166 (1958): 1191. On such tensions in the British context, see Flurin Condrau and Robert G. W. Kirk, "Negotiating Hospital Infections: The Debate between Ecological Balance and Eradication Strategies in British Hospitals, 1947–1969," *Dynamis* 31 (2011): 385–405.

40. W. Emory Burnett et al., "Program for Prevention and Eradication of Staphylococcic Infections," *JAMA* 166 (1958): 1183.

41. Robert I. Wise, Elizabeth A. Ossman, and Dwight R. Littlefield, "Personal

Reflections on Nosocomial Staphylococcal Infections and the Development of Hospital Surveillance," *Reviews of Infectious Diseases* 11 (1989): 1005–19; Elaine Larson, "A Retrospective on Infection Control. Part 2: Twentieth Century—The Flame Burns," *American Journal of Infection Control* 25 (1997): 340–49. On the earlier British development of hospital infection control in the context of resistant staphylococci, see Kathryn Hillier, "Babies and Bacteria: Phage Typing, Bacteriologists, and the Birth of Infection Control," *Bulletin of the History of Medicine* 80 (2006): 733–61, and Flurin Condrau, "Standardising Infection Control: Antibiotics and Hospital Governance in Britain, 1948–1960," in Christian Bonah, Anne Rasmussen, and Christoph Masutti, eds., *Harmonizing Drugs: Standards in 20th Century Pharmaceutical History* (Paris: Editions Glyphe, 2009), 327–39.

42. Burney, "Staphylococcal Disease," 3.

43. *Proceedings of the National Conference on Hospital-Acquired Staphylococcal Disease*, 147.

44. On the epidemics, see Bud, *Penicillin*, 119–20.

45. "Antibiotics and Influenza," *JAMA* 165 (1957): 58–59; "Antibiotics and Virus Diseases," *JAMA* 165 (1957): 53–54; Robert K. Plumb, "Use of Antibiotics in Flu Is Opposed," *New York Times* (Sept. 7, 1957): 16.

46. Morton Mintz, *The Therapeutic Nightmare* (Boston: Houghton Mifflin, 1965), 221. While such a comment does not appear in the official conference proceedings, it likely followed a paper by two clinicians from the Philippines supporting the up-front use of antibiotics in influenza. See A. A. Florentin and A. B. Sison, "The Role of Antibiotics in Asian Influenza," *Antibiotics Annual* (1957–1958): 941–47. For a similar suggestion to administer antibiotics to vulnerable influenza patients in Britain during the epidemic, see John T. MacFarlane and Michael Worboys, "The Changing Management of Acute Bronchitis in Britain, 1940–1970: The Impact of Antibiotics," *Medical History* 52 (2007): 59–60.

47. Leonard Engel, *Medicine Makers of Kalamazoo* (New York: McGraw-Hill, 1961), 170. For Pfizer's own aggressive response to the influenza epidemic, including the development of a "Pfizer Flu Emergency Plan" that consisted of an antibiotic ordering blank, see George Guess to [Pfizer] district managers, 8/9/57; "Excerpts from Minutes of Meeting of President's Staff Committee," 8/12/57; "Pfizer Flu Emergency Plan"; George Guess to [Pfizer] regional managers, 10/23/57, all in Box 247, FTC Dockets 7211, RG 122; and "Taming of the Flu" in the Pfizer *1957 Annual Report*, included as an advertising supplement to the *New York Times* (Mar. 23, 1958).

48. "Development of WHO Programme of Research in Antibiotics," 3/21/50, WHO/Antib/2; "Report on First Session," 4/19/50, WHO/Antib/8, both in WHOL. This "report" was reprinted as the *World Health Organization Technical Report Series no. 26: Expert Committee on Antibiotics, Report on the First Session, Geneva 11–15 April 1950*, Box 10, ff 6, SWP. See also Brock Chisholm to Selman Waksman, 1/22/51, Box 10, ff 6, SWP, and *The First Ten Years of the World Health Organization* (Geneva: World Health Organization, 1958), 407–8.

49. Selman A. Waksman, "Plan of Organization of an Antibiotics Programme," 3/16/50, WHO/Antib/1, WHOL, also in Box 10, ff 6, SWP; Dr. Grzegorzewski to Dr. Eliot, "Interoffice Memorandum," 3/16/50, TS3, WHOA; Director, Division of Therapeutic Substances, to Waksman, 8/26/53, Box 10, ff 6, SWP; Dir., Ths to DDG through ADG/OTS, "Expert Advisory Panel on Antibiotics," 11/1/56, A10/136/1, WHOA; Gunnar Löfström to Waksman, 3/6/57, A10/136/2, WHOA, also in Box 10, ff 6, SWP.

50. ADG/OTS to Dir., Ths, "Antibiotics/Expert Advisory Panel," 11/2/56, A10/136/1, WHOA.

51. P. Dorolle to Maxwell Finland, 10/8/58, Box 14, ff 11, MFP; Selman A. Waksman to R. Sansonnens, 3/11/59, A10/522/2, WHOA, also in Box 10, ff 6, SWP.

52. Henry Welch was originally considered for the panel; on Max Finland's copy of the proposal his name is crossed off, and he did not attend. See R. Sansonnens to Finland, 2/19/59, Box 14, ff 11, MFP.

53. "Report to the Director-General, Scientific Group on Antibiotics Research, Geneva, 26–30 May 1959," Box 10, ff 6, SWP; an earlier draft version appears in Box 14, ff 11, MFP.

54. Christoph Gradmann, "Sensitive Matters: The World Health Organization and Antibiotic Resistance Testing, 1945–1975," *Social History of Medicine* 26 (2013): 555–74.

55. "Programme Proposals for Research Activities in 1960/61—Antibiotics," 9/11/59, p. 2, ACMR1/21, WHOL.

56. Selman A. Waksman to R. Sansonnens, 3/11/59, A10/522/2, WHOA, also in Box 10, ff 6, SWP; Waksman, "Plans for the Establishment of an International Repository of Antibiotics and of Antibiotic-Producing Cultures of Micro-organisms," 5/31/60, A10/522/3, WHOA; see slightly later iteration, dated 6/10/60, in Box 10, ff 6, SWP. A conference concerning the establishment of such repositories was held in July 1960, but the WHO ultimately decided not to fund the project. See "Programme Proposals for Research Activities," pp. 2–4; "Scientific Group on Antibiotics Research Establishment of a Centre for Antibiotics and for Micro-Organisms Producing Them," A10/522/3, WHOA; R. Sansonnens to Selman Waksman, 3/31/60, Box 10, ff 6, SWP; Sansonnens to Waksman, 2/24/61; Waksman to Sansonnens, 3/1/61; Sansonnens to Waksman, 3/23/61; and Waksman to Sansonnens, 3/27/61, all in Box 1, ff 2, SWP. A centralized "Centre of Information on Antibiotics" was established in Liège, Belgium, in 1965, but it appears to have become dormant by the early 1970s; see "Report on the Activities of the International Centre of Information on Antibiotics from 1965 to 1970," in "Scientific Group on Microbic Sensitivity Testing of Antibiotics, Geneva 29 June–4 July 1970," Box 10, ff 6, SWP. On the dissolution of the "Expert Panel," see Chief, BS, to Director General, 9/3/71; F. T Perkins to D. Tejada-de-Rivero, 10/21/75, both in A10/136/3, WHOA.

57. Walsh McDermott, "The Problem of Staphylococcal Infection," *British*

Medical Journal 2 (1956): 839; see also McDermott, "Host Factors in Experimental Staphylococcal Infection: The Problem of Staphylococcal Infections," *Annals of the New York Academy of Sciences* 65 (1956): 58–66, and Wesley W. Spink to Maxwell Finland, 11/12/54, Box 5, ff 8, MFP, wherein Spink wondered if "Walsh is too immersed in the problem of tuberculosis."

58. McDermott, "Problem of Staphylococcial Infection," 840.

59. See, e.g., Walsh McDermott, "Microbial Persistance," *Yale Journal of Biology and Medicine* 30 (1958): 257–91, and René Dubos, "The Unknowns of Staphylococcal Infection," *Annals of the New York Academy of Sciences* 65 (1956): 243–46, and "The Micro-Environment of Inflammation or Metchnikoff Revisited," *Lancet* 269 (1955): 1–5. Dubos's primary interest, in this respect, concerned tuberculosis; see his "Unsolved Problems in Tuberculosis," *American Review of Tuberculosis* 70 (1954): 391–401.

60. Gordon Ethelbert Ward Wolstenholme and Cecilia M. O'Connor, *Ciba Foundation Symposium on Drug Resistance in Micro-Organisms: Mechanisms of Development* (Boston: Little, Brown, 1957); see also Angela N. H. Creager, "Adaptation or Selection? Old Issues and New Stakes in the Postwar Debates over Bacterial Drug Resistance," *Studies in the History and Philosophy of Biology and Biomedical Science* 38 (2007): 159–90.

61. American Hospital Association, "Prevention and Control of Staphylococcal Infections in Hospitals (Bulletin 1)," in *Proceedings of the National Conference on Hospital-Acquired Disease*, appendix B.

62. See, e.g., Perrin H. Long, "The Use and Abuse of Chemotherapeutic and Antibiotic Agents," *NEJM* 237 (1947): 837–39; "Streptomycin, Its Uses and Abuses," *Journal of the Omaha Mid-West Clinical Society* 8 (1947): 35–39; and "Antibiotics—Their Use and Abuse," *Journal of the Michigan State Medical Society* 59 (1960): 417–18.

63. Maxwell Finland, "Antibiotics—Whither?" *Journal of Pediatrics* 39 (1951): 641 [signed as M.F.].

64. Ibid., 643.

65. See, e.g., "Antibiotics as Prophylactic Agents" [unsigned editorial by Maxwell Finland], *NEJM* 252 (1955): 872–73, and Finland, "Antibiotic-Resistant Micrococcic Infections," *JAMA* 158 (1955): 188–190.

66. Maxwell Finland, "Emergence of Antibiotic Reistance," *NEJM* 253 (1955): 909–22, 969–79, 1019–28. Finland likewise wrote an unsigned accompanying editorial: "Resistance of Bacteria to Antibiotics," *NEJM* 253 (1955): 1040.

67. Maxwell Finland, "Clinical Uses of the Presently Available Antibiotics," *Antibiotics Annual* (1953–1954): 10, 25. For early demonstration of such dreams, even in the "use and abuse" literature, see William S. Hoffman, "Penicillin: Its Use and Possible Abuse," *Journal of the American Dental Association* 34 (1947): 99.

68. Maxwell Finland, "Antibacterial Agents: Uses and Abuses in Treatment and Prophylaxis," *Rhode Island Medical Journal* 43 (1960): 503.

69. Harry Dowling to Maxwell Finland, 12/18/62, Box 2, ff 9, MFP.

70. On Finland's demurring, see Finland to Dowling, 1/7/63; John R. Lewis to Finland, 3/28/63; and Finland to Dowling, 8/28/63, all in Box 2, ff 9, MFP. Two articles appeared in the July 27, 1963, issue of *JAMA*, but despite the promise in an accompanying editorial of subsequent contributions—with the hope that "perhaps it will become possible in the future to match organism, patient, and antimicrobial drug precisely"—very few articles appear to have followed. See William L. Hewitt, "The Penicillins: A Review of Strategy and Tactics," *JAMA* 185 (1963): 264–72; Edwin M. Ory and Ellard M. Yow, "The Use and Abuse of the Broad Spectrum Antibiotics," *JAMA* 185 (1963): 273–79; "Use and Misuse of Antibiotics," *JAMA* 185 (1963): 315; Richard H. Meade, "Treatment of Meningitis," *JAMA* 185 (1963): 1023–30; Robert G. Carney, "Topical Use of Antibiotics," *JAMA* 186 (1963): 646–48; Harry F. Dowling and Mark H. Lepper, "Hepatic Reactions to Tetracycline," *JAMA* 188 (1964): 307–9; and Rees B. Rees, "Cutaneous Reactions to Antibiotics," *JAMA* 189 (1964): 685–86.

71. Tsutomu Watanabe, "Infective Heredity of Multiple Drug Resistance in Bacteria," *Bacteriological Reviews* 27 (1963): 87–115.

72. See Tomoichiro Akiba, Kotaro Koyama, Yoshito Ishiki, Sadao Kimura, and Toshio Fukushima, "On the Mechanism of the Development of Multiple-Drug-Resistant Clones of Shigella," *Japanese Journal of Microbiology* 4 (1960): 219–27; Tsutomu Watanabe and T. Fukusawa, "'Resistance Transfer Factor,' an Episome in Enterobacteriaceae," *Biochemical and Biophysical Research Communications* 3 (1960): 660–65; and David L. Cowen and Alvin B. Segelman, eds., *Antibiotics in Historical Perspective* (Rahway, NJ: Merck Sharp & Dohme International, 1981), 63.

73. See Joshua Lederberg, "Cell Genetics and Hereditary Symbiosis," *Physiological Reviews* 32 (1952): 403–30.

74. Norton D. Zinder, "Infective Heredity in Bacteria," *Cold Spring Harbor Symposium on Quantitative Biology* 18 (1953): 261–69; Joshua Lederberg and Esther M. Lederberg, "Infection and Heredity," in Dorothea Rudnick, ed., *Cellular Mechanisms in Differentiation and Growth* (Princeton: Princeton University Press, 1956), 101–24.

75. Watanabe, "Infective Heredity of Multiple Drug Resistance in Bacteria," 108; "Infectious Drug Resistance," *NEJM* 275 (1966): 277. This unsigned editorial was not written by Max Finland.

76. See, e.g., G. O. Gale and J. S. Kiser, "Antibiotic Resistance—Theory and Practice," *Transactions of the New York Academy of Sciences* 29 (1967): 960–67; Maxwell Finland, "Changing Ecology of Bacterial Infections as Related to Antibacterial Therapy," *Journal of Infectious Diseases* 122 (1970): 430; Susumu Mitsuhashi, ed., *Transferable Drug Resistance Factor R* (Tokyo: University Park Press, 1971); John A. Osmundsen, "Are Germs Winning the War against People?" *Look* (October 18, 1966): 140–41; and "Resistance to Antibiotics," *JAMA* 203 (1968): 1132. Osmundsen's use of "superbugs" is the earliest use of the term to denote antibiotic-resistant bacteria that I have found.

77. "Infectious Drug Resistance," 277.

78. Wrote Henry D. Isenberg and James I. Berkman in 1971: "Hospital-acquired infectious disease was an expression not used in polite society until very recently. . . . The recent acceptance of this phrase is interpreted by some as evidence for an 'honesty transfer factor' passed from the younger to the older generations by a still unknown mechanism. More cynical observers see it as a politically socially expedient opportunity in a decade dedicated to environment and ecology." In "The Role of Drug-Resistant and Drug-Selected Bacteria in Nosocomial Disease," *Annals of the New York Academy of Sciences* 182 (1971): 52.

79. David Kinkela, *DDT and the American Century: Global Health, Environmental Politics, and the Pesticide that Changed the World* (Chapel Hill: University of North Carolina Press, 2011), esp. 106–35.

80. While such scrutiny would spark a ban on the use of certain antibiotics in Britain, it would spark only debate and committee formation in the United States throughout the late 1960s and 1970s. See Bud, *Penicillin*, 163–91, and Mark R. Finlay, "Reframing the History of Agricultural Antibiotics in the Postwar World: An International and Comparative Perspective" (talk given at the European Science Foundation DRUGS Research Network Programme on "Beyond the Magic Bullet: Reframing the History of Antibiotics," University of Oslo, 2011).

81. Richard Gleckman and Morton A. Madoff, "Environmental Pollution with Resistant Microbes," *NEJM* 281 (1969): 677.

82. J. C. Gould, "General Discussion," in Maxwell Finland, Walter Marget, and Karl Bartmann, eds., *Bacterial Infections: Changes in Their Causative Agents, Trends and Possible Basis* (Heidelberg: Springer-Verlag Berlin, 1971), 209. Finland was one of the program organizers at the conference at which this statement was made. On the increasing usage of antibiotics in the United States throughout this period, see Henry E. Simmons and Paul D. Stolley, "This Is Medical Progress? Trends and Consequences of Antibiotic Use in the United States," *JAMA* 227 (1974): 1023–28.

83. Maxwell Finland, in discussion following H. Knothe, "Epidemiological Investigations of R-Factor Bearing Enterobacteriaceae in Man and Animals in Germany," in Finland, Marget, and Bartmann, *Bacterial Infections*, 204–7.

84. Finland, "Changing Ecology of Bacterial Infections," 430. See also Finland, "Changing Patterns of Susceptibility of Common Bacterial Pathogens to Antimicrobial Agents," *Annals of Internal Medicine* 76 (1972): 1033.

85. Finland, "Changing Ecology of Bacterial Infections," 431.

86. Edward H. Kass and Katherine Murphey Hayes, "A History of the Infectious Diseases Society of America," *Reviews of Infectious Diseases* 10, suppl (1988): 4–9.

87. Ibid., 6, 12–13, 17–25, 43–51. Journal revenue would lead to a nearly thirty-fold increase in society assets throughout the 1970s, helping to propel the expansion of the society's mission and mandate. "IDSA Council Meeting Minutes," 10/4/79, Business and Council Meeting Minutes, 1977–1982, IDSA.

88. "IDSA Minutes of the Annual Business Meeting," 10/24/64, IDSA, 1964–1971, IDSA. See also Kass and Murphey, "History of the Infectious Diseases Society of America," 13.

89. Kass and Murphey, "History of the Infectious Diseases Society of America," 27. Part of the issue—particularly with respect to antibiotic resistance—may have concerned the contemporary attitude of the leaders of federal agencies. NIAID director Dorland Davis, in speaking before the Committee on Appropriations of the House of Representatives regarding the increasing prevalence of gram-negative infections, reassured his audience that "fortunately, the drug industry continues to come up with new antibiotics, which Institute grantees promptly test in well-controlled studies." His communication was passed along to the IDSA council. See "Statement by Director, NIAID, Public Health Services, on Allergy and Infectious Diseases," attached to "Minutes of the Meeting of Officers and Council," 5/3/66, IDSA, 1964–1971, IDSA.

90. Leighton E. Cluff, "Infectious Diseases: A Perspective," *Journal of Infectious Diseases* 129 (1974): 88.

91. Calvin M. Kunin, "Antibiotic Accountability," *NEJM* 301 (1979): 380.

92. Regarding irrational prescribing, see Mary Castle, Catherine M. Wilfert, Thomas R. Cate, and Suydam Osterhout, "Antibiotic Use at Duke University Medical Center," *JAMA* 237 (1977): 2819–22; Dennis G. Maki, "A Study of Antimicrobial Misuse in a University Hospital," *American Journal of the Medical Sciences* 275 (1978): 271–82; William Schaffner, Wayne A. Ray, and Charles F. Federspiel, "Surveillance of Antibiotic Prescribing in Office Practice," *Annals of Internal Medicine* 89 part 2 (1978): 796–99; Mervyn Shapiro, Timothy R. Townsend, Bernard Rosner, and Edward H. Kass, "Use of Antimicrobial Drugs in General Hospitals," *NEJM* 301 (1979): 351–55; and Gerald J. Jogerst and Stephen E. Dippe, "Antibiotic Use among Medical Specialties in a Community Hospital," *JAMA* 245 (1981): 842–46. Regarding audits and restrictive policies, see John E. McGowan and Maxwell Finland, "Effects of Monitoring the Usage of Antibiotics: An Interhospital Comparison," *Southern Medical Journal* 69 (1976): 193–95; Calvin M. Kunin and Herman Y. Efron, "Audits of Antimicrobial Usage: Guidelines for Peer Review," *JAMA* 237 (1977): 1001–2; William A. Craig et al., "Hospital Use of Antimicrobial Drugs: Survey at 19 Hospitals and Results of Antimicrobial Control Program," *Annals of Internal Medicine* 89 part 2 (1978): 793–95; and Rose A. Recco, "Antibiotic Control in a Municipal Hospital," *JAMA* 241 (1979): 2283–86.

93. Robert Edelman, "Introduction," *Annals of Internal Medicine* 89 part 2 (1978): 741.

94. Ibid., 741. Three years later, Krause would publish on the enduring relevance of microbial disease, despite the advent of antibiotics; see Richard M. Krause, *The Restless Tide: The Persistent Challenge of the Microbial World* (Washington, DC: National Foundation for Infectious Diseases, 1981).

95. Edelman, "Introduction," 741.

96. Robert Bud, *The Uses of Life: A History of Biotechnology* (New York: Cambridge University Press, 1993), 174.

97. Susan Wright, *Molecular Politics: Developing American and British Regulatory Policy for Genetic Engineering* (Chicago: University of Chicago Press, 1994); Donald S. Fredrickson, *The Recombinant DNA Controversy: Science, Politics, and the Public Interest, 1974–1981* (Washington, DC: ASM Press, 2001); C. K. J. Paniker and K. N. Vamela, "Transferable Chloramphenicol Resistance in *Salmonella typhi*," *Nature* 239 (1972): 109–10; E. S. Anderson, "The Problem and Implications of Chloramphenicol Resistance in the Typhoid Bacillus," *Journal of Hygiene* 74 (1975): 289–99; L. P. Elwell, J. M. Inamine, and B. H. Minshew, "Common Plasmid Specifying Tobramycin Resistance in Two Enteric Bacteria Isolated from Burn Patients," *Antimicrobial Agents and Chemotherapy* 13 (1978): 312–17; P. L. Sadowski, B. C. Peterson, D. N. Gerding, and P. P. Cleary, "Physical Characterization of Ten R Plasmids Obtained from an Outbreak of Nosocomial *Klebsiella pneumoniae* Infections," *Antimicrobial Agents and Chemotherapy* 15 (1979): 616–24; T. F. O'Brien et al., "Dissemination of an Antibiotic Resistance Plasmid in Hospital Patient Flora," *Antimicrobial Agents and Chemotherapy* 17 (1980): 537–43; L. S. Tomkins, J. J. Plorde, and S. Falkow, "Molecular Analysis of R-Factors from Multiresistant Nosocomial Isolates," *Journal of Infectious Diseases* 141 (1980): 625–36.

98. Wendell H. Hall, "The Abuse and Misuse of Antibiotics," *Minnesota Monthly* 35 (1952): 629; Hussar, "Proposed Crusade," 380; Erwin Neter, "Use and Abuse of Antibiotics," *Virginia Medical Monthly* 81 (1954): 362; C. Henry Kempe, "The Use of Antibacterial Agents: Summary of a Round Table Discussion," *Pediatrics* 15 (1955): 227; George R. Fisher, "The Use and Abuse of Antibiotics," *Journal of the Iowa State Medical Society* 50 (1960): 245. Most prominently, see Louis Weinstein, "Antibiotics: I. General Considerations," in Louis Goodman and Alfred Gilman, eds., *The Pharmacological Basis of Therapeutics*, 3rd ed. (New York: Macmillan, 1965), 1187.

99. Charlotte Muller, "The Overmedicated Society: Forces in the Marketplace for Medical Care," *Science* 176 (1972): 488–92.

100. "Antibiotic Sugar Pills," *NEJM* 273 (1965): 825–26. While this unsigned editorial bears all the hallmarks of a Max Finland piece, it does not appear in his collected reprints.

101. "Another Side of the Coin," *JAMA* 227 (1974): 1048. Such "top-down" declarations were occurring in the context of transformative notions of the patient as "consumer" during the same period. See Nancy Tomes, "Patients or Health-Care Consumers? Why the History of Contested Terms Matters," in Rosemary A. Stevens, Charles E. Rosenberg, and Lawton R. Burns, eds., *History and Health Policy in the United States: Putting the Past Back In* (New Brunswick, NJ: Rutgers University Press, 2006), 83–110.

102. "Another Side of the Coin," 1049.

103. Eugene J. Gangarosa, Leonardo J. Mata, David R. Perera, L. Barth Reller,

and Cesar Mendizabal Morris, "Shiga bacillus dysentery in Central America," in A. M. Davies, ed., *Uses of Epidemiology in Planning Health Services* (Belgrade: Savremena Administracija, 1973), 1:259–67; E. S. Anderson and H. R. Smith, "Chloramphenicol Resistance in the Typhoid Bacillus," *British Medical Journal* 3 (1972): 329–30; *Surveillance for the Prevention and Control of Health Hazards Due to Antibiotic-Resistant Enterobacteria: Report of a WHO Meeting*, World Health Organization Technical Report Series 624 (Geneva: World Health Organization, 1978), pp. 13–18, WHOL.

104. Anderson, "Problem and Implications of Chloramphenicol Resistance," 289.

105. Bud, *Penicillin*, 175–91.

106. Milton Silverman, *The Drugging of the Americas: How Multinational Drug Companies Say One Thing about Their Products to Physicians in the United States and Another to Physicians in Latin America* (Berkeley: University of California Presss, 1976), and Jeremy A. Greene, "When Did Medicines Become Essential?" *Bulletin of the World Health Organization* 88 (2010): 483, and "Making Medicines Essential: The Emergent Centrality of Pharmaceuticals in Global Health," *BioSocieties* 6 (2011): 10–33.

107. Lars L. Gustafsson and Katarina Wild, "Marketing of Obsolete Antibiotics in Central America," *Lancet* 1 (1981): 31–33; the authors excused trimethoprim-sulfamethoxazole as the one rational fixed-dose combination antibiotic at the time. See also "Drug Use in the Third World," *Lancet* 2 (1980): 1231–32; M. Moshadeque Hossain, Roger I. Glass, and M. R. Khan, "Antibiotic Use in a Rural Community in Bangladesh," *International Journal of Epidemiology* 11 (1982): 402–5; and Calvin M. Kunin, "Antibiotic Resistance—A World Health Problem We Cannot Ignore," *Annals of Internal Medicine* 99 (1983): 859–60.

108. *Control of Harmful Residues in Food for Human and Animal Consumption: The Public Health Aspects of Antibiotics in Feedstuffs: Report of a Working Group, Bremen, 1–5 October 1973* (Copenhagen: WHO Regional Office for Europe, 1974); *Public Health Aspects of Antibiotic-Resistant Bacteria in the Environment: Report on a Consultation Meeting, Brussels 9–12 December 1975* (Copenhagen: WHO Regional Office for Europe, 1976), both in WHOL; Theodore M. Brown, Marcos Cueto, and Elizabeth Fee, "The World Health Organization and the Transition from 'International' to 'Global' Health," *American Journal of Public Health* 96 (2006): 62–72.

109. *Public Health Aspects of Antibiotic-Resistant Bacteria in the Environment*, p. 13.

110. D. Barua, Y. Watanabe, and L. Houang, "Organization of Surveillance for Antibiotic Resistance in Enteric Bacteria," in *Meeting on Surveillance for Prevention and Control of Health Hazards due to Antibiotic-Resistant Enterobacteriaceae, Geneva 18–24 October 1977*, BAC/ENT/77.2, WHOL.

111. "Plasmid International Conference" [undated account of the meeting], SLP; Joshua Lederberg, foreword to David G. White, Michael N. Alekshun, and

Patrick F. McDermott, eds., *Frontiers in Antimicrobial Resistance: A Tribute to Stuart Levy* (Washington, DC: ASM, 2005), xiii–xiv; Ellen Koenig, "Commentary: The Birth of the Alliance for the Prudent Use of Antibiotics (APUA)," in ibid., 517–18.

112. "Plasmid Conference, Santo Domingo, D.R.: Outline of Prospectus for Fogarty," SLP. Tellingly, proposed titles for the conference included "Medical Implications of Bacterial Plasmids: Antibiotic Resistance, Gastroenteritis, and Recombinant DNA" and "The Problems and Potential Benefits of Bacterial Plasmids to Public Health."

113. Stuart B. Levy, "Survey of a Conference: Turista or Not Turista," in Stuart B. Levy, Royston C. Clowes, and Ellen L. Koenig, eds., *Molecular Biology, Pathogenicity, and Ecology of Bacterial Plasmids* (New York: Plenum, 1981), 676–77. Several participants sent Levy accounts of the antibiotic resistance profiles and "plasmid profiles" of their infecting strains (see SLP).

114. "Plasmid Conference, Santo Domingo"; Bud, *Penicillin*, 188–89.

115. "Statement Regarding Worldwide Antibiotic Misuse," in Levy, Clowes, and Koenig, *Molecular Biology, Pathogenicity, and Ecology of Bacterial Plasmids*, 679–81.

116. Levy served as president of the organization, and Mexico's Yankel Kupersztoch (who had chaired the session at which the "Statement Regarding Worldwide Antibiotic Misuse" was constructed and signed) served as vice president. Levy's explicit emphasis on antibiotic overuse and misuse, its scientific grounding in plasmid ecology, and notions of a rapidly worsening crisis were all evident from the first paragraph of the organization's quarterly newsletter: "Since the initial discovery of antibiotics, scientists were concerned about resistant strains and predicted their emergence. What was not anticipated was the rapidity and extent of the change in the ecology of resistance plasmids which is now evident in our world environment. This situation is the direct result of massive use of antibiotics and a lack of restraint in their prescription and utilization"; Levy, "APUA: History and Goals," *APUA Newsletter* 1 (Summer 1983): 1. On Kupersztoch's chairing of the session at which the statement was signed, see the untitled transcript of "Discussion on 'Appropriate Use of Antibiotics,'" 12/15/83, p.1, SLP.

117. Thomas F. O'Brien, Ralph L. Kent, and Antone A. Medeiros, "Computer-Generated Plots of Results of Antimicrobial-Susceptibility Tests," *JAMA* 210 (1969): 84–92; Thomas F. O'Brien et al., "International Comparison of Prevalence of Resistance to Antibiotics," *JAMA* 239 (1978): 1518–23; John F. Stelling and Thomas F. O'Brien, "Surveillance of Antibiotic Resistance: The WHONET Program," *Clinical Infectious Diseases* 24, suppl 1 (1997): S157–S168; Bud, *Penicillin*, 208, 278; Kathleen T. Young and Thomas F. O'Brien, "Alliance for the Prudent Use of Antibiotics: Scientific Vision and Public Health Mission," in *Frontiers in Antimicrobial Resistance*, 519–27.

118. "Antimicrobial Resistance," *Bulletin of the World Health Organization* 61 (1983): 383–94; "Control of Antibiotic-Resistant Bacteria: Memorandum from

a WHO Meeting," *Bulletin of the World Health Organization* 61 (1983): 423–33; M.T. Parker, "Antibiotic Resistance in Pathogenic Bacteria," *WHO Chronicle* 36 (1982): 191–96; "Surveillance of Antimicrobial Resistance: Report of a Consultation, Geneva 22–26 November, 1982," BVI/PHA/ANT/82.2, WHOL.

119. "Surveillance of Antimicrobial Resistance: Report of a Consultation," pp. 5, 12. The 1981 working group's drawing of attention to the ongoing need for individual hospital antibiotic policies likewise pointed to such links between the local and the global; see "Control of Antibiotic-Resistant Bacteria," 425.

120. Mark Lappé, *Germs That Won't Die: Medical Consequences of Antibiotics* (New York: Anchor, 1982), 161–85. On probiotics, see Scott H. Podolsky, "Metchnikoff and the Microbiome," *Lancet* 380 (2012): 1810–11.

121. Bud, *Penicillin*, 190.

122. Lappé, *Germs That Won't Die*, 118–20; "IDSA Council Meeting Minutes," 5/17/82, IDSA, 1977–1982, IDSA.

123. "IDSA Council Meeting Minutes," 5/18–19/82, in IDSA, 1977–1982, IDSA. In 1980, the traditionally inward-looking society hired its first government lobbyist as a "venture into public affairs." In "IDSA Board of Directors Meeting Minutes," 9/24/80, IDSA, 1977–1982, IDSA.

124. Calvin M. Kunin, "The Responsibility of the Infectious Disease Community," *Journal of Infectious Diseases* 151 (1985): 388–98.

125. Calvin M. Kunin and Stephen Chambers, "Responsibility of the Infectious Disease Community for Optimal Use of Antibiotics: Views of the Membership of the Infectious Diseases Society of America," *Reviews of Infectious Diseases* 7 (1985): 547–59. The mailed questionnaire elicited responses from 881 members and fellows.

126. "IDSA Council Meeting Minutes," 4/29–30/85, IDSA, 1982–1987, IDSA.

127. "Report from the Antimicrobial Agents Committee," *Journal of Infectious Diseases* 156 (1987): 700–705; J. Joseph Marr, Hugh L. Moffet, and Calvin M. Kunin, "Guidelines for the Use of Antimicrobial Agents in Hospitals: A Statement by the Infectious Diseases Society of America," *Journal of Infectious Diseases* 157 (1988): 869–76.

128. "Minutes of the IDSA Retreat," 2/7–8/86, IDSA, 1982–1987, IDSA; Calvin Kunin (chairman), Allen R. Ronald, Hugh Moffet, David N. Gilbert, John E. McGowan Jr., George A. Jacoby Jr., William Craig, and J. Joseph Marr, "Report of the Antibiotic Use Committee, IDSA, February 1986," Box 28, ff 48, EKP. See also Edward H. Kass to Calvin Kunin, 3/3/86, in ibid. Kass's markup of Kunin's draft is illustrative, variably substituting "guide the use," "improve," and "review" for "regulate," and substituting "examine carefully" for "control."

129. AR_1, pp. 353, 355.

130. Jennings cited the Rockefeller University's dean of graduate students, James D. Hirsch, in terms reminiscent of the "antibioticist" literature of the early 1950s: "For many years it was feared that microorganisms would find ways to evade the action of . . . antimicrobial agents and that they would become resistant

faster than new agents could be developed. Resistant forms did appear and this initial fear was reinforced, but in practice, the fear proved to be unfounded. New impressive antimicrobials are now emerging so rapidly that the physician can't keep up with them, much less the microbes!" (AR_1, p. 384). See also James G. Hirsch, "The Greatest Success Story in the History of Medicine," *Medical Times* 108 (Sept. 1980): 42.

131. AR_1, p. 306.

132. Untitled transcript of "Discussion on 'Appropriate Use of Antibiotics,'" 12/15/83, p. 5, SLP.

133. Ernst Freese to L. Houang, 4/25/80, SLP.

134. Ibid.; "International Meeting Program, Status Report," 3/11/81; Chief, Laboratory of Molecular Biology, IRP, NINCDS [Freese] to James [*sic*; listed as John in published writings] P. Burke, Fogarty International Center, 1/19/83, all in SLP.

135. "Justification for a Task Force on Public Health Problems Resulting from the Unlimited Use of Antibiotics in Humans and in Animal Foods," 10/20/80; "Prospectus for the Pan American Conference on Public Health Problems Caused by Antibiotics," spring 1982, both in SLP.

136. "Prospectus," 1/21/82, SLP. Richard Krause likewise appears to have helped advocate for the conference. See Ernst Freese to Stuart Levy, 7/6/81, SLP.

137. Earl C. Chamberlayne to Stuart B. Levy, 5/19/82, SLP.

138. See, e.g., untitled transcript of "Discussion on 'Appropriate Use,'" pp. 12–13.

139. John P. Burke and Stuart B. Levy, "Summary Report on Worldwide Antibiotic Resistance: International Task Forces on Antibiotic Use," *Reviews of Infectious Diseases* 7 (1985): 560–64; Stuart B. Levy, John P. Burke, and Craig K. Wallace, introduction [to the Task Force Reports], *Reviews of Infectious Diseases* 9 (1987): S231.

140. Nananda F. Col and Ronald W. O'Connor, "Estimating Worldwide Current Antibiotic Usage," *Reviews of Infectious Diseases* 9, suppl 3 (1987): S232–S243; Thomas F. O'Brien et al., "Resistance of Bacteria to Antibacterial Agents," in ibid., S244–S260; Harold J. Simon, Peter I. Folb, and Heonir Rocha, "Policies, Laws and Regulations Pertaining to Antibiotics," in ibid., S261–S269; Calvin M. Kunin et al., "Social, Behavioral, and Practical Factors Affecting Antibiotic Use Worldwide," in ibid., S270–S285; Jerry Avorn et al., "Information and Education as Determinants of Antibiotic Use," in ibid., S286–S296; Robert H. Liss and F. Ralph Batchelor, "Economic Evaluations of Antibiotic Use and Resistance—A Perspective," in ibid., S297–S312.

141. Acting Chief, ISB to Acting Director, FIC, "Trip Report[s], Meeting with Dr. T. Kereselidze and Dr. Georges Causse, Both of World Health Organization," 9/19/83, SLP.

142. Tim Beardsley, "NIH Retreat from Controversy," *Nature* 319 (1986): 611.

143. Sidney M. Wolfe to Margaret M. Heckler [secretary of HHS], 9/19/85; "Telephone Conversation with Jerry Avorn," 12/17/85, both in SLP. See also Philip R. Lee to Stuart B. Levy, 2/28/86, SLP.

144. Stuart B. Levy, "Antibiotic Resistance 1992–2002: A Decade's Journey," in Stacey L. Knobler, Stanley M. Lemon, Marjan Nafjafi, and Tom Burroughs, eds., *The Resistance Phenomenon in Microbes and Infectious Disease Vectors: Implications for Human Health and Strategies for Containment* [workshop summary] (Washington, DC: National Academies Press, 2003), 37. The same year, as Jeremy Greene points out, saw the FDA shy away from engagement with the WHO program concerning "the rational use of drugs" and cost-containment more broadly; see *The Rational Use of Drugs: Report of the Conference of Experts, Nairobi, 25–29 November 1985* (Geneva: World Health Organization, 1987), and Jeremy A. Greene, *Generic: The Unbranding of Modern Medicine* (Baltimore: Johns Hopkins University Press, 2014).

145. Ruth L. Berkelman and Phyllis Freeman, "Emerging Infections and the CDC Response," in Packard et al., *Emerging Illnesses and Society*, 356–57. The key factor cited at the time by NIH officials concerned the ongoing sticking-point regarding the inadequacy of the available scientific information on which to base judgments. See Craig D. Wallace to Stuart Levy, 12/9/85, and Donald Ian Macdonald to Sidney M. Wolfe, 1/6/86, both in SLP. For the view that such concerns (and Reagan-era budgetary stringency) superseded even the concerns of industry, see "Telephone Conversation with Ed Kass," 10/22/85, SLP.

146. With respect to antibiotic resistance, see Joshua Lederberg to Philip Lee, 6/9/77, Box 21, ff 70, JLP, and clippings in Box 185, ff 8–10, JLP. In July 1967, though, Lederberg had written a fascinating note to FDA commissioner James Goddard regarding the potential use of antibiotics as growth promoters for infants and children in developing nations, well aware that this could entail a tricky calculus of costs and benefits; Lederberg to Goddard, 7/10/67, Box 18, ff 168, JLP. I am grateful to the late Mark Finlay for having drawn my attention to this letter.

147. Joshua Lederberg, "Mankind Had a Near Miss from a Mystery Pandemic," *Washington Post* (Sept. 7, 1968): A19; see also Lederberg, "The Infamous Black Death May Return to Haunt Us," *Washington Post* (Aug. 31, 1968): A13, and "Yellow Fever Still Survives in Jungles of Africa, Brazil," *Washington Post* (Mar. 14, 1970): A17. On the resemblance of Lederberg to the character of Dr. Stone in Michael Crichton's 1969 thriller, *The Andromeda Strain*, see Bud, *Uses of Life*, 174.

148. Bud, *Penicillin*, 196–97. See also Joshua Lederberg to James Hughes, 7/2/87, Box 228, ff 91, JLP; Lederberg to Disque Deane, 4/4/87, Box 54, ff 2, JLP; "Notice of Closing of 68th Street Avery Gate Secondary to ACT UP Demonstration," Box 54, ff 5, JLP; Lederberg to Andrew Marshall, 7/30/87, Box 54, ff 5, JLP.

149. Joshua Lederberg to Paul and Ellen Growald, 5/7/86, Box 183, ff 21, JLP. On Lederberg's concerns regarding airborne spread, see Lederberg to Anthony Fauci, 2/17/88, Box 39, ff 10, JLP, and Joshua Lederberg, "'Pandemic' as a

Natural Evolutionary Phenomenon," in Arien Mack, ed., *In Time of Plague: The History and Social Consequences of Lethal Epidemic Disease* (New York: New York University Press, 1991), 35–36.

150. Joshua Lederberg to Andrew Marshall, 12/10/86, Box 183, ff 21, JLP.

151. Elie Wiesel to Joshua Lederberg, 10/26/87, Box 54, ff 9, JLP.

152. Joshua Lederberg, "Medical Science, Infectious Disease, and the Unity of Humankind," *JAMA* 260 (1988): 684.

153. Joshua Lederberg to Gustav Nossal, 1/25/88, Box 54, ff 11, JLP. Ironically, Macfarlane Burnet had famously stated in 1972: "If for the present we retain a basic optimism and assume no major catastrophes occur and that any wars are kept at the 'brush fire' level, the most likely forecast about the future of infectious disease is that it will be very dull. There may be some wholly unexpected emergence of a new and dangerous infectious disease, but nothing of the sort has marked the last fifty years"; Macfarlane Burnet and David White, *Natural History of Infectious Disease*, 4th ed. (London: Cambridge University Press, 1972), 263.

154. Bud, *Penicillin*, 197; Joshua Lederberg to Stephen Morse, 2/27/88, Box 54, ff 12, JLP; Joshua Lederberg, "Viruses and Humankind: Intracellular Symbiosis and Evolutionary Competition," in Stephen S. Morse, ed., *Emerging Viruses* (New York: Oxford University Press, 1993), 3–9. See also Morse, preface to *Emerging Viruses*, vii–xi; Llewellyn J. Legters, Linda H. Brink, and Ernest T. Takafuji, "Are We Prepared for a Viral Epidemic Emergency?" in ibid., 269–82; Donald A. Henderson, "Surveillance Systems and Intergovernmental Cooperation," in ibid., 283–89. On Morse, see also Nicholas B. King, "The Scale Politics of Emerging Diseases," *Osiris* 19 (2004): 64–66.

155. Joshua Lederberg to Purnell Choppin, 5/13/88, Box 54, ff 15, JLP; Lederberg to Donald A. Henderson, 7/11/89, Box 55, ff 1, JLP. See also Institute of Medicine, *The Future of Public Health* (Washington, DC: National Academy Press, 1988); Joshua Lederberg and Robert E. Shope, preface, in Lederberg, Shope, and Stanley C. Oaks Jr., eds., *Emerging Infections: Microbial Threats to Health in the United States* (Washington, DC: National Academy Press, 1992), vi.

156. Lederberg, Shope, and Oaks, *Emerging Infections*, 34.

157. Frank M. Snowden, "Emerging and Reemerging Diseases: A Historical Perspective," *Immunological Reviews* 225 (2008): 9–26.

158. Nancy Tomes, "The Making of a Germ Panic, Then and Now," *American Journal of Public Health* 90 (2000): 191–98.

159. Lederberg, Shope, and Oaks, *Emerging Infections*, 159–60.

160. Stuart B. Levy, *The Antibiotic Paradox: How Miracle Drugs are Destroying the Miracle* (New York: Plenum, 1992). The second edition (Cambridge: Perseus, 2002) was subtitled *How the Misuse of Antibiotics Destroys Their Curative Powers*.

161. Berkelman and Freeman, "Emerging Infections and the CDC Response," 350–87; see also Robert J. Howard, "Perspective: Media Coverage of Emerg-

ing and Re-Emerging Diseases behind the Headlines," *Statistics in Medicine* 20 (2001): 1357–61.

162. Bud, *Penicillin*, 199; Laurie Garrett, *The Coming Plague: Newly Emerging Diseases in a World Out of Balance* (New York: Farrar, Straus & Giroux, 1994). On Lederberg's media savvy, see, e.g., Joshua Lederberg to Stuart Levy, 8/1/97, Box 248, ff 61, JLP.

163. See, e.g., Mitchell L. Cohen, "Epidemiology of Drug Resistance: Implications for a Post-Antimicrobial Era," *Science* 257 (1992): 1050–55; Harold C. Neu, "The Crisis in Antibiotic Resistance," *Science* 257 (1992): 1064–72; Mike Toner, "When Bugs Fight Back," *Atlanta Journal-Constitution* (15-part series running Aug. 23–Oct. 16, 1992); Sharon Begley, "The End of Antibiotics?" *Newsweek* (Mar. 7, 1994): 63; Geoffrey Cowley, John F. Lauerman, Karen Springen, Mary Hager, and Pat Wingert, "Too Much of a Good Thing" [internal story capsule], *Newsweek* (Mar. 28, 1994): 50–51; and Michael D. Lemonick, "The Killers All Around," *Time* (Sept. 12, 1994): 62–69 ["Revenge" on cover].

164. On contemporary media interest, see, e.g., "Conversation with [ABC's] Terri Lichstein," 3/1/94, SLP. See also Barbara Rosenkrantz, "Appendix A: Coverage of Antibiotic Resistance in the Popular Literature, 1950 to 1994," in U.S. Congress, Office of Technology Assessment, *Impacts of Antibiotic-Resistant Bacteria* (Washington, DC: U.S. Government Printing Office, 1995).

165. *Impacts of Antibiotic Resistant Bacteria*; American Society for Microbiology, *Report of the ASM Task Force on Antibiotic Resistance* (1995), produced as a supplement to *Antimicrobial Agents and Chemotherapy*, available at www.asm.org/images/PSAB/ar-report.pdf.

166. Centers for Disease Control and Prevention, National Center for Infectious Diseases, *Addressing Emerging Infectious Disease Threats: A Prevention Strategy for the United State*s (Atlanta: Centers for Disease Control, 1994); Berkelman and Freeman, "Emerging Infections and the CDC," 356–61, 370–79.

167. Berkelman and Freeman, "Emerging Infections and the CDC," 363.

168. Stuart Levy to Joshua Lederberg, 8/16/93; Lederberg to Levy, 8/1/97, both in Box 248, ff 61, JLP. For an example of explicit attempts to form a research "network" concerning antibiotic resistance, see the mention of the original RFP by Anthony Fauci (director of NIAID) for the "Network on Antimicrobial Resistance in Staphylococcal aureus (NARSA)," in AR_2, p. 45.

169. Berkelman and Freeman, "Emerging Infections and the CDC," 360, 374–75.

170. *EI*, title page.

171. See Snowdon, "Emerging and Reemerging Diseases," 14–15. Laurie Garrett's follow-up to *Coming Plague* would be *Betrayal of Trust: The Collapse of Global Public Health* (New York: Hyperion, 2000).

172. Lederberg, "Medical Science, Infectious Disease, and the Unity of Humankind," 685.

173. Lederberg, "Pandemic," 31; cf. the dating of the plague conference with Bud, *Penicillin*, 197.

174. For earlier positings of this tension (with respect to the risks and benefits of prophylaxis), see R. Bradley Sack, "Prophylactic Antibiotics? The Individual versus the Community," *NEJM* 300 (1979): 1107–8. On persisting clinician resistance to such a reformulation, see Joshua P. Metlay, Judy A. Shea, Linda B. Crossette, and David A. Asche, "Tensions in Antibiotic Prescribing: Pitting Social Concerns against the Interests of Individual Patients," *Journal of General Internal Medicine* 17 (2002): 87–94; Richard S. Saver, "In Tepid Defense of Population Health: Physicians and Antibiotic Resistance," *American Journal of Law and Medicine* 34 (2008): 431–91. On the ethical challenges inherent in such a reformulation, see Kevin R. Foster and Hajo Grundmann, "Do We Need to Put Society First? The Potential for Tragedy in Antimicrobial Resistance," *PLOS Medicine* 3 (2006): 177–80.

175. King, "Scale Politics of Emerging Diseases," 62–76.

176. *WHO Scientific Working Group on Monitoring and Management of Bacterial Resistance to Antimicrobial Agents* (Geneva: World Health Organization, 1994); *The Medical Impact of the Use of Antimicrobials in Food Animals. Report of a WHO Meeting* (Geneva: World Health Organization, 1997); *The Current Status of Antimicrobial Resistance Surveillance in Europe: Report of a WHO Workshop held in Collaboration with the Italian Associazione Culturale Microbiologia Medica* (Geneva: World Health Organization, 1998).

177. *Report of the Select Committee on Science and Technology of the House of Lords. Resistance to Antibiotics and Other Antimicrobial Agents* (London: Stationery Office, 1998); D. Greenwood, "Lords-a-Leaping: The Soulsby Report on Antimicrobial Drug Resistance," *Journal of Medical Microbiology* 47 (1998): 749–50; Bud, *Penicillin*, 201–6. The European Surveillance of Antimicrobial Consumption project was likewise initiated in 2001.

178. Moysis Lelekis and Panos Gargalianos, "The Influence of National Policies on Antibiotic Prescribing," in Ian M. Gould and Jos W.M. van der Meer, eds., *Antibiotic Policies: Theory and Practice* (New York: Kluwer Academic/Plenum, 2005), 545–66; Aníbal Sosa, "Antibiotic Policies in Developing Countries," in ibid., 593–616; *WHO Global Strategy for Containment of Antibiotic Resistance* (Geneva: World Health Organization, 2001).

179. Polly F. Harrison and Joshua Lederberg, eds., *Antimicrobial Resistance: Issues and Options* (Washington, DC: National Academy Press, 1998), 71.

180. AR_2, p. 81.

181. Ibid., p. 7.

182. King, "Scale Politics of Emerging Diseases," 74–76.

183. As an example, much of the research on the impact of patients (or parents of pediatric patients) on antibiotic-prescribing habits dates from this era onward. For a contemporary overview and study of parental pressures, see How-

ard Bauchner, Stephen I. Pelton, and Jerome O. Klein, "Parents, Physicians, and Antibiotic Use," *Pediatrics* 103 (1999): 395–401. This notion of commodification borrows from Keith Wailoo, *Dying in the City of the Blues: Sickle Cell Anemia and the Politics of Race and Health* (Chapel Hill: University of North Carolina Press, 2001), 17–18.

184. Interagency Task Force on Antimicrobial Resistance, "A Public Health Action Plan to Combat Antimicrobial Resistance," 2001, available at www.cdc.gov/drugresistance/actionplan/aractionplan.pdf.

185. John G. Bartlett, Neil O. Fishman, and Robert J. Guidos, "Statement of the Infectious Diseases Society of America before the Food and Drug Administration Part 15 Hearing Panel on Antimicrobial Resistance, April 28, 2008," available at www.idsociety.org/US_AR_Efforts; "Inventory of Projects. Progress Report: Implementation of a Public Health Action Plan to Combat Antimicrobial Resistance, Progress Through 2007," available at www.cdc.gov/drugresistance/actionplan/2007_report/ann_rept.pdf; N. Kent Peters, Dennis M. Dixon, Steven M. Holland, and Anthony S. Fauci, "The Research Agenda of the National Institute of Allergy and Infectious Diseases for Antimicrobial Resistance," *Journal of Infectious Diseases* 197 (2008): 1087–93; *ES*, pp. 3–11, 67–73; Brad Spellberg, "Testimony of the Infectious Diseases Society of America: Antibiotic Resistance: Promoting Critically Needed Antibiotic Research and Development and Appropriate Use ('Stewardship') of These Precious Drugs," in *PDAEJU*, pp. 74–102.

186. On the AMA, see, e.g., "Statement of the American Medical Association" [presented by Sandra Adamson Fryhofer], in *PDAEJU*, pp. 103–11.

187. "Facilitators Report, Participants in Spring Long Range Planning Retreat [4/24–25/93]," 6/22/93, independent file, IDSA. On IDSA-FDA collaboration, see David N. Gilbert, Thomas R. Beam Jr., and Calvin M. Kunin, "The Path to New FDA Guidelines for Clinical Evaluation of Anti-Infective Drugs," *Reviews of Infectious Diseases* 13, suppl 10 (1991): S890–S894; Thomas R. Beam Jr., David N. Gilbert, and Calvin M. Kunin, "General Guidelines for the Clinical Evaluation of Anti-Infective Drug Products," *Clinical Infectious Diseases* 15, no. S1 (1992): S5–S32.

188. "IDSA Council Meeting," 1/9–10/99, IDSA Council 1996–1999, IDSA.

189. Ibid.

190. "IDSA Council and Committee Chairs Meeting," 11/12/98, IDSA Council 1996–1999, IDSA.

191. "IDSA Council Meeting," 10/4/00; "IDSA Council Meeting," 7/25/01; "2002 IDSA Member Needs Survey," all in IDSA Board of Directors/Education and Research Foundation Executive Committee, 2000–2002, IDSA.

192. "IDSA Executive Committee Conference Call," 12/5/01; "IDSA Executive Committee Conference Call," 1/9/02, both in IDSA Board of Directors/Education and Research Foundation Executive Committee, 2000–2002, IDSA; David M. Shlaes and Robert C. Moellering, "The United States Food and Drug Administration and the End of Antibiotics," *Clinical Infectious Diseases* 34 (2002): 420–22.

193. David N. Gilbert and John E. Edwards Jr., "Is There Hope for the Prevention of Future Antimicrobial Shortages?" *Clinical Infectious Diseases* 35 (2002): 215–16; David M. Shlaes, "Reply," *Clinical Infectious Diseases* 35 (2002): 216–17; "IDSA Executive Committee Conference Call," 12/4/02, Board of Directors /Education and Research Foundation Executive Committee, 2000–2002, IDSA.

194. "IDSA Council Meeting," 3/21–23/03, IDSA Board of Directors/Education and Research Foundation Executive Committee, 2003–2006, IDSA.

195. Ibid.

196. "IDSA Executive Committee Conference Call," 5/7/03; "IDSA Council Meeting," 6/6–8/03; "IDSA Executive Committee Conference Call," 9/3/03; "IDSA Executive Committee Conference Call," 11/5/03; "IDSA Board Meeting," 3/12–13/04, all in IDSA Board of Directors/Education and Research Foundation Executive Committee, 2003–2006, IDSA; Brad Spellberg, John H. Powers, Eric P. Brass, Loren G. Miller, and John E. Edwards Jr., "Trends in Antimicrobial Drug Development: Implications for the Future," *Clinical Infectious Diseases* 38 (2004): 1279–86.

197. For its work to inform antibiotic stewardship programs, see David M. Shlaes et al., "Society for Healthcare Epidemiology of America and Infectious Diseases Society of America Joint Committee on the Prevention of Antimicrobial Resistance: Guidelines for the Prevention of Antimicrobial Resistance in Hospitals," *Infection Control and Hospital Epidemiology* 18 (1997): 275–91; Timothy H. Dellit et al., "Infectious Diseases Society of America and the Society for Healthcare Epidemiology of America Guidelines for Developing an Institutional Program to Enhance Antimicrobial Stewardship," *Clinical Infectious Diseases* 44 (2007): 159–77. For its public focus on the drying pharmaceutical pipeline, see, e.g., the congressional testimony of John E. Edwards Jr., on behalf of the IDSA, in *PB*, pp. 105–8. It became clear early in the formulation of the task force that it would "focus on the development of new drugs (especially for resistant pathogens)," given that other committees within the IDSA were working at the time on antimicrobial resistance. By 2004, the name of the task force would be changed to the Task Force on Antimicrobial Availability, subsequently the AATF. For early discussion regarding the purview of the task force, see "IDSA Executive Committee Conference Call," 2/15/03, IDSA Board of Directors/Education and Research Foundation Executive Committee, 2003–2006, IDSA.

198. "Special Conference Call of the IDSA Board of Directors," 5/5/04, IDSA Board of Directors/Education and Research Foundation Executive Committee, 2003–2006, IDSA.

199. Infectious Diseases Society of America, *Bad Bugs, No Drugs: As Antibiotic Discovery Stagnates . . . A Public Health Crisis Brews* (Alexandria, VA: IDSA, 2004). See also George H. Talbot, John Bradley, John E. Edwards Jr., David Gilbert, Michael Scheld, and John G. Bartlett, "Bad Bugs Need Drugs: An Update on the Development Pipeline from the Antimicrobial Availability Task Force of the Infectious Diseases Society of America," *Clinical Infectious Diseases* 42 (2006): 657–68.

200. *Bad Bugs, No Drugs*, 22. See also Brad Spellberg et al., "The Epidemic of Antibiotic-Resistant Infections: A Call to Action for the Medical Community from the Infectious Diseases Society of America," *Clinical Infectious Diseases* 46 (2008): 155–64; Calvin M. Kunin, "Why Did It Take the Infectious Diseases Society of America So Long to Address the Problem of Antibiotic Resistance?" *Clinical Infectious Diseases* 46 (2008): 1791–92.

201. Helen W. Boucher et al., "Bad Bugs, No Drugs: No ESKAPE! An Update from the Infectious Diseases Society of America," *Clinical Infectious Diseases* 48 (2009): 1–12; "The 10 x '20 Initiative: Pursuing a Global Commitment to Develop 10 New Antibacterial Drugs by 2020," *Clinical Infectious Diseases* 50 (2010): 1081–83.

202. Brad Spellberg et al., "Reply to Kunin: Infectious Diseases Society of America's Efforts to Contain Antibiotic Resistance," *Clinical Infectious Diseases* 46 (2008): 1793.

203. "IDSA Board Meeting," 3/2–3/07; "IDSA Board Meeting," 10/3–4/07, both in IDSA Board of Directors/Education and Research Foundation Executive Committee, 2007–2009, IDSA.

204. In 2012, IDSA formalized such an integrated mandate and began efforts to combine the three committees into a single Antimicrobial Resistance Committee, under the leadership of UCSF's Henry Chambers. See "IDSA Board Meeting," 6/22–23/12; "IDSA Executive Committee," 9/12/12, both in IDSA Board of Directors/Education and Research Foundation Executive Committee, 2010–2012, IDSA.

205. See, e.g., *PDAEJU*, pp. 74–102; "The STAAR Act Coalition," in AR_4, pp. 106–8; "Statement of the Infectious Diseases Society of America presented at the Interagency Task Force on Antimicrobial Resistance (ITFAR) Meeting" (Washington, DC, Nov. 15, 2012), available at www.idsociety.org/US_AR_Efforts/.

206. IDSA, "Combating Antimicrobial Resistance: Policy Recommendations to Save Lives," *Clinical Infectious Diseases* 52, suppl 5 (2011): S397–S428. The IDSA advocated, e.g., for $1.7 billion for the assistant secretary for preparedness and response's Biomedical Advanced Research and Development Authority (BARDA), an additional $500 million for the NIAID, $50 million for the CDC, and $40 million for the FDA's Center for Drug Evaluation and Research.

207. This derived from the GAIN (Generating Antibiotic Incentives Now) Act and passed as part of FDASIA (the Food and Drug Administration Safety and Innovation Act) in 2012. It permitted five years of extended market exclusivity for qualifying products.

208. As a brief but representative sampling, drawn from the last four months of 2013 alone, see the CDC's *Antibiotic Resistance Threats in the United States, 2013*, available at www.cdc.gov/drugresistance/threat-report-2013/; David E. Hoffman, "Waking up to a Medical 'Nightmare,'" *Washington Post* (Oct. 22, 2013): A17; Charlotte Alter, "Experts Warn of Antibiotics 'Crisis': Fears of a 'Superbug,'" healthland.time.com (Nov. 18, 2013); Megan McArdle, "The Post-

Antibiotic World Would Have Fewer Miracles," *Bloomberg Opinion* (Nov. 24, 2013); Mike Wereschagin, "Time Bomb," *Pittsburgh Tribune Review* (Dec. 8, 2013): A1, A6.

209. For still earlier use of "doomsday," see Sebastian G. B. Amyes, *Magic Bullets, Lost Horizons: The Rise and Fall of Antibiotics* (London: Taylor & Francis, 2001), 171–97. For "crying wolf," see Jim Matheson, "Regarding Introduction of the Strategies to Address Antimicrobial Resistance (STAAR) Act," 155 Cong. Rec. (May 13, 2009): H12445. On such "boy who cries wolf" fatigue more generally, see Tomes, "Making of a Germ Panic," 197.

210. See, e.g., Kate Rawson, "GAIN Begins (Part 1): FDA Completes Turnaround on Antibiotic R&D," *RPM Report* (Feb. 20, 2013).

211. "Docket No. FDA-20120N-1248, Statement of the Infectious Diseases Society of America (IDSA)," available at www.idsociety.org/uploadedFiles/IDSA/Policy_and_Advocacy/Current_Topics_and_Issues/Advancing_Product_Research_and_Development/Bad_Bugs_No_Drugs/Statements/IDSA%20LPAD%20Statement%20to%20FDA.March%201%202013.pdf.

212. See, e.g., the discussion in "A New Pathway for Antibiotic Innovation: Exploring Drug Development for Limited Populations" [meeting organized by the PEW Charitable Trusts], 1/31/13, transcript available at www.pewhealth.org/uploadedFiles/PHG/Content_Level_Pages/Other_Resource/antibiotics-transcript.pdf. On Dec. 12, 2013, the ADAPT Act (Antibiotic Development to Advance Patient Treatment Act of 2013), H.R. 3742, "to provide for approval of certain drugs and biological products indicated for use in a limited population of patients in order to address increases in bacterial and fungal resistance to drugs and biological products, and for other purposes," was brought forth by bipartisan supporters in the House of Representatives. More such activity is expected as this book goes to press.

213. This is not just antiquarian musing. Of nearly 200 antibiotic advertisements from the year 2000, none advocated conservative use or mentioned antibiotic resistance. See Jacob Gilad, Lia Moran, Francisc Schlaeffer, and Abraham Borer, "Antibiotic Drug Advertising in Medical Journals," *Scandinavian Journal of Infectious Diseases* 37 (2005): 910–12.

214. For the back-and-forth between Outterson and colleagues, on the one hand, and Brad Spellberg, on the other, see Kevin Outterson, Julie Balch Samora, and Karen Keller-Cuda, "Will Longer Antimicrobial Patents Improve Global Public Health?" *Lancet Infectious Diseases* 7 (2007): 559–66; Brad Spellberg, "Antibiotic Resistance and Antibiotic Development," *Lancet Infectious Diseases* 8 (2008): 211–12; Outterson, "Author's Reply," ibid., 212–13; Spellberg, *Rising Plague*, 198–206; Aaron S. Kesselheim and Kevin Outterson, "Fighting Antibiotic Resistance: Marrying New Financial Incentives to Meeting Public Health Goals," *Health Affairs* 29 (2010): 1689–96; Kevin Outterson, John H. Powers, Ian M. Gould, and Aaron S. Kesselheim, "Questions about the 10 x '20 Initiative," *Clinical Infectious Diseases* 51 (2011): 751–52; and Outterson and Kesselheim, "Improving

Antibiotic Markets for Long Term Sustainability," *Yale Journal of Health Policy, Law, and Ethics* 11 (2011): 103–67.

215. Tamar F. Barlam and Margarita Divall, "Antibiotic-Stewardship Practices at Top Academic Centers throughout the United States and at Hospitals throughout Massachusetts," *Infection Control and Hospital Epidemiology* 27 (2006): 695, 700. Among academic hospitals, the figure was 89%.

216. Ibid., 697. "As one program director noted, it is 'unpleasant' to be involved with antibiotic oversight" (ibid.).

Conclusion

Epigraphs. Selman A. Waksman, "Antibiotics" [unpublished interview with R. Liles], 10/1/63, Box 31, ff 3, SWP; Sally C. Davies, *Annual Report of the Chief Medical Officer, vol. 2, 2011: Infections and the Rise of Antimicrobial Resistance* (London: Department of Health, Mar. 2013).

1. Selman A. Waksman, *The Antibiotic Era: A History of Antibiotics and of Their Role in the Conquest of Infectious Diseases and in Other Fields of Human Endeavor* (Tokyo: Waksman Foundation of Japan, 1975). Waksman reportedly first uttered the "ancient saying" that "From the Earth Shall Come Thy Salvation" upon being notified of his receipt of the 1952 Nobel Prize; Rutgers then used the phrase on the cover of the publicity folder announcing Waksman's award (Box 33, ff 10, SWP). One assumes that Waksman was referring to the text of Isaiah 45:8, which comes closest to his phraseology.

2. Mark L. Nelson [of Paratek Pharmaceuticals], in AR_3, p. 74.

3. AR_3, pp. 61–62. One can, of course, likewise support present directions by invoking more problematic pasts, as has happened when those describing the benefits of the unbridled antibiotic era have referred back to the era of difficult-to-treat mastoiditis, empyemas, etc.

4. See, e.g., Davies, *Annual Report*, 14; *Antibiotic Resistance Threats in the United States, 2013* (Atlanta: Centers for Disease Control and Prevention, 2013), 5, available at www.cdc.gov/drugresistance/threat-report-2013/.

5. Harry M. Marks, *The Progress of Experiment: Science and Therapeutic Reform in the United States, 1900–1990* (New York: Cambridge University Press), 1997.

6. John Harley Warner, *The Therapeutic Perspective: Medical Practice, Knowledge, and Identity in America, 1820–1885* (Cambridge, MA: Harvard University Press, 1986), and *Against the Spirit of System: The French Impulse in Nineteenth-Century American Medicine* (Princeton: Princeton University Press, 1998); Charles S. Bryan and Scott H. Podolsky, "Dr. Holmes at 200—The Spirit of Skepticism," *NEJM* 361 (2009): 846–47; James Whorton, "'Antibiotic Abandon': The Resurgence of Therapeutic Rationalism," in John Parascandola, ed., *The History of Antibiotics: A Symposium* (Madison, WI: American Institute of the History of Pharmacy, 1980), 125–36; Jeremy A. Greene, David S. Jones, and Scott H. Po-

dolsky, "Therapeutic Evolution and the Challenge of Rational Medicine," *NEJM* 367 (2012): 1077–82.

7. Sydney Wolfe, in AR_I, p. 398.

8. Daniel J. Shapiro, Lauri A. Hicks, Andrew T. Pavia, and Adam L. Hersh, "Antibiotic Prescribing for Adults in Ambulatory Care in the USA, 2007–09," *Journal of Antimicrobial Chemotherapy* 69 (2014): 234–40; Michael L. Barnett and Jeffrey A. Linder, "Antibiotic Prescribing to Adults with Sore Throat in the United States, 1997–2010," *JAMA Internal Medicine* 174 (2014): 138–40.

9. Jeanne Daly, *Evidence-Based Medicine and the Search for a Science of Clinical Care* (Berkeley: University of California Press, 2005), 89. For a McMaster study on "rational" and "irrational" antibiotic therapy, in which the authors thanked Sackett for his encouragement and advice, see M. R. Achong et al., "Changes in Hospital Antibiotic Therapy after a Quality-of-Use Study," *Lancet* 2 (1977): 1118–22.

10. Stefan Timmermans and Marc Berg, *The Gold Standard: The Challenge of Evidence-Based Medicine and Standardization in Health Care* (Philadelphia: Temple University Press, 2003); George Weisz, "From Clinical Counting to Evidence-Based Medicine," in Gérard Jorland, Annick Opinel, and George Weisz, eds., *Body Counts: Medical Quantification in Historical and Sociological Perspective* (Montreal: McGill-Queen's University Press, 2005), 377–93.

11. Michael Misocky, "The Epidemic of Antibiotic Resistance: A Legal Remedy to Eradicate the 'Bugs' in the Treatment of Infectious Diseases," *Akron Law Review* 30 (1996–1997): 745–46; see also Scott B. Markow, "Penetrating the Walls of Drug-Resistant Bacteria: A Statutory Prescription to Combat Antibiotic Misuse," *Georgetown Law Journal* 87 (1998–1999): 531–62.

12. Tamar F. Barlam and Margarita Divall, "Antibiotic-Stewardship Practices at Top Academic Centers throughout the United States and at Hospitals throughout Massachusetts," *Infection Control and Hospital Epidemiology* 27 (2006): 695–703.

13. For a general introduction to this literature, see Moyssis Lelekis and Panos Gargalianos, "The Influence of National Policies on Antibiotic Prescribing," in Ian M. Gould and Jos W. M. van der Meer, eds., *Antibiotic Policies: Theory and Practice* (New York: Kluwer Academic/Plenum Publishers, 2005), 545–66.

14. Zdeněk Modr, "Antibiotic Policy in Czechoslovakia," *Journal of Antimicrobial Chemotherapy* 4 (1978): 305–8, and "Statutory Control of Antibiotic Use in Man versus Voluntary Restriction," in Sir Charles H. Stuart-Harris and David M. Harris, eds., *The Control of Antibiotic-Resistant Bacteria* (London: Academic Press, 1982), 211–26. In the latter article, Modr directly cited Max Finland, along with Britain's Mary Barber, as inspiration for the invoking of antibiotic policies. For the shifts in antibiotic usage (but reappropriation of the "antibiotic centres") attendant to the political shift in the former Czechoslovakia since the fall of the Communist regime, see V. Jindrák et al., "Improvements

in Antibiotic Prescribing by Community Paediatricians in the Czech Republic," *Eurosurveillance* 13 (2008): 1–5.

15. Flemming Hald Steffensen et al., "Changes in Reimbursement Policy for Antibiotics and Prescribing Patterns in General Practice," *Clinical Microbiology and Infection* 3 (1997): 653–57; Thomas Lund Sørensen and Dominique Monnet, "Control of Antibiotic Use in the Community: The Danish Experience," *Infection Control and Hospital Epidemiology* 21 (2000): 387–89; Gaia Lusini et al., "Antibiotic Prescribing in Paediatric Populations: A Comparison between Viareggio, Italy and Funen, Denmark," *European Journal of Public Health* 19 (2009): 434–38.

16. S. Mölstad et al., "Sustained Reduction of Antibiotic Use and Low Bacterial Resistance: 10-Year Follow-up of the Swedish Strama Programme," *Lancet Infectious Diseases* 8 (2008): 125–32.

17. Helena Seppälä et al., "The Effect of Changes in the Consumption of Macrolide Antibiotics on Erythromycin Resistance in Group A Streptococci in Finland," *NEJM* 337 (1997): 441–46. The article had been cited over 750 times as of August 2013, as per Web of Science. Iceland, in response to increasing pneumococcal penicillin resistance, utilized broad-based public and clinician educational measures by "key opinion leaders" in the 1990s to lead to decreased antibiotic use (especially in children) and an apparently correlated drop in pneumococcal resistance. See Karl G. Kristinsson, "Modification of Prescribers' Behavior: The Icelandic Approach," *Clinical Microbiology and Infection* 5 (1999): 4S43–4S47. Norway, with a drug regulatory system based on a "clause of need," has long used antibiotics at a rate far lower than its European counterparts. For more recent educational efforts in Norway to reduce unnecessary prescribing, see Svein Gjelstad et al., "Improving Antibiotic Prescribing in Acute Respiratory Tract Infections: Cluster Randomized Trial from Norwegian General Practice (Prescription Peer Academic Detailing (Rx-Pad) Study)," *British Medical Journal* 347 (2013): f4403, available at http://dx.doi.org/10.1136/bmj.f4403.

18. Mölstad et al., "Sustained Reduction of Antibiotic Use and Low Bacterial Resistance," 129–30. For the similar absence of such adverse consequences from reduced antibiotic rates, see Dilip Nathwani, Jacqueline Sneddon, Andrea Patton, and William Malcolm, "Antimicrobial Stewardship in Scotland: Impact of a National Program," *Antimicrobial Resistance and Infection Control* 1 (2012): 7, doi:10.1186/2047-2994-1-7. The formation of the European Union has provided the opportunity for more easily facilitated between-country comparisons. See, e.g., Otto Cars, Sigvard Mölstad, and Arne Melander, "Variation in Antibiotic Use in the European Union," *Lancet* 357 (2001): 1851–53; Herman Goosens, Matus Ferech, Robert Vander Stichele, and Monique Elseviers, "Outpatient Antibiotic Use in Europe and Association with Resistance: A Cross-National Database Study," *Lancet* 365 (2005): 579–87; and Germaine Hanquet et al., "Surveillance of Invasive Pneumococcal Disease in 30 EU Countries: Towards a European System?" *Vaccine* 28 (2010): 3920–28.

19. See, e.g., *The Rational Use of Drugs: Report of the Conference of Experts, Nairobi, 25–29 November 1985* (Geneva: World Health Organization, 1987), and *WHO Global Strategy for Containment of Antibiotic Resistance* (Geneva: World Health Organization, 2001).

20. For the similar drawing of attention to such historical structural factors in Great Britain, see John T. MacFarlane and Michael Worboys, "The Changing Management of Acute Bronchitis in Britain, 1940–1970: The Impact of Antibiotics," *Medical History* 52 (2007): 47–72.

21. Both talks were delivered at the European Science Foundation DRUGS Research Network Programme on "Is This the End? The Eclipse of the Therapeutic Revolution" (University of Zurich, Oct. 4–6, 2012). See also Kristin Peterson, *Speculative Markets: Drug Circuits and Derivative Life in Nigeria* (Durham, NC: Duke University Press, 2014).

22. C. M. Chukwuani, M. Onifade, and K. Sumonu, "Survey of Drug Use Practices and Antibiotic Prescribing Pattern at a General Hospital in Nigeria," *Pharmacy World & Science* 24 (2002): 188–95; Joseph O. Fadare and Igbiks Tamuno, "Antibiotic Self-Medication among University Medical Undergraduates in Northern Nigeria," *Journal of Public Health and Epidemiology* 3 (2011): 217–20; C. E. Vialle-Valentin, R. F. LeCates, F. Zhang, A. T. Testa, and D. Ross-Degnan, "Predictors of Antibiotic Use in African Communities: Evidence from Medicines Household Surveys in Five Countries," *Tropical Medicine and International Health* 17 (2012): 211–22; Olanike O. Kehinde and Babatunde E. Ogunnowo, "The Pattern of Antibiotic Use in an Urban Slum in Lagos State, Nigeria," *West African Journal of Pharmacy* 24 (2013): 49–57.

23. They have likewise drawn attention to the limitations of the evaluations themselves of such measures. See K. A. Holloway, V. Ivanovska, A. K. Wagner, C. Vialle-Valentin, and D. Ross-Degnan, "Have We Improved Use of Medicines in Developing and Transitional Countries and Do We Know How To? Two Decades of Evidence," *Tropical Medicine and International Health* 18 (2013): 656–64.

24. Salmaan Keshavjee and Paul E. Farmer, "Tuberculosis, Drug Resistance, and the History of Modern Medicine," *NEJM* 367 (2012): 931–36.

25. René and Jean Dubos, *The White Plague: Tuberculosis, Man, and Society* (Boston: Little, Brown, 1952); Paul Farmer, "Social Scientists and the New Tuberculosis," *Social Science and Medicine* 44 (1997): 347–58, and *Infections and Inequalities: The Modern Plagues* (Berkeley: University of California Press, 1999); Paul Farmer, Jaime Bayona, and Mercedes Becerra, "Multidrug-resistant Tuberculosis and the Need for Biosocial Perspectives," *International Journal of Lung Disease* 5 (2001): 885–86.

26. C. A. Hart and S. Kariuki, "Antimicrobial Resistance in Developing Countries," *British Medical Journal* 317 (1998): 647–50; Iruka N. Okeke, Adebayo Lamikanra, and Robert Edelman, "Socioeconomic and Behavioral Factors Leading to Acquired Bacterial Resistance to Antibiotics in Developing Countries," *Emerging Infectious Diseases* 5 (1999): 18–27; Iruka N. Okeke et al., "Anti-

microbial Resistance in Developing Countries. Part I: Recent Trends and Current Status," *Lancet Infectious Diseases* 5 (2005): 481–93; Margaret B. Planta, "The Role of Poverty in Antimicrobial Resistance," *Journal of the American Board of Family Medicine* 20 (2007): 533–39; Iruka N. Okeke, "Poverty and Root Causes of Resistance in Developing Countries," in Aníbal de J. Sosa et al., eds., *Antimicrobial Resistance in Developing Countries* (New York: Springer, 2010), 27–35.

27. Aryanti Radyowijati and Hilbrand Haak, "Improving Antibiotic Use in Low-Income Countries: An Overview of Evidence on Determinants," *Social Science and Medicine* 57 (2003): 733–44.

28. Henry Welch, "Opening Remarks," *Antibiotics Annual* (1956–1957): 1–2.

29. Iruke N. Okeke et al., "Antimicrobial Resistance in Developing Countries. Part II: Strategies for Containment," *Lancet Infectious Diseases* 5 (2005): 568–80; Hilbrand Haak and Aryanti Radyowijati, "Determinants of Antimicrobial Use: Poorly Understood—Poorly Researched," in *Antimicrobial Resistance in Developing Countries*, 283–300.

30. Calvin M. Kunin, "The Responsibility of the Infectious Disease Community for the Optimal Use of Antimicrobial Agents," *Journal of Infectious Diseases* 151 (1985): 389.

INDEX

Page numbers in italics indicate figures.